# Fault-Zone Properties and Earthquake Rupture Dynamics

This is Volume 94 in the
INTERNATIONAL GEOPHYSICS SERIES
A series of monographs and textbooks
Edited by RENATA DMOWSKA, DENNIS HARTMANN, and H. THOMAS ROSSBY
A complete list of books in this series appears at the end of this volume.

# Fault-Zone Properties and Earthquake Rupture Dynamics

**Eiichi Fukuyama**
National Research Institute for Earth Science and
Disaster Prevention
Tsukuba, Japan

AMSTERDAM • BOSTON • HEIDELBERG • LONDON
NEW YORK • OXFORD • PARIS • SAN DIEGO
SAN FRANCISCO • SINGAPORE • SYDNEY • TOKYO

ELSEVIER

Academic Press is an imprint of Elsevier

Elsevier Academic Press
30 Corporate Drive, Suite 400, Burlington, MA 01803, USA
525 B Street, Suite 1900, San Diego, California 92101-4495, USA
84 Theobald's Road, London WC1X 8RR, UK

This book is printed on acid-free paper. ∞

Library of Congress Cataloging-in-Publication Data
Application Submitted

British Library Cataloguing in Publication Data
A catalogue record for this book is available from the British Library

ISBN 13: 978-0-12-374452-4

For all information on all Elsevier Academic Press publications
visit our Web site at www.elsevierdirect.com

Printed and bound by CPI Group (UK) Ltd, Croydon, CR0 4YY

Transferred to digital print 2012

# Contents

## 4 Fault Zone Structure and Deformation Processes along an Exhumed Low-Angle Normal Fault: Implications for Seismic Behavior    69

*Cristiano Collettini, Robert E. Holdsworth and Steven A. F. Smith*

## 5 Pseudotachylytes and Earthquake Source Mechanics    87

*Giulio Di Toro, Giorgio Pennacchioni and Stefan Nielsen*

## 6. The Critical Slip Distance for Seismic and Aseismic Fault Zones of Finite Width — 135

## 7. Scaling of Slip Weakening Distance with Final Slip during Dynamic Earthquake Rupture — 163

## 8. Rupture Dynamics on Bimaterial Faults and Nonlinear Off-Fault Damage — 187

## 9. Boundary Integral Equation Method for Earthquake Rupture Dynamics  217

*Taku Tada*

In April 2004, I attended the Annual Meeting of the Seismological Society of America (SSA) held at Palm Springs, CA, U.S.A. My main motivation to join the meeting was to attend the luncheon where an award ceremony for the Harry Fielding Reid Medal was held; Professor Raul Madariaga was the recipient of the medal. Since he took care of me during my post-doc period at Institut de Physique du Globe de Paris (IPGP) in 1991-1992, I wanted to join the ceremony. During the SSA meeting, I had an opportunity to have a dinner with David D. Oglesby. On the way back to the hotel after the dinner, we discussed the future direction on the study of dynamic rupture propagation and we agreed to propose a special session at the upcoming American Geophysical Union (AGU) Fall Meeting. I believe this was the starting point of this book. We then proposed a session entitled "Fault Structure, Friction, Stress, and Dynamics" for the 2004 AGU Fall Meeting, where we tried to include the marginal subjects related to earthquake rupture dynamics. The special session was quite successful, having 65 contributions. Through the session, we were convinced that it is important that the study of earthquake rupture dynamics interacts with structural geology where direct observations of fault structures are being made in the field. We felt it necessary to continue to propose a session in the next AGU Meeting. Since David became a member of the programming committee of the AGU Fall Meeting at that time, I instead invited Douglas S. Dreger, Paola Vannucchi, César R. Ranero, and Harold J. Tobin to hold a special session entitled "Fault-zone Properties and Earthquake Rupture Dynamics" in the 2005 AGU Fall Meeting. This session was focused more clearly on the marginal research subjects between structural geology and seismology for the understanding of the dynamics of earthquake faulting. The session was again very successful with 88 contributions. I was convinced again that the research on dynamic rupture propagation should be developed along with geological observation in the field, as well as the laboratory experiments in addition to the seismological analysis of earthquake sources. Therefore, I started to organize a book project to cover this marginal research field between seismology and structural geology on the earthquake faulting.

In October 2006, I attended a workshop for celebrating the retirement of Professor Takeshi Mikumo from Universidad Nacional Autonoma de Mexico (UNAM) after his 15 years career of very active research in Mexico. I saw Renata Dmonska there and she asked me if there is a review work in my field because she has been working as an editor at Elsevier. Thus I contacted

Renata after the workshop and she kindly guided me in the publication process, which made it quite efficient to organize the book project.

During the project, I received much encouragement that I will never forget from my colleagues including Renata Dmonska, Raul Madariaga, Takeshi Mikumo, Kojiro Irikura, David D. Oglesby, Toshi Shimamoto, Paola Vannucchi, Douglas S. Dreger, Michel Bouchon, César R. Ranero and Harold J. Tobin. I would like to acknowledge all the reviewers of the manuscript who spent their valuable time to improve the chapters in the book significantly. And I would like to thank Linda Versteeg, Sara Pratt and the staff at Elsevier for their dedicated and patient assistance.

Finally, I sincerely hope that this book helps those who work on earthquake faulting, especially for young researchers who are going to study earthquake sources. Because the research on earthquake sources includes various aspects in understanding the earthquake faulting, one should not stick to a small research area where one started to study. Development of a new marginal research area is highly required. I would be very glad if the book helps such people.

*Eiichi Fukuyama*
December 2008
Tsukuba, Japan

## Foreword

Earthquake dynamics is one of the most pluridisciplinary endeavors in Earth Science, encompassing most disciplines in Geology, Geophysics, Rock Mechanics, and Fracture Mechanics. These studies range from field observations of faults and seismic ruptures through experimental rupture mechanics, fracture mechanics, and seismic radiation from earthquake faulting. Such a broad field attracts researchers from many horizons, using different scientific approaches and diverse technical experiences. The collection of papers selected by Eiichi Fukuyama and his colleagues is a very ample cross-section of these different approaches, ranging from field observations of faulting—both recent and old—to the most advanced techniques for simulating seismic ruptures in the laboratory and in computers, as well as the use of seismic data to constrain these models. These different approaches are required in order to understand and eventually use this information for the scientific prediction of ground motion and, perhaps one day, predicting earthquakes.

Over the years earthquake models have evolved from very simple, point-like, double couple source models, into full fledged fracture mechanical models of rupture propagation under the control of friction, taking into account the material properties of the rocks surrounding the fault. Many interesting phenomena of seismic ruptures and the properties of earthquakes were discovered using seismological models of earthquakes; we understand, for instance, that earthquakes scale with one, or at most, two variables: their fault area and stress drop. New methods of modeling seismic ruptures under realistic conditions are being developed in order to invert near field strong motion recordings. The current challenge is to do full-size dynamic inversions taking into account realistic models of friction and geometry; this is one of the biggest challenges in computational geophysics.

Studying present day earthquakes is very difficult because interesting events are rare and do not always occur under the best observational conditions. Moreover, current earthquakes are usually inaccessible to direct observation being buried under tens of km of rock. Drilling provides essential information, but is limited to small sections of faults because the drills are smaller than the main scales that control rupture propagation during large earthquakes. Scaling from the local fault structure to the full size of an earthquake is a major current challenge. The study of fossil fault zones has opened new ways of understanding the details of friction and the way strain energy is partitioned at the source between seismic wave generation and the energy dissipated into fracture energy, heat, etc. Earthquake slips produce very particular structures around the fault zone that are the subject of an intensive research effort, especially older facture zones, eroded, uplifted, and exhumed. The study of these areas will help in understanding the energy balance of earthquakes, especially the partition between fracture energy and heat.

Until very recently, laboratory fracture experiments were limited to very slow speeds that were several orders of magnitude too small with respect to slip rates during earthquakes. Recent development in servo-controlled experimental techniques for rock and fracture mechanics have opened the way to new experimental techniques at slip rates that approach those observed in major earthquakes. These experiments have shown that at least at low confining pressures, velocity weakening of friction through melting, thermal pressurization, and phase transformations can strongly reduce friction facilitating the very large slips and slip rates observed during the largest mega earthquakes like the giant Sumatra earthquake in 2004. The full impact of these new experiments in seismic source studies is a subject of intense research because of the numerous unknown parameters and material properties that need to be elucidated before these models can be effectively applied to understand natural friction at mid-crustal depths.

Seismic wave generation is of course the ultimate objective of many of these studies because large earthquakes produce strong seismic waves and tsunamis in the oceans. We expect that the details of rupture propagation, variations in the speed of rupture, and the heterogeneity of the elastic structure are at the origin of the intricate details of strong ground motion that causes damage to structures. These large waves can wreak havoc in costal cities and cities situated in soft sedimentary basins or narrow river valleys. Earthquake engineers have developed a substantial knowledge of strong ground motions, based mainly on empirical methods developed after years of accumulation of ground motion recordings. There is strong evidence that these empirical methods do not cover the entire gamut of possible ground motions, in particular extreme ground shaking. For instance, a few months ago, vertical accelerations of up to 4g were observed during the Iwate-Miyagi Nairiku Earthquake of 13 June 2008 in Japan, and a broad region of China was heavily damaged by the recent Wenchuan earthquake of 12 May 2008.

The present volume provides a very broad selection of chapters dealing with many different aspects of earthquake studies, so that the reader will find an entry point to most of the problems currently debated in the scientific literature. The excellent introduction to the volume by Eiichi Fukuyama provides an up-to-date survey of most of the problems posed in earthquake studies, including a very well documented selection of the most essential references.

*Raul Madariaga*
Laboratoire de Géologie, Ecole Normale Supérieure,
24 rue Lhomond, 75231 Paris Cedex 05, France.
madariag@geologie.ens.fr

# List of Contributors

**Massimo Cocco** (cocco@ingv.it) Istituto Nazionale di Geofisica e Vulcanologia, 00143 Rome, Italy, Coauthors: **Elisa Tinti, Chris Marone and Alessio Piataanesi**

**Cristiano Collettini** (colle@unipg.it) Dipartimento di Scienze della Terra, Università degli Studi di Perugia, 06123 Perugia, Italy, Coauthors: **Robert E. Holdsworth and Steven A. F. Smith**

**Giulio Di Toro** (giulio.ditoro@unipd.it) Dipartimento di Geoscienze, Università degli Studi di Padova, 35137 Padova, Italy, Coauthors: **Giorgio Pennacchioni and Stefan Nielsen**

**Eiichi Fukuyama** (fuku@bosai.go.jp) National Research Institute for Earth Science and Disaster Prevention, Tsukuba, 305-0006, Japan

**Aiming Lin** (slin@ipc.shizuoka.ac.jp) Graduate School of Science and Technology, Shizuoka University, Shizuoka, 422-8529, Japan

**Chris J. Marone** (cjm@geosc.psu.edu) Istituto Nazionale di Geofisica e Vulcanologia, 00143 Rome, Italy, Permanent address: Department of Geosciences, Pennsylvania State University, University Park, PA 16802, USA, Coauthors: **Massimo Cocco, Eliza Richardson and Elisa Tinti**

**Taku Tada** (kogutek@ni.aist.go.jp) National Institute of Advanced Industrial Science and Technology, Tsukuba, 305-8567, Japan

**Paola Vannucchi** (paola.vannucchi@unifi.it) Dipartimento di Scienze della Terra, Università degli Studi di Firenze, 50121 Firenze, Italy, Coauthors: **Francesca Remitti, Jason Phipps-Morgan and Giuseppe Bettelli**

**Teruo Yamashita** (tyama@eri.u-tokyo.ac.jp) Earthquake Research Institute, the University of Tokyo, Tokyo, 113-0032, Japan

## Chapter 1

# Introduction

## Fault-Zone Properties and Earthquake Rupture Dynamics

**Eiichi Fukuyama**
*National Research Institute for Earth Science and Disaster Prevention, Tsukuba, Japan*

An earthquake is a phenomenon that ruptures the crust to release the stress accumulated by the tectonic loading. Once an earthquake occurs, slip offsets sometimes appear on the surface as a fault outcrop, and the repeated occurrence of earthquakes forms a fault zone. These features have been recognized for many years (e.g., *Omori*, 1910; *Reid*, 1910). However, recent advances in the studies of earthquake faulting in view of both structural geology and seismology reveal that an earthquake faulting process is not simple but a quite complicated phenomenon at various scales. We shall look into these features in more detail in this book.

Research on earthquake faulting and rupture dynamics has developed together with earthquake prediction research programs. In the 1970s and 1980s, one of the main research targets in seismology was to predict the occurrence of large earthquakes, which had been considered feasible if very dense and accurate observations were conducted (e.g., *Mogi*, 1992). At that time, earthquake prediction had been considered an ultimate approach to reducing the damage from large earthquakes. However, as research developed, everybody started to realize that earthquakes are not so simple but very complicated phenomena (e.g., see www.nature.com/nature/debates/earthquake). *Kanamori* (2003, p. 1213) said "Despite the progress made in understanding the physics of earthquakes, the predictions of earthquake activity we can make today are inevitably very uncertain, mainly because of the highly complex nature of earthquake process." We surely admit that earthquake prediction could save a number of lives if the prediction is accurate enough to evacuate the public. Actually, there exist a few examples that succeeded in sending an appropriate warning before the earthquake, which could tremendously reduce the damages (e.g., the 1975 Haicheng, China, earthquake; *Scholz*, 1977). But it should be pointed out that such predictions are rare, and most disastrous earthquakes occurred without such warnings even if they occurred inside a densely distributed seismological observation network (e.g., *Geller*, 1997; *Bakun et al.*, 2005). Recent research focus is migrating to early warning (e.g., *Kanamori et al.*, 1997), which is different from earthquake prediction in view of its timing. In the early warning system, a warning is issued by observing the first P-wave arrival emitted from a large earthquake close to the epicenter. Thus, an early warning is made after the earthquake occurrence but as early as possible. Because the speed of wave propagation is slower than that of information transmission, one can get a warning slightly before the strong shaking arrives. This has been achieved due to recent rapid advances in information technology.

In such prediction-oriented research, understanding of earthquake source mechanics is crucial, because one needs a model for prediction, and the model should be based on scientific background. Therefore, the investigation of earthquake rupture dynamics became an important target and it significantly progressed. The most important question for prediction is how an earthquake initiates. Two end-member models are proposed: the *cascade* model and the *preslip* model (*Ellsworth and Beroza*, 1995). In the *cascade* model, initial small slips occur randomly in time and space close to the hypocenter, and due to the accumulation of small slips, the main rupture is triggered and starts to propagate. Thus, in this model, no information on the earthquake size is included in the initial behavior of the faulting, so that prediction of earthquake occurrence becomes impossible based on this model. In contrast, in the *pre-slip* model, before the large earthquake occurs, a preslip region appears at the hypocenter, whose size is proportional to the final earthquake size. Thus, in this model, the nucleation process includes the information on the final size of the earthquake. *Shibazaki and Matsu'ura* (1998) investigated two Japanese earthquakes and obtained the scaling relations, which support the *preslip* model and might be useful for the prediction. On the contrary, *Bakun et al.* (2005) reported that no crustal deformation was observed before the 2004 Parkfield earthquake, suggesting that either the *preslip* signal was too tiny to be detected or the *preslip* model was not appropriate for this earthquake.

These nucleation modelings are also related to early warning study, because in the early warning system, the final earthquake size has to be estimated as early as possible using the very beginning part of waveforms observed at near distance stations (e.g., *Olson and Allen*, 2005; *Rydelek and Horiuchi*, 2006). Note that source duration time for large earthquakes is not negligible; the rupture duration of magnitude 8 earthquakes exceeds 1 minute. Thus, if earthquake nucleation follows the *cascade* model, we need to wait until the rupture terminates after slipping over the entire slip region. But if it follows the *preslip* model, we do not have to wait for more than 1 minute and can make a warning before the termination of the rupture.

To discriminate these models, we need to understand the physical properties of the earthquake source, where a rupture initiates and propagates. Most large earthquakes occur along the preexisting discontinuities including plate

boundaries, shear zones, and active faults, whose geometry is not always planar but has many kinks, joints, jogs, steps, and branches (e.g., *Yeats et al.,*
1997). We also notice that these features can be seen at various scales from
millimeters to tens of kilometers (e.g., *Ben-Zion and Sammis,* 2003). These
fault complexities may be related to the earthquake rupture mechanics.

To investigate the properties of fault zone directly, there have been several
drilling projects, including the Nojima Fault in Japan (*Ando,* 2001; *Ikeda,*
2001), the Chelungpu Fault in Taiwan (*Ma et al.,* 2006), the Aigion Fault in
Greece (*Cornet et al.,* 2004), and the San Andreas Fault in the United States
(*Hickman et al.,* 2004). In addition to these inland drilling projects, offshore
subduction drilling projects are planned and have started, such as the Nankai
Trough in southwest Japan (*Park et al.,* 2002; *Tobin and Kinoshita,* 2006) and
the Central America Trench off southwest Costa Rica (*DeShon et al.,* 2003;
*Ranero et al.,* 2008). These drilling projects generated important information
on the fault zone materials (for the Nojima Fault, e.g., *Boullier et al.,* 2001;
*Kobayashi et al.,* 2001; *Lin et al.,* 2001; *Ohtani et al.,* 2001; *Tanaka et al.,*
2001a, 2001b; for the Chelungpu Fault, *Hirono et al.,* 2007, 2008; for the
Aigion Fault, *Rettenmaier et al.,* 2004; for the San Andreas Fault, *Solum
et al.,* 2006) as well as the *in situ* stress conditions (for the Nojima Fault,
*Ikeda et al.,* 2001; *Tsukahara et al.,* 2001; *Yamashita et al.,* 2004; for the
Chelungpu Fault, *Wu et al.,* 2007; *Hung et al.,* 2008; for the Aigion Fault,
*Sulem,* 2007; for the San Andreas Fault, *Hickman and Zoback,* 2004). There
remain, however, some problems on the drilling project strategy. The information is obtained only at a spot of the fault zone, so it is difficult to generalize the
obtained features. It is even difficult to identify the most recent slip surface
from the drilled core samples (e.g., *Hirono et al.,* 2007; *Tanaka et al.,* 2007).
In addition, the depth is still shallow ($\sim$3 km at maximum) compared to
the depth of the seismogenic zone (5 to 10 km), so that the pressure-
temperature condition might be different.

To overcome these problems, field investigations have been conducted on
exhumed faults, on which earthquakes occurred many years ago at seismogenic depth and now are exposed on the surface. From the observation of
exhumed faults, one can recognize that fault slip occurred within a very thin
region on the fault (e.g., *Chester and Chester,* 1998; *Di Toro and Pennacchioni,* 2005), whereas the slip lasts for tens of kilometers along the fault.
At shallow depth, fault slip zone is surrounded by cataclasite, breccia, and
deformed host rocks. Sometimes, pseudotachylyte (*Sibson,* 1975), which is
considered to be a product caused by melting as a result of a sudden increase
and decrease in temperature (*Di Toro et al.,* 2009, for details), is found at the
slip zone. At deep depth (deeper than 10 km, i.e., temperature higher than
350°C), no clear slip zone is found at the fault core and the mylonite zone
is observed instead of the cataclasite zone (e.g., *Shigematsu and Yamagishi,*
2002). This complicated fault zone structure is related to the fault slip behavior during earthquakes.

To interpret these field observations, we need some theoretical considerations. In the traditional framework of fracture mechanics (e.g., *Freund*, 1990), a fault is considered to be a zero-thickness planar surface, and shear stress drops immediately at the crack tip when the rupture front passes by. However, shear stress does not drop immediately but requires a certain amount of slip during a real earthquake. This slip is called the slip weakening distance (*Ida*, 1972; *Palmar and Rice*, 1973). *Marone and Kilgore* (1993) suggested that the thickness of the slip zone is related to the slip weakening distance. This fault weakening process affects the dynamic rupture propagation significantly. There are several causes of slip weakening behavior. During the high-speed slip, gauges are created on the sliding surfaces, which weaken the strength of the fault (e.g., *Matsu'ura et al.*, 1992). If the slip rate is much higher, fault materials start to melt due to friction heating, which weakens the fault strength and produces pseudotachylyte (e.g., *Hirose and Shimamoto*, 2003; *Di Toro et al.*, 2004). Another possible mechanism is thermal pressurization: when there is fluid on the fault surface, due to the thermal expansion of fluid caused by frictional heating, effective normal stress reduces, which makes the fault strength weak (*Mase and Smith*, 1987; *Rice*, 2006).

In such a slip weakening model, fracture energy becomes an important parameter and can be estimated from the observed data (e.g., *Kanamori and Heaton*, 2000; *Cocco and Tinti*, 2008). Fracture energy is defined as the integration of the traction drop history along slip displacement until the slip weakening distance (*Kanamori and Heaton*, 2000). The fracture energy is related to earthquake rupture dynamics and controls the rupture propagation velocity. In Griffith fracture theory (*Griffith*, 1920), the fracture advances if the external energy applied at the crack tip is equal to or greater than the energy consumed at the crack tip as a work at each time interval. Because the fracture energy is considered to be the work consumed at the rupture front (*Tinti et al.*, 2005, 2008), rupture velocity can be controlled by this parameter. *Burridge* (1973) suggested that in the case of in-plane rupture (mode II), rupture propagates with a velocity that is less than Rayleigh wave speed or between S- and P-wave speeds (supershear rupture). *Andrews* (1976b) confirmed this feature numerically. *Rosakis et al.* (1999) observed supershear rupture propagation in the laboratory experiments. It should be noted that supershear rupture can occur only when the in-plane rupture and maximum rupture velocity is S-wave velocity for antiplane rupture. In most earthquakes, rupture propagates with subshear rupture velocity, but there are some reports on supershear rupture propagation such as the 1979 Imperial Valley, California, earthquake (*Archuleta*, 1984); the 1992 Landers, California, earthquake (*Olsen et al.*, 1997); the 1999 Izumit, Turkey, earthquake (*Bouchon et al.*, 2001, 2002); the 2001 Kunlun, Tibet, earthquake (*Bouchon and Vallée*, 2003); and the 2002 Denali, Alaska, earthquake (*Dunham and Archuleta*, 2004; *Ellsworth et al.*, 2004). It should be noted that all these earthquakes were strike-slip earthquakes with surface breaks. This is because in strike-slip earthquakes, in-plane rupture dominates and free surface promotes the rupture propagation near the free surface (*Aagaard et al.*, 2001; *Dunham and Archuleta*, 2004).

In addition to the weakening process of faulting, detailed fault geometry affects the earthquake rupture. For example, during the 1992 Landers, California, earthquake, due to the existence of the Kickapoo Fault, which is a small fault segment connecting the Johnson Valley and Homestead Valley faults, the rupture did not continue to propagate along the Johnson Valley Fault; instead it switched to the rupture along the Homestead Valley, then continued to rupture along the Emerson and Camp Rock faults (*Hart et al.*, 1993; *Sieh et al.*, 1993; *Wald and Heaton*, 1994). These rupture transitions can be explained by considering the fault geometry and stress field applied to the fault together with appropriate constitutive relation on the fault (*Aochi and Fukuyama*, 2002; *Poliakov et al.*, 2002; *Aochi et al.*, 2003; *Kame et al.*, 2003). Because of the complex geometry of the fault system, inelastic deformation is generated as off-fault damage to compensate the geometrical mismatch and to release the local stress concentration. These features are observed in the field (e.g., *Chester et al.*, 1993) and can be reproduced by the modeling (*Yamashita*, 2000; *Dalguer et al.*, 2003; *Andrews*, 2005; *Ando and Yamashita*, 2007). Such off-fault damage is enhanced when the fault has bi-material—that is, both sides of the fault wall have different elastic constants (e.g., *Andrews and Ben-Zion*, 1997; *Harris and Day*, 1997). In addition, off-fault shear branches decelerate the rupture propagation and sometimes make the rupture terminate (*Kame and Yamashita*, 1999). Supershear rupture propagation could enhance the off-fault damages by generating tensile cracks as a result of Mach wave cones (*Bhat et al.*, 2007).

Recent notable progress on numerical modeling of earthquake rupture dynamics is a result of the development of numerical techniques such as finite differences, finite elements, spectral elements, boundary elements, and boundary integral equation method, in addition to the rapid advances of computer resources. Increases in the detailed observations in the field also accelerate the development of numerical computation method. Since the end of the 1970s, many computation techniques have been proposed (*Andrews*, 1976a, 1976b, 1985; *Madariaga*, 1976; *Das and Aki*, 1977; *Archuleta and Frazer*, 1978; *Mikumo and Miyatake*, 1978; *Day*, 1982a, 1982b). All of these methods assumed planar fault geometry. More recently, nonplanar fault geometry has been modeled (e.g., *Harris et al.*, 1991; *Kase and Kuge*, 1998; *Oglesby et al.*, 1998, 2000; *Aochi et al.*, 2000; *Aagaard et al.*, 2001, 2004; *Dalguer et al.*, 2001; *Ando et al.*, 2004; *Cruz-Atienza and Virieux*, 2004; *Kase and Day*, 2006; *Ely et al.*, 2008). Because the fault zone structure in nature is quite complicated, these features should be included in more detail. Stress field is another important feature in the modeling of nonplanar fault systems, because under the Coulomb friction law, friction is controlled by both normal and shear stresses under a predefined coefficient of friction. It is reasonable to assume that the background stress field is formed by remotely applied tectonic loading. Thus, the change in fault plane orientation alters the amount of accumulated shear stress. Strength also changes because it depends on the amount of normal stress under the Coulomb friction law.

I will give a brief overview of the topics discussed in the following chapters. In Chapter 2, *Lin* (2009) precisely describes the macroscopic fault slip of the 2001 Kunlun, Tibet, earthquake based on the satellite image analysis with field observations. Coseismic surface ruptures are recognized as distinct shear faults, echelon extensional cracks, and mole tracks within the width of a few meters to a half-kilometer and extending the length of 450 km. This is one of the largest intracontinental earthquakes ever reported. *Lin* (2009) demonstrated that fault trace appears in a complicated form on the surface even from the macroscopic view. In Chapter 3, *Vannucchi et al.* (2009) discuss the characteristic feature of the subduction plate boundary of erosive and accretionary margins in relation to the seismic-aseismic transition. They review the subduction process that relates to the seismogenesis in various aspects. They comment that the seismic-aseismic transition is not the smectite-illite cray mineral transformation but the behavior of fluid at depth. They also compare the fossil examples of Northern Apennines (erosive margin) and Shimanto Belt (accretionary margin) with the observations conducted in the subduction zone. In Chapter 4, *Collettini et al.* (2009) discuss the low-angle normal fault paradox. It is considered that normal faulting is difficult to slip at low dip angles because of the Coulomb stress criteria (*Sibson*, 1994). However, they interpret the observation of low-angle normal faults as resulting from the existence of $CO_2$-rich fluid, which decreases the effective normal stress on the fault and enables it to slip based on the geological field survey of the exhumed Zuccale low-angle normal fault in the isle of Elba in central Italy. In Chapter 5, *Di Toro et al.* (2009) review pseudotachylytes observed in the field as the product of high-velocity slippage. They try to explain the field observation with the laboratory experiments with a high velocity rotary shear apparatus. They also interpret the field observation based on a theoretical consideration of the generation of pseudotachylyte.

In Chapter 6, *Marone et al.* (2009) propose a simple model for the slip weakening in the finite width fault zone using several parallel faults that obey rate- and state-friction law. In their model, effective dynamic slip weakening distance, which is measured outside the fault zone, is proportional to the fault zone width. In Chapter 7, *Cocco et al.* (2009) discuss the slip weakening distance and fracture energy estimated by the slip evolution history obtained by the seismic waveform inversions. They point out that to investigate the physical interpretation of these parameters, these estimations should be made carefully with the assistance of numerical modeling of dynamic rupture. In Chapter 8, *Yamashita* (2009) reviews recent progress on the effect of off-fault damage and bi-material faults. He points out that on bi-material planar faults, off-fault damage should be developed to avoid the ill-posed stress condition. Generation of the tensile cracks and shear fault branches forms the off-fault damages, and bi-material planar interface enhances the generation of the damages. In Chapter 9, *Tada* (2009) describes the boundary integral equation method used to analyze dynamic rupture propagation. He explains how to derive the boundary integral equations and their discretized forms. He further describes how to handle complicated fault geometries, which is required to model the fault realistically. In Chapter 10, *Fukuyama* (2009) applies the boundary integral equation method to the disastrous 1995 Kobe earthquake. Under the Coulomb friction condition, complicated fault geometry is tightly linked to the stress field around the fault. From the numerical experiments, the initial stress field is found to be rotated from the tectonic stress field.

Finally, I do really hope that the interconnected research between the field observation of fault zones and the physical modeling of earthquakes will give us a way to proceed for the further understanding of these complicated earthquake generation systems.

## REFERENCES

Aagaard, B. T., G. Anderson, and K. W. Hudnut, (2004), Dynamic rupture modeling of the transition from thrust to strike-slip motion in the 2002 Denali Fault earthquake, Alaska, *Bull. Seismol. Soc. Am.*, **94**(6B), S190-S201.

Aagaard, B. T., T. H. Heaton, and J. F. Hall, (2001), Dynamic earthquake ruptures in the presence of lithostatic normal stresses: Implications for friction models and heat production, *Bull. Seismol. Soc. Am.*, **91**(6), 1765-1796.

Ando, M., (2001), Geological and geophysical studies of the Nojima fault from drilling: An outline of the Nojima Fault Zone probe, *Island Arc*, **10**(3-4), 206-214,

Ando, R., T. Tada, and T. Yamashita, (2004), Dynamic evolution of a fault system through interactions between fault segments, *J. Geophys. Res.*, **109**, B05303, doi:10.1029/2003JB002665.

Ando, R., and T. Yamashita, (2007), Effects of mesoscopic-scale fault structure on dynamic earthquake ruptures: Dynamic formation of geometrical complexity of earthquake faults, *J. Geophys. Res.*, **112**, B09303, doi:10.1029/2006JB004612.

Andrews, D. J., (1976a), Rupture propagation with finite stress in antiplane strain, *J. Geophys. Res.*, **81**(20), 3575-3582.

Andrews, D. J., (1976b), Rupture velocity of plane strain shear cracks, *J. Geophys. Res.*, **81**(32), 5679-5687.

Andrews, D. J., (1985), Dynamic plane-strain shear rupture with a slip-weakening friction law calculated by a boundary integral method, *Bull. Seismol. Soc. Am.*, **75**(1), 1-21.

Andrews, D. J., (2005), Rupture dynamics with energy loss outside the slip zone, *J. Geophys. Res.*, **110**, B01307, doi:10.1029/2004JB003191.

Andrews, D. J. and Y. Ben-Zion, (1997), Wrinkle-like slip pulse on a fault between different materials, *J. Geophys. Res.*, **102**, 553-571.

Aochi, H. and E. Fukuyama, (2002), Three-dimensional nonplanar simulation of the 1992 Landers earthquake, *J. Geophys. Res.*, **107**(B2), 2035, doi:10.1029/2000JB000061.

Aochi, H., E. Fukuyama, and M. Matsu'ura, (2000), Spontaneous rupture propagation on a nonplanar fault in 3-D elastic medium, *Pure Appl. Geophys.*, **157**, 2003-2037.

Aochi, H., R. Madariaga, and E. Fukuyama, (2003), Constraint of fault parameters inferred from nonplanar fault modeling, *Geochem. Geophys. Geosyst.*, **4**(2), 1020, doi:10.1029/2001GC000207.

Archuleta, R. J., (1984), A faulting model for the 1979 Imperial Valley earthquake, *J. Geophys. Res.* **89**, 4559-4585.

Archuleta, R. J. and G. A. Frazer, (1978), Three-dimensional numerical simulations of dynamic faulting in a half-space, *Bull. Seismol. Soc. Am.*, **68**(3), 541-572.

Bacun, W. H., B. Aagaard, B. Dost, W. L. Ellsworth, J. L. Hardebeck, R. A. Harris, C. Ji, M. J. S. Johnston, J. Langbein, J. J. Lienkaemper, A. J. Michael, J. R. Murray, R. M. Nadeau, P. A. Reasenberg, M. S. Reichle, E. A. Roeloffs, A. Shakal, R. W. Simpson, and F. Waldhauser, (2005), Implications for prediction and hazard assessment from the 2004 Parkfield earthquake, *Nature*, **437**, 969-974, doi:10.1038/nature04067.

Ben-Zion, Y. and C. G. Sammis, (2003), Characterization of Fault Zones, *Pure Appl. Geophys.*, **160**, 677-715

Bhat, H. S., R. Dmowska, G. C. P. King, Y. Klinger, and J. R. Rice, (2007), Off-fault damage patterns due to supershear ruptures with application to the 2001 Mw 8.1 Kokoxili (Kunlun) Tibet earthquake, *J. Geophys. Res.*, **112**, B06301, doi:10.1029/2006JB004425.

Bouchon, M., M. P. Bouin, H. Karabulut, M. N. Toksöz, M. Dietrich, and A. Rosakis, (2001), How fast is rupture during an earthquake? New insights from the 1999 Turkey earthquakes, *Geophys. Res. Lett.* **28**, 2723-2726.

Bouchon, M., M. N. Toksöz, H. Karabulut, M.-P. Bouin, M. Dietrich, M. Aktar, and M. Edie (2002), Space and time evolution of rupture and faulting during the 1999 Izmit (Turkey) earthquake, *Bull. Seismol. Soc. Am.*, **92**(1), 256-266, doi:10.1785/0120000845.

Bouchon, M. and M. Vallée, (2003), Observation of long supershear rupture during the magnitude 8.1 Kunlunshan earthquake, *Science* **301**, 824-826.

Boullier, A.-M., T. Ohtani, K. Fujimoto, H. Ito, and M. Dubois, (2001), Fluid inclusions in pseudotachylytes from the Nojima fault, Japan, *J. Geophys. Res.*, **106**(B10), 21965-21977.

Burridge, R., (1973), Admissible speeds for plane-strain self-similar shear cracks with friction but lacking cohesion, *Geophys. J. Roy. astr. Soc.*, **35**, 439-455.

Chester, F. M. and J. S. Chester, (1998), Ultracataclasite structure and friction processes of the Punchbowl fault, San Andreas System, California, *Tectonophys.*, **295**(1-2), 199-221.

Chester, F. M., J. P. Evans, and R. L. Biegel, (1993), Internal structure and weakening mechanisms of the San Andreas fault, *J. Geophys. Res.*, **98**(B1), 771-786.

Cocco, M. and E. Tinti, (2008), Scale dependence in the dynamics of earthquake propagation: Evidence from seismological and geological observations, *Earth Planet. Sci. Lett.*, **273**(1-2), 123-131.

Cocco, M., E. Tinti, C. Marone, and A. Piatanesi, (2009), Scaling of slip weakening distance with final slip during dynamic earthquake rupture, In: *Fault-zone Properties and Earthquake Rupture Dynamics*, edited by E. Fukuyama, *International Geophysics Series*, **94**, 163-186, Elsevier.

Collettini, C., R. E. Holdsworth, and S. A. F. Smith, (2009), Fault zone structure and deformation process along an exhumed low-angle normal fault, In: *Fault-zone Properties and Earthquake Rupture Dynamics*, edited by E. Fukuyama, *International Geophysics Series*, **94**, 69-85, Elsevier.

Cornet, F. H., M. L. Doan, I. Moretti, and G. Borm, (2004), Drilling through the active Aigion Fault: The AIG10 well observatory, *C. R. Geoscience*, **336**, 395-406.

Cruz-Atienza, V. M. and J. Virieux, (2004), Dynamic rupture simulation of non-planar faults with a finite-difference approach, *Geophys. J. Int.*, **158**, 939-954.

Dalguer, L. A., K. Irikura, J. D. Riera, and H. C. Chiu, (2001), Fault dynamic rupture simulation of the hypocenter area of the thrust fault of the 1999 Chi-Chi (Taiwan) earthquake, *Geophys. Res. Lett.*, **28**(7), 1327-1330.

Dalguer L. A., K. Irikura, and J. D. Riera, (2003), Generation of new cracks accompanied by the dynamic shear rupture propagation of the 2000 Tottori (Japan) earthquake, *Bull. Seismol. Soc. Am.*, **93**(5), 2236-2252.

Das, S. and K. Aki, (1977), A numerical study of two-dimensional spontaneous rupture propagation, *Geophys. J. Roy. astr. Soc.*, **50**, 643-668.

Day, S. M., (1982a), Three-dimensional finite difference simulation of fault dynamics: Rectangular faults with fixed rupture velocity, *Bull. Seismol. Soc. Am.*, **72**(3), 705-727.

Day, S. M., (1982b), Three-dimensional simulation of spontaneous rupture: The effect of nonuniform prestress, *Bull. Seismol. Soc. Am.*, **72**(6), 1851-1902.

DeShon, H. R., S. Y. Schwartz, S. L. Bilek, L. M. Dorman, V. Gonzalez, J. M. Protti, E. R. Flueh, and T. H. Dixon, (2003), Seismogenic zone structure of the southern Middle America Trench, Costa Rica, *J. Geophys. Res.*, **108**(B10), 2491, doi:10.1029/2002JB002294.

Di Toro, G., D. L. Goldsby, and T. E. Tullis, (2004), Friction falls towards zero in quartz rock as slip velocity approaches seismic rates, *Nature*, **427**, 436-439.

Di Toro, G. and G. Pennacchioni, (2005), Fault plane processes and mesoscopic structure of a strong-type seismogenic fault in tonalites (Adamello batholith, Southern Alps), *Tectonophysics*, **402**, 54-79.

Di Toro, G., G. Pennacchioni, and S. Nielsen, (2009), Pseudotachylytes and earthquake source mechanics, In: *Fault-zone Properties and Earthquake Rupture Dynamics*, edited by E. Fukuyama, *International Geophysics Series*, **94**, 87-113, Elsevier.

Dunham, E. M. and R. J. Archuleta, (2004), Evidence for a supershear transient during the 2002 Denali Fault earthquake, *Bull. Seismol. Soc. Am.*, **94**(6B), S256-S268, doi:10.1785/0120040616.

Ellsworth, W. L. and G. C. Beroza, (1995), Seismic evidence for an earthquake nucleation phase, *Science*, **268**, 851-855.

Ellsworth, W. L., M. Celebi, J. R. Evans, E. G. Jensen, R. Kayen, M. C. Metz, D. J. Nyman, J. W. Roddick, P. Spudich, and C. D. Stephens, (2004), Near-field ground motion of the 2002 Denali Fault, Alaska, earthquake recorded at Pump Station 10, *Earthq. Spectra*, **20**(3), 597-615, doi:10.1193/1.1778172.

Ely, G. P., S. M. Day, and J.-B. Minster, (2008), A support-operator method for viscoelastic wave modeling in 3-D heterogeneous media, *Geophys. J. Int.*, **172**(1), 331-344.

Freund, L. B., (1990), *Dynamic Fracture Mechanics*, Cambridge University Press, Cambridge.

Fukuyama, E., (2009), Dynamic rupture propagation of the 1995 Kobe, Japan, earthquake, In: *Fault-zone Properties and Earthquake Rupture Dynamics*, edited by E. Fukuyama, *International Geophysics Series*, **94**, 269-283, Elsevier.

Geller, R. J., (1997), Earthquake prediction: A critical review, *Geophys. J. Int.*, **131**, 425-450.

Griffith, A. A., (1920), The phenomenon of rupture and flow in solids, *Phil. Trans., Roy. Soc. London*, **A221**, 163-198.

Harris, R. A., R. J. Archuleta, and S. M. Day, (1991), Fault steps and the dynamic rupture process: 2-D numerical simulations of a spontaneously propagating shear fracture, *Geophys. Res. Lett.*, **18**(5), 893-896.

Harris, R. A. and S. M. Day, (1997), Effects of a low velocity zone on a dynamic rupture, *Bull. Seismol. Soc. Am.*, **87**, 1267-1280.

Hart, E. W. W. A. Bryant, and J. A. Treiman, (1993), Surface faulting associated with the June 1992 Landers earthquake, *California, Calif. Geol.*, **46**(Jan./Feb.), 10-16.

Hickman, S. and M. Zoback, (2004), Stress orientations and magnitudes in the SAFOD pilot hole, *Geophys. Res. Lett.*, **31**, L15S12, doi:10.1029/2004GL020043.

Hickman, S., M. Zoback, and W. Ellsworth, (2004), Introduction to special section: Preparing for the San Andreas Fault Observatory at Depth, *Geophys. Res. Lett.*, **31**, L12S01, doi:10.1029/2004GL020688.

Hirono, T., E.-C. Yeh, W. Lin, H. Sone, T. Mishima, W. Soh, Y. Hashimoto, O. Matsubayashi, K. Aoike, H. Ito, M. Kinoshita, M. Murayana, S.-R. Song, K.-F. Ma, J.-H. Hung, C.-Y. Wang, Y.-B. Tsai, T. Kondo, M. Nishimura, S. Moriya, T. Tanaka, T. Fujiki, L. Maeda, H. Muraki, T. Kuramoto, K. Sugiyama, and T. Sugawara, (2007), Nondestructive continuous physical property measurements of core samples recovered from hole B, Taiwan Chelungpu-Fault Drilling Project, *J. Geophys. Res.*, **112**, B07404, doi:10.1029/2006JB004738.

Hirono, T., M. Sakaguchi, K. Otsuki, H. Sone, K. Fujimoto, T. Mishima, W. Lin, W. Tanikawa, M. Tanimizu, W. Soh, E.-C. Yeh, and S.-R. Song, (2008), Characterization of slip zone associated with the 1999 Taiwan Chi-Chi earthquake: X-ray CT image analyses and microstructural observations of the Taiwan Chelungpu fault, *Tectonophys.*, **449**, 63-84.

Hirose, T. and T. Shimamoto, (2003), Fractal dimension of molten surfaces as a possible parameter to infer the slip-weakening distance of faults from natural pseudotachylytes, *J. Struct. Geol.*, **25**, 1569-1574.

Hung, J.-H., K.-F. Ma, C.-Y. Wang, H. Ito, W. Lin, and E.-C. Yeh, (2008), Subsurface structure, physical properties, fault-zone characteristics and stress state in scientific drill holes of Taiwan Chelungpu Fault Drilling Project, *Tectonophys.*, doi:10.1016/j.tecto.2007.11.014, *in press.*

Ida, Y., (1972), Cohesive force across the tip of a longitudinal-shear crack and Griffith's specific surface energy, *J. Geophys. Res.*, **77**, 3796-3805.

Ikeda, R., (2001), Outline of the fault drilling project by NIED in the vicinity of the 1995 Hyogo-ken Nanbu earthquake, Japan, *Island Arc*, **10**(3-4), 199-205.

Ikeda, R., Y. Iio, and K. Omura, (2001), In situ stress measurements in NIED boreholes in and around the fault zone near the 1995 Hyogo-ken Nanbu earthquake, Japan, *Island Arc*, **10**(3-4), 252-260.

Kame, N. and T. Yamashita, (1999), A new light on arresting mechanism of dynamic earthquake faulting, *Geophys. Res. Lett.*, **26**(13), 1997-2000.

Kame, N., J. R. Rice, and R. Dmowska, (2003), Effects of prestress state and rupture velocity on dynamic fault branching, *J. Geophys. Res.*, **108**(B5), 2265, doi:10.1029/2002JB002189.

Kanamori, H., (2003), Earthquake prediction: An overview, In: *International Handbook of Earthquake and Engineering Seismology*, edited by Lee, W. H. K., H. Kanamori, P. C. Jennings, and C. Kisslinger, *International Geophysics Series*, **81B**, 1205-1216, Academic Press.

Kanamori, H. and T. H. Heaton, (2000), Microscopic and macroscopic physics of earthquakes, In: *Geocomplexity and the Physics of Earthquakes*, edited by Rundle, J., D. L. Turcotte, and W. Klein, *Geophysical Monograph Series*, **120**, 147-163, American Geophysical Union, Washington, D.C.

Kanamori, H., E. Hauksson, and T. Heaton, (1997), Real-time seismology and earthquake hazard mitigation, *Nature*, **390**, 461-464, doi:10.1038/37280.

Kase, Y. and S. M. Day, (2006), Spontaneous rupture processes on a bending fault, *Geophys. Res. Lett.*, **33**, L10302, doi:10.1029/2006GL025870.

Kase, Y. and K. Kuge, (1998), Numerical simulation of spontaneous rupture process on two non-coplanar faults: The effect of geometry on fault interaction, *Geophys. J. Int.*, **135**, 911-922.

Kobayashi, K., S. Hirano, T. Arai, R. Ikeda, K. Omura, H. Sano, T. Sawaguchi, H. Tanaka, T. Tomita, N. Tomida, T. Matsuda, and A. Yamazaki, (2001), Distribution of fault rocks in the fracture zone of the Nojima Fault at a depth of 1140m: Observations from the Hirabayashi NIED drill core, *Island Arc*, **10**(3-4), 411-421.

Lin, A., (2009), Geometry and slip distribution of co-seismic surface ruptures produced by the 2001 Kunlun, northern Tibet, earthquake, In: *Fault-zone Properties and Earthquake Rupture Dynamics*, edited by E. Fukuyama, *International Geophysics Series*, **94**, 15-36, Elsevier.

Lin, A., T. Shimamoto, T. Maruyama, M. Shigetomi, T. Miyata, K. Takemura, H. Tanaka, S. Uda, and A. Murata, (2001), Comparative study of cataclastic rocks from a drill core and outcrops of the Nojima Fault zone on Awaji Island, Japan, *Island Arc*, **10**(3-4), 368-380.

Ma, K.-F., H. Tanaka, S.-R. Song, C.-Y. Wang, J.-H. Hung, Y.-B. Tsai, J. Mori, Y.-F. Song, E.-C. Yeh, W. Soh, H. Sone, L.-W. Kuo, and H.-Y. Wu, (2006), Slip zone and energetics of a large earthquake from the Taiwan Chelungpu-fault Drilling Project, *Nature*, **444**, 473-476, doi: 10.1038/nature05253.

Madariaga, R., (1976), Dynamics of an expanding circular fault, *Bull. Seismol. Soc. Am.*, **66**(3), 639-666.

Marone, C. and B. Kilgore, (1993), Scaling of the critical slip distance for seismic faulting with shear strain in fault zones, *Nature*, **362**, 618-621.

Marone, C., M. Cocco, E. Richardson, and E. Tinti, (2009), The critical slip distance for seismic and aseismic fault zones of finite width, In: *Fault-zone Properties and Earthquake Rupture Dynamics*, edited by E. Fukuyama, *International Geophysics Series*, **94**, 135-162, Elsevier.

Mase, C. W. and L. Smith, (1987), Effects of frictional heating on the thermal, hydrologic, and mechanical response of a fault, *J. Geophys. Res.*, **92**(B7), 6249-6272.

Matsu'ura, M., H. Kataoka, and B. Shibazaki, (1992), Slip-dependent friction law and nucleation processes in earthquake rupture, *Tectonophys.*, **211**(1-4), 135-148.

Mikumo, T. and T. Miyatake, (1978), Dynamical rupture process on a three-dimensional fault with non-uniform frictions and near-field seismic waves, *Geophys. J. Roy. Astr. Soc.*, **54**, 417-438.

Mogi, K., (1992), Opening address, *Tectonophys.*, **211**(1-4), xi-xii.

Oglesby, D. D., R. J. Archuleta, and S. Nielsen, (1998), Earthquakes on dipping faults: The effects of broken symmetry, *Science*, **280**, 1055-1059.

Oglesby, D. D., R. J. Archuleta, and S. B. Nielsen, (2000), The three-dimensional dynamics of dipping faults, *Bull. Seismol. Soc. Am.*, **90**(3), 616-628.

Ohtani, T., H. Tanaka, K. Fujimoto, T. Higuchi, N. Tomida, and H. Ito, (2001), Internal structure of the Nojima Fault zone from the Hirabayashi GSJ drill core, *Island Arc*, **10**(3-4), 392-400.

Olsen, K. B., R. Madariaga, and R. J. Archuleta, (1997), Three-dimensional dynamic simulation of the 1992 Landers earthquake, *Science*, **278**, 834-838.

Olson, E. L. and R. M. Allen, (2005), The deterministic nature of earthquake rupture, *Nature*, **438**, 212-215, doi:10.1038/nature04214.

Omori, F., (1910), Note on the great Mino-Owari earthquake of Oct. 28th, 1891, *Publ. Earthq. Inv. Comm.*, **4**, 13-24.

Palmer, A. C. and J. R. Rice, (1973), The growth of slip surfaces in the progressive failure of over-consolidated clay, *Proc. Roy. Soc. London* Ser. A, **332**, 527-548.

Park, J. O., T. Tsuru, S. Kodaira, P. R. Cummins, and Y. Kaneda, (2002), Splay fault branching along the Nankai subduction zone, *Science*, **297**, 1157-1160.

Poliakov, A. N. B., R. Dmowska, and J. R. Rice, (2002), Dynamic shear rupture interactions with fault bends and off-axis secondary faulting, *J. Geophys. Res.*, **107**(B11), 2295, doi:10.1029/2001JB000572.

Ranero, C. R., I. Grevemeyer, H. Sahling, U. Barckhausen, C. Hensen, K. Wallmann, W. Weinrebe, P. Vannucchi, R. von Huene, and K. McIntosh, (2008), Hydrogeological system of erosional convergent margins and its influence on tectonics and interplate seismogenesis, *Geochem. Geophys. Geosyst.*, **9**, Q03S04, doi:10.1029/2007GC001679.

Rettenmaier, D., V. Giurgea, H. Hötzl, and A. Förster, (2004), The AIG10 drilling project (Aigion, Greece): Interpretation of the litho-log in the context of regional geology and tectonics, *C. R. Geosciences*, **336**(4-5), 415-423.

Reid, H. F., (1910), The mechanism of the earthquake, In: *The California Earthquake of April 18, 1906, Report of the State Earthquake Investigation Commission*, **2**, 1-192, Carnegie Institutions, Washington, D.C.

Rice, J. R., (2006), Heating and weakening of faults during earthquake slip, *J. Geophys. Res.*, **111**, B05311, doi:10.1029/2005JB004006.

Rosakis, A. J., O. Samudrala, and D. Coker, (1999), Cracks faster than the shear wave speed, *Science*, **284**, 1337-1340.

Rydelek, P. and S. Horiuchi, (2006), Is earthquake rupture deterministic?, *Nature*, **442**, E5-E6, doi:10.1038/nature04963.

Scholz, C. H., (1977), A physical interpretation of the Haicheng earthquake prediction, *Nature*, **267**, 121-124, doi:10.1038/267121a0.

Shibazaki, B. and M. Matsu'ura, (1998), Transition process from nucleation to high-speed rupture propagation: Scaling from stick-slip experiments to natural earthquakes, *Geophys. J. Int.*, **132**, 14-30.

Shigematsu, N. and H. Yamagishi, (2002), Quartz microstructures and deformation conditions in the Hatagawa shear zone, north-eastern Japan, *Island Arc*, **11**, 45-60.

Sibson, R. H., (1975), Generation of pseudotachylyte by ancient seismic faulting, *Geophys. J. Roy. Astr. Soc.*, **43**(3), 775-794.

Sibson, R. H., (1994), An assessment of field evidence for "Byerlee" friction, *Pure Appl. Geophys.*, **142**, 645-662.

Sieh, K., L. Jones, E. Hauksson, K. Hudnut, D. Eberthart-Philips, T. Heaton, S. Hough, H. Kanamori, A. Lilje, S. Lindvall, S. F. McGill, J. Mori, C. Rubin, J. A. Spotila, J. Stock, H. K. Thio, J. Treiman, B. Wernicke, J. Zachariasen, (1993), Near-field investigations of the Landers earthquake sequence, April to July 1992, *Science*, **260**, 171-176.

Solum, J. G., S. H. Hickman, D. A. Lockner, D. E. Moore, B. A. van der Pluijm, A. M. Schleicher, and J. P. Evans, (2006), Mineralogical characterization of protolith and fault rocks from the SAFOD Main Hole, *Geophys. Res. Lett.*, **33**, L21314, doi:10.1029/2006GL027285.

Sulem J., (2007), Stress orientation evaluated from strain localisation analysis in Aigion Fault, *Tectonophys.*, **442**, 3-13.

Tada, T., (2009), Boundary integral equation method for earthquake rupture dynamics, In: *Fault-zone Properties and Earthquake Rupture Dynamics*, edited by E. Fukuyama, *International Geophysics Series*, **94**, 217-267, Elsevier.

Tanaka, H., K. Fujimoto, T. Ohtani, and H. Ito, (2001a), Structural and chemical characterization of shear zones in the freshly activated Nojima fault, Awaji Island, southwest Japan, *J. Geophys. Res.*, **106**(B5), 8789-8810.

Tanaka, H., S. Hinoki, K. Kosaka, A. Lin, K. Takemura, A. Murata, and T. Miyata, (2001b), Deformation mechanisms and fluid behavior in a shallow, brittle fault zone during coseismic and interseismic periods: Results from drill core penetrating the Nojima Fault, Japan, *Island Arc*, **10**(3-4), 381-391.

Tanaka, H., K. Omura, T. Matsuda, R. Ikeda, K. Kobayashi, M. Murakami, and K. Shimada, (2007), Architectural evolution of the Nojima fault and identification of the activated slip layer by Kobe earthquake, *J. Geophys. Res.*, **112**, B07304, doi:10.1029/2005JB003977.

Tinti, E., P. Spudich, and M. Cocco, (2005), Earthquake fracture energy inferred from kinematic rupture models on extended faults, *J. Geophys. Res.*, **110**, B12303. doi:10.1029/ 2005JB003644.

Tinti, E., P. Spudich, and M. Cocco, (2008), Correction to "Earthquake fracture energy inferred from kinematic rupture models on extended faults," *J. Geophys. Res.*, **113**, B07301, doi:10.1029/ 2008JB005829.

Tobin, H. J. and M. Kinoshita, (2006), NanTroSEIZE: The IODP Nankai Trough seismogenic zone experiment, *Sci. Drilling*, **2**, 23-27.

Tsukahara, H., R. Ikeda, and K. Yamamoto, (2001), In situ stress measurements in a borehole close to the Nojima Fault, *Island Arc*, **10**(3-4), 261-265.

Vannucchi, P., F. Remitti, J. Phipps-Morgan, and G. Bettelli, (2009), Aseismic-seismic transition and fluid regime along subduction plate boundaries and a fossil example from the Northern Apennines of Italy, In: *Fault-zone Properties and Earthquake Rupture Dynamics*, edited by E. Fukuyama, *International Geophysics Series*, **94**, 37-68, Elsevier.

Wald, D. J. and T. H. Heaton, (1994), Spatial and temporal distribution of slip for the 1992 Landers, California, earthquake, *Bull. Seismol. Soc. Am.*, **84**(3), 668-691.

Wu, H.-Y., K.-F. Ma, M. Zoback, N. Boness, H. Ito, J.-H. Hung, and S. Hickman, (2007), Stress orientations of Taiwan Chelungpu-Fault Drilling Project (TCDP) hole-A as observed from geophysical logs, *Geophys. Res. Lett.*, **34**, L01303, doi:10.1029/2006GL028050.

Yamashita, T., (2000), Generation of microcracks by dynamic shear rupture and its effects on rupture growth and elastic wave radiation, *Geophys. J. Int.*, **143**, 395-406.

Yamashita, T., (2009), Rupture dynamics on bi-material faults and non-linear off-fault damage, In: *Fault-zone Properties and Earthquake Rupture Dynamics*, edited by E. Fukuyama, *International Geophysics Series*, **94**, 187-216, Elsevier.

Yamashita, F., E. Fukuyama, and K. Omura, (2004), Estimation of fault strength: Reconstruction of stress before the 1995 Kobe earthquake, *Science*, **306**, 261-263.

Yeats, R. S., K. Sieh, and C. R. Allen, (1997), *The Geology of Earthquakes*, Oxford University Press, New York.

# Geometry and Slip Distribution of Coseismic Surface Ruptures Produced by the 2001 Kunlun, Northern Tibet, Earthquake

**Aiming Lin**

*Graduate School of Science and Technology, Shizuoka University, Shizuoka, Japan*

This chapter presents a case study of deformation structures of the coseismic surface ruptures produced by the 2001 $M_w$ 7.8 Kunlun earthquake along the strike-slip Kunlun Fault in northern Tibet. Field investigations, seismic data, and interpretations of high-resolution remote sensing images reveal the geometric and deformational characteristics and strike-slip offset distribution of the 2001 Kunlun coseismic surface ruptures. The coseismic surface ruptures are mainly composed of numerous distinct shear faults, echelon extensional cracks, and mole tracks, which are concentrated on a deformation zone ranging from a few meters up to ~500 m in width and extending for 450 km along the western segment of preexisting active Kunlun Fault striking E-W to WNW-ESE. Structural features and focal mechanism solutions reveal that the earthquake had a nearly pure strike-slip mechanism. The 2001 coseismic strike-slip offsets measured in field and from 1-m-resolution IKONOS and 0.61-m-resolution QuickBird images range from 2 m up to ~16 m, generally 3 to 8 m. The spatial distribution pattern of strike-slip offsets observed immediately after the 2001 earthquake in the field coincides well with that obtained from high-resolution remote sensing images and seismic inversion results. Both the rupture length and maximum displacement are the largest among the coseismic surface rupture zones ever reported in intracontinental earthquakes. Field evidence, seismic inversion results, and interpretations of high-resolution remote sensing images demonstrate that the geometric characteristics and the temporal and spatial displacement distributions of the coseismic surface ruptures are constrained by the preexisting geological structures of the active strike-slip Kunlun Fault.

## 1. INTRODUCTION

The magnitude $M_w$ 7.8 Kunlun earthquake occurred on November 14, 2001, in the Kunlun mountain area of northern Tibet (Figure 1), and it produced extensive surface rupturing over a distance of ~450 km along the preexisting strike-slip Kunlun Fault (Figure 1). The coseismic surface ruptures are dominated by strike-slip faulting with a maximum left-lateral displacement up to ~16 m distributed in a wide zone of ~500 m (*Lin et al.*, 2002, 2003; *Lin and Nishikawa*, 2007). Both the rupture length and maximum strike-slip displacement are the largest among the coseismic surface rupture zones produced by intracontinental earthquakes ever reported. The large rupture length and displacement are of great importance as key parameters in estimating maximum earthquake magnitude and seismic moment and assessing individual seismic fault shear zone structure for seismogenic zone. The 2001 Kunlun earthquake, therefore, provides us with an unusual opportunity to understand the rupture mechanism and process along a large continental strike-slip fault.

The deformation characteristics and strike-slip offset distribution of the coseismic surface rupture zone revealed by field investigations, seismic data, and satellite images have been described in a series of previous papers (*China Seismological Bureau*, 2002; *Lin et al.*, 2002, 2003, 2004, 2006; *Xu et al.*, 2002, 2006; *Fu and Lin*, 2003; *Klinger et al.*, 2005; *Lasserre et al.*, 2005; *Li et al.*, 2005; *Lin and Nishikawa*, 2007; *Lin and Guo*, 2008b, 2009). On the basis of the satellite remote sensing data including SPOT and IKONOS images and field investigations, both the eastern and western end extensions (N90°15'~N94°50') of the 2001 coseismic surface rupture were detected immediately after the earthquake, and the total rupturing length is first estimated to be up to ~400 km, that is a straight distance between the eastern and western ends of the surface rupture zone measured from 1:1,000,000 topographic map (*Lin et al.*, 2002, 2003; *Fu and Lin*, 2003; *Lin and Nishikawa*, 2007). The length of surface rupture trace measured directly from 1:100,000 topographic map and ETM images, and high-resolution IKONOS and Quick-Bird images is up to ~450 km (*Lin and Nishikawa*, 2007). Meanwhile, it is also reported that large coseismic strike-slip offsets of >10 m and maximum up to ~16 m in several locations along the surface rupture zone in the field (*Lin et al.*, 2002, 2003) and by interpretations of 1-m-resolution IKONOS and 0.61-m-resolution QuickBird images (*Lin and Nishikawa*, 2007). However, some previous studies showed that the total rupture length is uncertain, which probably ranges from 350 to 426 km (*Xu et al.*, 2002) and no offset amount over 10 m (*Xu et al.*, 2002, 2006; *King et al.*, 2005; *Klinger et al.*, 2005). The primary field works carried out immediately after the earthquake were reported by *Lin et al.* (2002, 2003, 2004), *China Seismological Bureau* (2002), and *Xu et al.* (2002), which present the coseismic offsets measured at only 30 to 40 locations along the 450-km-long surface rupture zone. Although there are some differences in the field-measured offset amounts

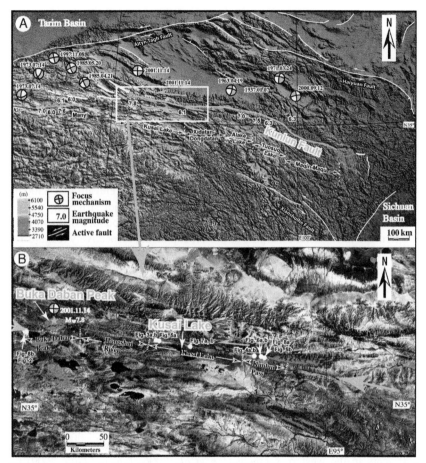

**FIGURE 1**    (a) Shuttle Radar Topography Mission (SRTM, 90-m resolution) color-shaded relief map showing the topographic features and distribution of the major active faults in the northern Tibet plateau and (b) Landsat TM image showing the 2001 coseismic surface rupture (indicated by red arrows) in the study area. (a) The Kunlun Fault is divided into six main segments, from the west to east: Kusai Lake, Xidatan-Dongdatan, Alake Lake, Tuosuo Lake, and Maqing-Maqu segments, respectively (*Seismological Bureau of Qinghai Province and Institute of Crustal Dynamics*, 1999). Red lines indicate the Kunlun Fault. Star indicates the epicenter of the 2001 earthquake. Beach balls show the focal mechanisms of large historic earthquakes of M ≥ 6.0 from 1904 to 2001 (*International Seismological Centre*, 2001). (b) The 2001 Kunlun surface rupture (indicated by red arrows) is divided into four main segments, from the west to east: Buka Daban Peak, Hongshui River, Kusai Lake, and Kunlun Pass segment, respectively (*Lin et al.*, 2003). Solid white cycles show the locations of figures (Figures 2 to 8). (Part a modified from Lin and Guo, 2008b.) **(See Color Plate 1.)**

reported, all these primary works demonstrate that the temporal and spatial displacement distributions and the rupture process are restricted by the preexisting geological structures of the Kunlun Fault.

This chapter focuses on the topic related with the geometric and deformational characteristics and strike-slip distribution of coseismic surface ruptures produced by the 2001 $M_w$ 7.8 Kunlun earthquake, which occurred on the pre-existing strike-slip Kunlun Fault, by reviewing the previous studies.

## 2. TECTONIC SETTING

The Kunlun Fault, striking E-W to WNW-ESE for >1200 km, is located in the Kunlun mountains of north Tibet at an average elevation > 4500 m (Figure 1). This fault is considered one of the three major strike-slip active faults, including the Altyn Tagh and Haiyuan faults that bounded on the northwestern and northeastern margins of the Tibet plateau, respectively (Figure 1), to accommodate the eastward extrusion of Tibet by the ongoing collision between the Indian and Eurasian plate (e.g., *Molnar and Tapponnier*, 1975; *Tapponnier et al.*, 2001). The western segment of the fault (90°E~95°E), which triggered the 2001 $M_w$ 7.8 Kunlun earthquake (*Lin et al.* 2002, 2003; *Xu et al.*, 2002), cuts through south-sloping alluvial fans and bajadas and is visible on satellite images as a straight lineament trending E-W to WNW-ESE (Figure 1b). A Pleistocene-Holocene average slip rate is estimated to be 10 to 20 mm/yr for this segment (*Kidd and Molnar*, 1988; *van der Woerd et al.*, 2002; *Lin et al.*, 2006), which decreases gradually to a slip-rate of 2 to 4 mm/yr at the eastern end (~102°E) of the fault (*Kirby et al.*, 2007; *Lin and Guo*, 2008b) with an average gradient of 1 mm/100 km from the west to the east along the Kunlun Fault (*Lin and Guo*, 2008b). Paleoseismic studies show that the average recurrence interval of large earthquakes ranges from 300 to 400 years in the western segment (*Lin et al.*, 2006; *Lin and Guo*, 2009) to 1000 years in the eastern segment of the fault (*Lin and Guo*, 2008b). These studies indicate that the slip rate is nonuniform along the strike-slip Kunlun Fault and that the characteristic recurrence interval of large earthquakes becomes longer gradually from the western to the eastern segments along the fault.

More than 10 moderate-large earthquakes of M > 6.0 occurred on the Kunlun Fault, which generally show a main focal mechanism of left-lateral strike-slip (Figure 1a), but there are no historical records of large earthquakes before the 2001 earthquake on the 450-km-long western segment of the Kunlun Fault. There are three large historic earthquakes of M ≥ 7.0 before the 2001 $M_w$ 7.8 Kunlun earthquake occurred in the past century on the central-eastern segments along the Kunlun Fault, producing 40 to 180-km-long surface ruptures (Figure 1a). One is the 1937 M 7.5 earthquake that ruptured the 150- to 180-km-long Tuosuo Lake segment (*Jia et al.*, 1988; *Guo et al.*, 2007). The second is the 1963 M 7.0 earthquake that produced a 40-km-long surface rupture zone along the Alake Lake segment (*Guo et al.*, 2007).

The third one is the 1997 Manyi $M_w$ 7.6 earthquake that produced a 120-km-long surface rupture zone (*Xu*, 2000). These historic earthquakes indicate that the Kunlun Fault is currently active as a large earthquake source fault.

## 3. DEFORMATION CHARACTERISTICS OF THE 2001 COSEISMIC SURFACE RUPTURE

The 2001 $M_w$ 7.8 Kunlun earthquake ruptured the western segment of the Kunlun Fault and produced a 450-km-long coseismic surface rupture zone, which is composed of numerous shear faults, extensional cracks, and mole track structures (*Lin et al.*, 2002, 2003, 2004). Field investigations and seismic data reveal that this earthquake had a nearly pure strike-slip mechanism and that the temporal and spatial strike-slip displacement distribution pattern and rupture process are controlled by the preexisting geological structures of the Kunlun Fault (*Lin et al.*, 2002, 2003, 2004, 2006; *Lin and Nishikawa*, 2007). The details are summarized next.

### 3.1. Geometric Distribution and Deformational Structure

The 2001 Kunlun surface rupture zone occurred on the western segment of the Kunlun Fault, extending from the east of Kunlun Pass (~95°E) throughout the northern side of the Kusai Lake and terminating on the west of Buka Daban Peak along the western segment of the Kunlun Fault (Figure 1). The western termination is located in the western side of the ice-capped Buka Daban Peak around ~90.5°E (Figure 1), where the ruptures splay into several extensional cracks and mole track structures on alluvial fans, which can be recognized from high-resolution remote sensing images and by field investigations (Figure 2) (*Lin et al.*, 2003). The eastern termination of the surface rupture zone is characterized by some extensional cracks along which no distinct displacement can be observed. The surface rupture zone is divided into four segments: Buka Daban Peak, Hongshui River, Kusai, and Kunlun Pass segments (Figure 1), which mostly followed the preexisting active Kunlun Fault trace along which many fault outcrops are observed (Figure 3). This indicates that the deformational characteristics of surface ruptures are restricted by the preexisting geological structures. The coseismic surface ruptures are mainly composed of distinct shear faults, echelon extensional cracks showing a right-hand stepping geometric pattern, and mole tracks along the preexisting Kunlun Fault, which are distributed in a zone with a width ranging from a few meters to 10 km, but generally from 5 to 50 m (*Lin et al.*, 2003).

The distinct shear faults offset sinistrally the south-sloping alluvial fans, gullies and river channels, present-day ice-borne deposits, glaciers, and moraines (*Lin et al.*, 2002, 2003) that extend for a few tens to a few hundreds of meters in an individual fault trace (Figures 4 through 6). The fault scarps, up to 2 to 3 m high, immediately adjacent to the fault trace alternate along a

**FIGURE 2**   Representative example of extensional cracks that occurred in the western termination area of the 2001 surface rupture zone. See Figure 1b for detail location. **(See Color Plate 2.)**

strike from south facing to north facing, showing a scissoring structural pattern (Figure 4). Such scissoring structure of fault scarp is often observed across the south-sloping alluvial fans along the surface rupture zone and is best explained as a result of purely left-lateral strike-slip offset of the convex alluvial fans (*Lin et al.*, 2003). The distinct shear fault planes generally strike N75°W to E-W and dip 75° to 90°S parallel to the general trend of the surface rupture zone, on which the horizontal slickenside striations and fault steps were well observed at several locations. The striations are marked by parallel lineations in the grained materials with groves several millimeters to 4 to 5 cm wide and fault steps, showing a plunging angle of 0° to 10° ESE (Figure 3b). These coseismic deformational characteristics of surface markers and striations on the shear fault planes reveal that the earthquake had a mostly pure left-lateral strike-slip mechanism at surface.

The extensional cracks concentrated in the surface rupture zone generally show a right-hand stepping echelon pattern and are oblique to the general trend of the rupture zone with 40° to 70° angle, indicating a left-lateral shear sense (Figure 6) (*Lin et al.*, 2003; *Lin and Nishikawa*, 2007). Locally, the surface ruptures are composed of several parallel-oriented extensional crack zones (Figure 7a). In the central segment near 93.5°E where the displacements of 8 to 16 m were observed, the surface ruptures are distributed in a wide zone up to 10 km with an average interval of at least one crack every 10 m on the bajada in the southern side of the rupture zone (Figure 7b) (*Lin et al.*, 2003; *Lin and Nishikawa*, 2007). This

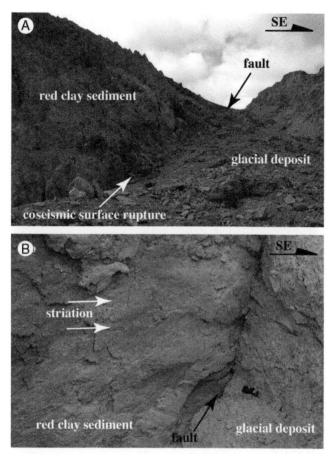

**FIGURE 3**    Fault outcrop of the 2001 rupture zone. (a) Red clay sediments (left site) are bounded with the glacial deposits composed of unconsolidated sand-gravel (right site) by fault along which the 2001 coseismic ruptures occurred. (b) Numerous striations are developed on the fault plane shown in (a), which indicate a major horizontal movement. See Figure 1b for detail location. **(See Color Plate 3.)**

wide distribution of cracks may be caused by the strong ground motion on the unconsolidated alluvial deposits (*Lin et al.*, 2003).

The mole tracks are widely developed on the alluvial fans, bajada, and frozen stream channels along the Kunlun rupture zone (*Lin et al.*, 2004). The sites that contain mole track structures record between 5 to 8 m of sinistral displacement, with no observable vertical offset. The mole tracks are divided into two types by the structural characteristics, angular-ridge, and bulge patterns, which form a linked row with the distinct shear faults (Figure 8a) (*Lin et al.*, 2004). The mole tracks are typically 50 cm to 1 m in height,

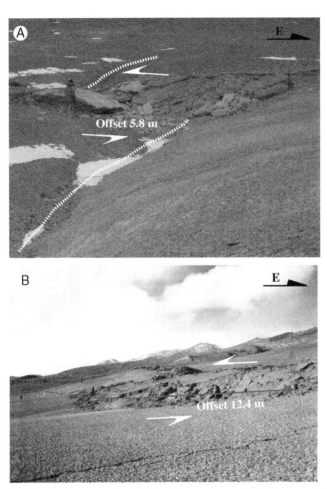

**FIGURE 4**  Typical outcrops of the 2001 surface ruptures. (a) Gully was offset 5.8 m.
(b) Alluvial fan was offset about 12.4 m. The south-facing and north-facing fault scarps are
developed on these alluvial fans, which are interpreted as a scissoring structure caused by pure
strike-slip faulting (*Lin et al.*, 2003). See Figure 1b for detail location. (**See Color Plate 4.**)

1 to 10 m in width, and 2 to 15 m in length. The trends of the mole track are
generally oblique to the general trend of the rupture zone by angles of 30° to
60°. The angular-ridge type of mole track structures resemble an empty trian-
gle-angular shaped frame in cross section and consist of opposing ramped-up
rigid plates (Figure 8b). Along the axis of the mole track, the topsoil layer was
generally separated from the underlying coarse-grained deposits and locally
offset by faulting. This type of mole track generally occurred in alluvial
deposits of interbedded clay, fine-grained sand, sandy-gravel, and gravel
and also within current stream channels. The topsoil layer and river water

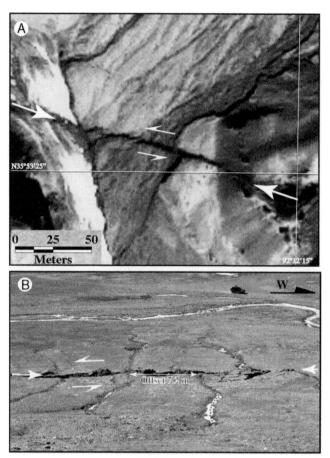

**FIGURE 5**  1-m-resolution IKONOS image (a) and field photograph (b) showing typical topographic features of strike-slip offset on gullies along the 2001 coseismic surface ruptures (indicated by large white arrows). The river channel (white part in the left side of the image) is displaced by 2.3 m during the 2001 earthquake (a). See Figure 1b for detail location. **(See Color Plate 5.)**

frozen as rigid plates of 30 to 50 cm thick covering the loose (unfrozen) alluvial deposits were ruptured and pushed up as that of bedded rocks. Bulge-shaped mole tracks were developed within unconsolidated–weak consolidated and nonfrozen coarse-grained alluvial deposits (*Lin et al.*, 2004). The bulges are conspicuous as small mountain ranges in the bajada and are easy to recognize in the field. Similar bulge-shaped mole track structures have also been described in many earthquakes such as the 1891 M 8.0 Nobi earthquake, Japan (*Koto*, 1893), the 1967 M 7.1 Mudurnu Valley earthquake, west Anatolia, Turkey (*Ambraseys and Nátopek*, 1969), and the 1973 M 7.6 Luhuo earthquake, China (*Deng et al.*, 1986). Such bulge-shaped ridges contain numerous

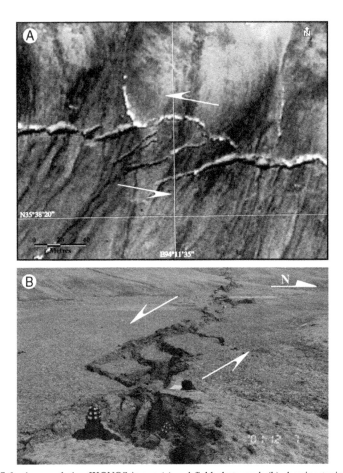

**FIGURE 6**  1-m-resolution IKONOS image (a) and field photograph (b) showing typical right-stepping echelon ruptures. The geometric pattern indicates a left-lateral movement of the 2001 rupture zone. See Figure 1b for detail location. **(See Color Plate 6.)**

extensional cracks within the hinge region of the bulges and are formed within unconsolidated or weak-consolidated deposits.

## 3.2. Coseismic Slip Distribution

### 3.2.1. Measurement Method of Strike-Slip Offset

The left-lateral displacements are marked by some distinguishable surface markers, such as present-day glaciers, moraine, roads, stream channels, gullies, and terraces, which are generally perpendicular to the surface ruptures (*Lin et al.*, 2002, 2003, 2004). Such straight-linear surface markers are used to measure the strike-slip offsets in the field (*Lin et al.*, 2002, 2003) and from

**FIGURE 7**   Photographs showing the parallel surface rupture zones (indicated by white arrows) (a) and extensional ruptures distributed in a wide zone (b). See Figure 1b for detail location. **(See Color Plate 7.)**

1-m-resolution IKONOS and 0.61-m-resolution QuickBird images (*Lin and Nishikawa*, 2007). The surface ruptures, along which strike-slip offsets were measured, are generally composed of a lot of echelon shear faults and cracks as stated previously (Figures 6 and 7). Therefore, total offset at one location within the surface rupture zone is measured in a cross section perpendicular to the fault trace and is calculated by adding all offsets measured along each individual shear fault or crack (Figure 9). As shown in Figure 9c, the total strike-slip offset (D) is obtained by adding d1, d2, and d3 measured along individual F1, F2, and F3 shear faults, respectively.

**FIGURE 8** Photographs showing the mole track structures developed along the 2001 coseismic rupture zone. (a) Mole tracks are linked along the rupture zone. (b) Typical angular-type of mole track developed on the unconsolidated alluvial deposits. See Figure 1b for detail location. **(See Color Plate 8.)**

## 3.2.2. Field Observations

The displacements observed in the field vary from several to a few tens of centimeters at both the eastern and western termination areas and from several to ~16 m at the central segment of the rupture zone (Figure 10) (*Lin et al.,* 2002, 2003). The obvious displacements observed at both the eastern and western end sites of the surface rupture zone indicate that the surface rupture length is up to 450 km along the Kunlun Fault (Figure 1). Both the rupture length and maximum displacement presented here are the largest in amount within the coseismic surface rupture zones produced by intracontinental earthquakes ever reported worldwide. The largest strike-slip offset of coseismic rupture zones produced by intracontinental earthquakes that has been reported in literature to date is 14.8 m caused by the 1931 M 8.0 Fuyun (Northwest China) earthquake, which produced a 180-km-long surface rupture zone

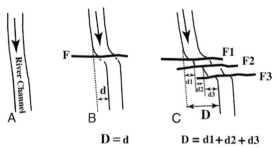

$$D = d \qquad\qquad D = d1 + d2 + d3$$

F1-F3: coseismic surface ripture    d: displacement

**FIGURE 9**    Sketch map showing the strike-slip offset of linear surface marker of river channel along an individual coseismic shear fault and surface rupture zone composed of several shear faults. (a) A linear surface marker of the river channel before the earthquake. (b) Strike-slip offset (D) measured along an individual coseismic shear fault (F). (c) Total strike-slip offset (D) measured along a coseismic surface rupture zone composed of shear faults F1 through F3, which is calculated by D = d1 + d2 + d3. (Modified from *Lin and Nishikawa*, 2007.)

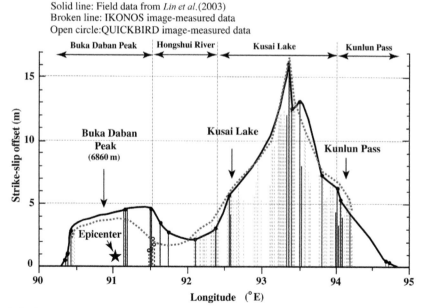

**FIGURE 10**    Diagram showing the details of left-lateral strike-slip offsets measured from the images (indicated by broken lines) and field data (indicated by solid lines) along the Kunlun rupture zone. The curves are connected by using the largest amounts in its related area. There is a similar distribution pattern along the surface rupture zone between the image- and field-measured strike-slip offsets. The largest offsets are distributed in the Kusai Lake segment. (Modified from *Lin et al.*, 2003, and *Lin and Nishikawa*, 2007.)

(*Ding*, 1985; *Lin and Lin*, 1998). The longest coseismic rupture zone is reported in the 1905 M 8.2 Bulnay (Mongolia) earthquake, which is 375 km with a maximum horizontal offset of 11 m (*Yeats et al.*, 1997). Some normal throw components on the extensional cracks are observed in the field and remote sensing images, which are explained to be caused by slip partitioning along the coseismic surface rupture (*Klinger et al.*, 2005).

### 3.2.3. Analysis of High-Resolution Remote Sensing Images

These strike-slip offsets were measured in the field immediately after the 2001 earthquake and were obtained from only 30 to 40 locations along the 450-km-long surface rupture zone (*Lin et al.*, 2002, 2003; *Xu et al.*, 2002) due to bad weather and high-mountain area where it is difficult to access and to work for a long time. For compensating the lack of field data and confirming the field-measured offset amounts, we also measured the strike-slip offsets using 1-m-resolution IKONOS and 0.61-m-resolution QuickBird images, which were acquired immediately after the earthquake (*Lin and Nishikawa*, 2007). The coseismic offsets were measured at 147 locations from the images covering a 300-km-long rupture zone (*Lin and Nishikawa*, 2007). The offsets were measured by using large-scale images of up to 1:1000, where the longitude and latitude were determined and shown on the images by using the ER-Mapper software, which is commercially developed for processing the digital remote sensing data. Each offset along an individual surface rupture was measured three times, and their average value is used as a measured offset with a standard deviation (*Lin and Nishikawa*, 2007).

The strike-slip offsets measured from the images vary from ~1 up to 16.7 m, and typically they are between 3 and 8 m, which is consistent with the field-measured offsets (Figure 10) (*Lin and Nishikawa*, 2007). There is also a coincidence in the distribution pattern of offsets along the surface rupture between the field- and the image-measured data (Figure 10). The image-measured data confirm the distribution pattern presumed by the related field data in some segments where there are no field-measured data. The largest offset up to 16.7 m was measured from one location along six individual shear faults distributed in a rupture zone of up to ~500 m wide (*Lin and Nishikawa*, 2007), which is comparable with the maximum offset (16.3 m) observed in the field (*Lin et al.*, 2002, 2003). Large offsets of >8 to 10 m measured at more than 10 locations along individual shear faults, which were also observed immediately after the earthquake in the field (Figure 10) (*Lin et al.*, 2002, 2003). These IKONOS imagery data were acquired in March and October 2002 immediately after the 2001 earthquake, and therefore the coseismic offsets could be preserved with little erosion and could be measured accurately from the images. This is a main reason that there is a coincidence between the field-observed and image-measured offsets. The high-resolution remote sensing images show that the ground deformational features of the

2001 coseismic surface rupture occurred in a wide area, which make it possible to measure the coseismic strike-slip offsets across a wide zone up to ~500 m and to observe the ground deformational features along a long surface rupture zone up to ~450 km in the remote and high mountain region.

### 3.2.4. Seismic Inversion Results

For understanding the seismic rupture process and mechanism of the 2001 Kunlun earthquake, we first used the broadband seismograph records at teleseismic distances retrieved from the Data Management Center, Incorporated Research Institutions for Seismology (IRIS-DMC) through Spyder (online near real-time data service) for inversion analysis of the Kunlun earthquake (*Lin et al.*, 2003). A unit fault cell of 20 km × 10 km was used in the inversion analysis. The inversion results show that focal mechanisms are nearly pure strike-slip and remain almost unchanged during the source process with a focal depth of 17 km and that the strike, dip, and rake of the total source obtained from focal mechanisms are 94°, 88°, and 0°, respectively. These results are consistent with those observed from the field investigations as stated earlier.

The source rupture process and the spatial distribution of fault displacements obtained from the inversion are shown in Figure 11. The inversion results show that (1) the left-lateral strike-slip motion is predominant over all of the fault plane, (2) a rupture started near the hypocenter area bilaterally and rapidly extended to east in a unilateral manner, and (3) the total length amounts to 400 km: 370 km to the east and 30 km to the west from the relocated epicenter (36.01°N, 91.10°E). The spatial resolution is about 10 km for the horizontal direction and 5 km for the dip direction, corresponding to the time resolution of 1 s (*Lin et al.*, 2003). Taking this resolution into account,

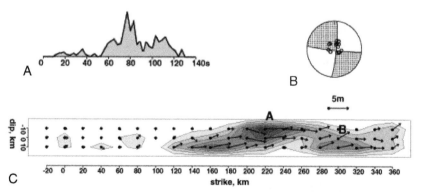

**FIGURE 11**  Results of waveform inversion. (a) Moment rate function. (b) Focal mechanism of the total source. (c) Spatial distribution of fault slip. The origin is located at the epicenter of 36.01°N, 91.10°E, and at a depth of 17 km. The vertical axis indicates the dip direction of the plane. Arrows indicate the slip vectors of the hanging wall relative to the opposite block. (Modified from *Lin et al.*, 2003.)

the pattern of moment-release distribution is very well consistent with that observed from the field. There are two distinct, large slip areas (asperities), A and B (Figure 11c), each having a spatial extent of about 60 km (a span of three grid points) and a time duration of 20 s. The largest displacement was 5.8 m on asperity A, about 220 km east from the relocated epicenter. It should be kept in mind that this displacement is an average value over a unit fault cell of 20 km × 10 km. *In situ* dislocation observed in the field may be locally twice or more than this average value. In fact, near the eastern side of the surface seismic fault, a large displacement up to ∼16 m was observed in the field. The site of this largest displacement should correspond to asperity A (Figure 11). Adjacent to it, asperity B occupies a deeper region. This is the reason why a large displacement was not observed in the field. The largest displacement averaged over a unit fault cell is 5.8 m.

## 4. DISCUSSION

### 4.1. Relationship between the Coseismic Surface Rupture and Preexisting Fault

Deformational characteristics of coseismic surface ruptures not only reflect the surface morphology of earthquake source fault but also show the structural characteristics at depth and preexisting tectonic environment (*King*, 1986; *Lin and Uda*, 1996; *Lin et al.*, 2001, 2003). The focal mechanism solutions show that the fault plane strikes nearly E-W with a steep angle (85°S), which coincides with that observed from the surface ruptures in the field, indicating that the Kunlun surface rupture zone is almost followed on the preexisting active Kunlun Fault trace (*Lin et al.*, 2003). The inversion results also have a good agreement with the field observations in rupture length, slip vector, and spatial displacement distribution pattern along the rupture zone as stated previously. These results clearly show that the deformational features of the surface ruptures reflect the rupturing characteristics on the preexisting Kunlun Fault at depth.

Topographically, the Kusai segment forms a boundary between the northern ice-capped high mountains and southern Kusai Lake (basin) where the south-sloping alluvial fans and bajadas are widely developed (*Lin et al.*, 2003). This boundary is also a geological boundary between the pre-Quaternary basement and Quaternary alluvial deposits that are widely distributed in the southern side of the fault trace (*Lin et al.*, 2003). This wide distribution of surface rupture is also observed in the western end segment, the Buka Daban Peak segments where the surface ruptures are distributed in a wide zone of 2 km on alluvial deposits, which seems to be related to a normal throw component inferred from the teleseismic and InSAR data modeling (*Antolik et al.*, 2004; *Ozacar and Beck*, 2004; *Lasserre et al.*, 2005; *Tocheport et al.*, 2006).

In contrast, the Hongshui River segment appears along the nearly E-W trending intermontane long valley where the basement outcrops, along which the displacements are generally smaller than that distributed in other segments (Figure 10). In the Kunlun pass segment, the eastern end segment, the surface ruptures are concentrated in a narrow zone of 2 to 5 m on the south-sloping mountain ridge where the basement rocks are exposed at the surface. Based on these geological structures, tectonic landform features, and the displacement distributions, it is suggested that the Hongshui River segment is a fault asperity that was not completely broken during this earthquake (*Lin et al.*, 2003).

As documented previously, the fact that the ruptures are mostly followed on the preexisting fault trace shows clearly that the deformational characteristics and displacement distributions of the 2001 Kunlun coseismic surface ruptures are restricted by the preexisting geological structures of the active strike-slip Kunlun Fault. This relationship between the coseismic surface ruptures and preexisting geological structures has also been reported in many other large earthquakes that have occurred in recent decades worldwide (e.g., *Unruh et al.*, 1994; *Lin and Uda*, 1996; *Lin et al.*, 2001; *Lin and Guo*, 2008a).

## 4.2. Coseismic Strike-Slip Displacement

The field investigations, analytical results of high-resolution remote sensing images, and seismic inversion results show that the displacement distribution is highly heterogeneous along the 450-km-long surface rupture zone and on the fault plane in depth. As documented earlier, large left-lateral displacements of >8 to 10 m have been detected at many locations in the Kusai segment in the field (*Lin et al.*, 2002, 2003) and from high-resolution IKONOS images (*Lin and Nishikawa*, 2007). However, some groups reported that there are no offsets of >8 m occurred along the coseismic surface rupture zone based on the preliminary field observation (*Xu et al.*, 2002, 2006) and analysis of local IKONOS images (*Klinger et al.*, 2005). What caused this difference? For the field measurement in the case of echelon surface ruptures, the displacements shown in this study were measured along all ruptures across the whole rupture zone of up to ~500m in width. However, it is possible that the offsets reported by other groups were measured along only one or two individual ruptures, not all the echelon ruptures across the whole rupture zone. *Xu et al.* (2002, 2006) argued that it is difficult to measure the lateral displacement at the sites where the coseismic surface ruptures show a complicated echelon pattern. However, the coseismic surface ruptures generally are composed of numerous echelon fractures showing a right-hand stepping geometric pattern indicating a left-lateral displacement sense, which reflects the deformational characteristics of the earthquake source fault. The strike-slip offsets are generally partitioned on these echelon fractures at an individual site. The displacement amounts shown in Figure 10, which are obtained by adding the offset amounts measured on each fracture across the profiles perpendicular to the rupture zone, are therefore considered the characteristic offsets related to the slip amounts that occurred on the surface in the measured sites.

In another way, in the case of narrow rupture zone of <10 m wide, the field-measured results reported by previous studies showed a good agreement (e.g., *Lin et al.*, 2002, 2003; *Xu et al.*, 2002). For instance, the maximum offset of 7.6 m, measured by *Xu et al.* (2002) at the location (93°19.37E, 35°46.053) where the rupture zone is concentrated in a narrow zone of <8 m, is similar to that observed in the field by *Lin et al.* (2002, 2003) and from the 1-m-resolution IKONOS images (Loc. No. 62 shown in *Lin and Nishikawa*, 2007). Therefore, the field measurement method is considered as a main account for the difference between our results and other groups' results and for the underestimation of the total lateral offsets in the field. Although *Xu et al.* (2002) reported that there are no offsets over 10 m on the basis of their preliminary field observations, they did not measure the offsets at the sites where the large offsets of >9 to 10 m were observed by *Lin et al.* (2002, 2003) and *Lin and Nishikawa* (2007). *Xu et al.* (2006) argued, without additional field-measured data, that the large displacements of >8 to 10 m reported in *Lin et al.* (2002, 2003) could be the cumulative displacements that were probably produced by multiple previous earthquakes. The large displacements of >8 m were measured immediately in December 2001 after the 2001 earthquake by using the current river channels and the lowest terrace risers as displacement markers where the fresh rupture structures were observed, as shown in many field photographs that have been published (*Lin et al.*, 2002, 2003; *Lin and Nishikawa*, 2007). Therefore, the field observations carried out immediately after the earthquake confirmed that these larger offsets of >8 m were produced only by the 2001 earthquake but not the accumulative amounts caused by previous earthquakes.

Furthermore, the large displacements of >8 to 10 m have also been detected by the analytical results of IKONOS and QuickBird images (Figure 10) (*Lin and Nishikawa*, 2007). *Klinger et al.* (2005) also analyzed the local IKONOS images near the Kusai Lake but showed that there are no offsets of >10 m. In their study, *Klinger et al.* (2005) did not measure the lateral offsets in a single location along an individual surface rupture using a single straight surface offset marker, as shown in Figure 10, but measured the area offset by restoring a local area topography along a wide zone of the surface ruptures to fit a pre-earthquake surface topography (Figure 4 in *Klinger et al.*, 2005). This method is generally used for obtaining an average offset amount of local or large area for active faults, but it cannot be used for measuring *in-situ* offset of coseismic surface rupture in centimeters to meters-scale. The lateral offset measured in this way is not an *in-situ* offset that occurred on a single rupture but may be a local area averaged offset amount. It is well known that the displacement distribution is not uniform along the Kunlun surface rupture even in a short distance of <10 m (*Lin et al.*, 2002, 2003, 2004; *Xu et al.*, 2002). For example, the offset is larger in the center than that observed in the ends of one single rupture of <10 m. Therefore, we cannot compare the offsets reported by *Klinger et al.* (2005) with the displacement amounts observed on individual ruptures in the field and from IKONOS images because of different measurement methods.

In another way, the seismic inversion results for the 2001 Kunlun earthquake show that the maximum averaged slip amount varies 5.8 (*Lin et al.*, 2003) to 7.5 m (*Antolik et al.*, 2004). Fault modeling results using InSAR data also show that the maximum slip amount is 8 to 10 m, with ~1 m of EEW dipping normal throw in the westernmost segment (*Lasserre et al.*, 2005). As stated earlier, the field data show that numerous extensional cracks developed on the south-facing slope at the westernmost segment (*Lin et al.*, 2002, 2003; *Fu and Lin*, 2003). The extensional cracks generally form locally in a dilatational area, in which a normal throw component might occur on the extensional cracks. This may explain the normal component detected by the InSAR data. It should also be kept in mind that these displacements obtained from the seismic and InSAR data inversion are the averaged slip amounts obtained from a gridded area that is 20 km wide × 10 km deep (*Lin et al.*, 2003) and 5 km wide × 5 km deep (*Antolik et al.*, 2004; *Lasserre et al.*, 2005), respectively. As stated previously, the lateral offsets are heterogeneous even along a short rupture of <10 m; therefore, it is difficult to compare directly the averaged value of offsets within an area of 25 to 200 km$^2$ of the fault plane with the field-measured offsets obtained from the surface ruptures at meter-scale sites.

Summarily, although it is difficult to compare the coseismic displacements obtained from different measurement methods, the field observations, seismic inversion results, and analytical results of high-resolution remote sensing images show that there is a good coincidence in the strike-slip displacement distribution pattern along the 450-km-long coseismic surface rupture zone. This fact also demonstrates that the deformation characteristics of coseismic surface rupture reflect the subsurface rupture structures of the seismic fault zone.

## 5. CONCLUSIONS

On the basis of the field observations, analytical results of high-resolution remote sensing images, and seismic inversion results of the coseismic rupture produced by the 2001 $M_w$ 7.8 Kunlun earthquake along the Kunlun Fault in northern Tibet, the following conclusions are obtained:

1. The coseismic surface ruptures produced by the 2001 $M_w$ 7.8 Kunlun earthquake occurred along the western segment of the Kunlun Fault for nearly 450 km, which are composed of numerous shear faults, echelon extensional cracks, and mole track structures concentrated in a zone of a few meters up to 500 m.

2. The coseismic strike-slip offsets of the surface ruptures produced by the 2001 $M_w$ 7.8 Kunlun earthquake range from 2 to ~16 m, generally 3 to 8 m.

3. The field observations, analytical results of high-resolution remote sensing images, and seismic inversion results show that there is a good coincidence in rupture length, slip vector, and spatial displacement distribution pattern along the 2001 Kunlun surface rupture zone.

4. The deformational characteristics and displacement distributions of the coseismic surface ruptures are controlled by the preexisting geological structures of the strike-slip Kunlun Fault.

## ACKNOWLEDGMENTS

I would like to express my sincere thanks to the Seismological Bureau of Xinjiang Province and Dr. G. Dang for providing the field photographs. Thanks are also due to graduate students J. Guo and M. Nishikawa for their assistance in remote sensing image processing. This work was supported by the Science Project (Project No. 18340158 for A. Lin) of the Ministry of Education, Culture, Sports, Science and Technology of Japan.

## REFERENCES

Ambraseys, N. N. and A. Natopek, (1969), The Mudurnu Valley, West Annatonia, Turkey, earthquake of 22 July 1967, *Bull. Seismol. Soc. Am.*, **59**, 521-589.

Antolik, M., R. E. Abercrombie, and G. Ekstrom, (2004), The 14 November 2001 Kokoxili (Kunlunshan), Tibet earthquake: Rupture transfer through a large extensional step-over, *Bull. Seismol. Soc. Am.*, **94**, 1173-1194.

China Seismological Bureau, (2002), *Album of the Kunlun Pass $M_s$ 8.1 Earthquake*, Beijing, Seismological Press, 105 pp. (in Chinese).

Deng, Q., D. Wu, P. Zhang, and S. Chen, (1986), Structure and deformational characteristics of strike-slip fault zones, *Pure Appl. Geophys.*, **124**, 203-223.

Ding, G., (1985), Some problems related in earthquake study, In: *The Fuyun Earthquake Fault Zone in Xinjiang, China*, edited by Ding, G., Seismological Press, 187-201 (in Chinese).

Fu, B. and A. Lin, (2003), Spatial distribution of the surface rupture zone associated with the 2001 $M_s$ 8.1 Central Kunlun earthquake, northern Tibet, revealed by satellite remote sensing data, *Int. J. Remote Sensing*, **24**, 2191-2198.

Guo, J., A. Lin, G. Su, and J. Zheng, (2007), The 1937 M7.5 earthquake and paleoseismicity along the Tuosuo lake segment of the Kunlun fault, northern Tibet, China, *Bull. Seismol. Soc. Am.*, **97**, 474-496.

International Seismological Centre, (2001) On-line Bulletin, http://www.isc.ac.uk/search/ (last accessed June 2007).

Jia, Y., H. Dai, and X. Su, (1988), Tuosuo Lake earthquake fault in Qinghai province, In: *Research on Earthquake Faults in China*, edited by Xinjiang Seismological Bureau, Xinjiang People Press, Urumqi, China, 66-71 (in Chinese).

Kidd, W. S. F. and P. Molnar, (1988), Quaternary and active faulting observed on the 1985 Academia-Sinica Royal-Society Geotraverse of Tibet, *Phil. Trans. Roy. Soc., London*, **327**, 337-363.

King, G., (1986), Speculations on the geometry of the initiation and termination process of earthquake rupture and relation to morphology and geological structures, *Pure Appl. Geophys.*, **124**, 567-685.

King, G., Y. Klinger, D. Bowman, and P. Tapponnier, (2005), Slip-partitioned surface breaks for the $M_w$ 7.8 2001 Kokoxili earthquake, China, *Bull. Seismol. Soc. Am.*, **95**, 731-738.

Kirby, E., N. Harkins, E. Wang, X. Shi, C. Fan, and D. Burbank, (2007), Slip rate gradients along the eastern Kunlun fault, *Tectonics*, **26**, TC2010, doi:10.1029/2006TC002033.

Klinger, Y., X. Xu, P. Tapponnier, J. van der Woerd, C. Lasserre, and G. King, (2005), High-resolution satellite imagery mapping of the surface rupture and slip distribution of the $M_w$ 7.8, November 2001 Kokoxili earthquake, Kunlun Fault, northern Tibet, China, *Bull. Seismol. Soc. Am.*, **95**, 1970-1987.

Koto, B., (1893), On the cause of the great earthquake in central Japan, 1891, *Journal of College Sciences, Imperial University*, **5**(4), 296-353.

Lasserre, C., G. Peltzer, F. Crampe, Y. Klinger, J. van der Woerd, and P. Tapponnier, (2005), Coseismic deformation of the 2001 Mw=7.8 Kokoxili earthquake in Tibet, measured by synthetic aperture radar interferometry, *J. Geophys. Res.*, **110**, B12408, doi:10.1029/2004JB003500.

Li, H., J. van der Woerd, P. Tapponnier, Y. Klinger, X. Qi, J. Yang, and Y. Zhu, (2005), Slip rate on the Kunlun fault at Hongshui Gou, and recurrence time of great events comparable to the 14/11/2001, Mw~7.9 Kokoxili earthquake, *Earth Planet. Sci. Lett.*, **237**, 285-299.

Lin, A. and J. Guo, (2008a), Co-seismic surface ruptures produced by the 2005 Pakistan $M_w$7.6 earthquake in the Muzaffarabad area, revealed by QuickBird imagery data, *Int. J. Remote Sensing*, **29**, 235-246.

Lin, A. and J. Guo, (2008b), Non-uniform slip rate and millennial recurrence interval of large earthquakes along the eastern segment of the Kunkun Fault, northern Tibet, *Bull. Seismol. Soc. Am.*, **98**, 2866-2878, doi:10.1785/0120070193.

Lin, A. and J. Guo, (2009), Prehistoric large earthquake inferred from liquefaction along the Kusai Hu segment of the Kunlun fault, northern Tibet, *J. Geol. Soc. London*, in press.

Lin, A. and S. Lin, (1998), Tree damage and the surface displacement: 1931 M8 Fuyun earthquake, *J. Geol.*, **106**, 749-755.

Lin, A. and M. Nishikawa, (2007), Coseismic lateral offsets of surface rupture zone produced by the 2001 Mw 7.8 Kunlun earthquake, Tibet from the IKONOS and QuickBird imagery, *Int. J. Remote Sensing*, **28**, 2431-2445, doi:10.1080/01431160600647233.

Lin, A. and S. Uda, (1996), Morphological characteristics of the earthquake surface ruptures occurred on Awaji Island, associated with the 1995 Southern Hyogo Prefecture Earthquake, *Island Arc*, **5**, 1-15.

Lin, A., B. Fu, J. Guo, Q. Zeng, G. Dang, W. He, and Y. Zhao, (2002), Coseismic strike-slip and rupture length produced by the 2001 Ms 8.1 Central Kunlun earthquake, *Science*, **296**, 2015-2017.

Lin, A., J. Guo, and B. Fu, (2004) Co-seismic mole-track structures produced by the 2001 Ms 8.1 Central Kunlun earthquake, China, *J. Struct. Geol.*, **26**, 1511-1519.

Lin, A., J. Guo, Y. Awata, and K. Kano, (2006), Average slip rate and recurrence interval of large magnitude earthquakes on the western segment of the Kunlun fault, northern Tibet, *Bull. Seismol. Soc. Am.*, **96**, 1597-1611.

Lin, A., M. Kikuchi, and B. Fu, (2003), Rupture segmentation and process of the 2002 Mw 7.8 Central Kunlun earthquake, China, *Bull. Seismol. Soc. Am.*, **93**, 2477-2492.

Lin, A., T. Ouchi, A. Chen, and T. Maruyama, (2001), Co-seismic displacements, folding and shortening structures along the Chelungpu surface rupture zone occurred during the 1999 Chi-Chi (Taiwan) earthquake, *Tectonophysics*, **330**, 225-244.

Molnar, P. and P. Tapponnier, (1975), Cenozoic tectonics of Asia: Effects of a continental collision, *Science*, **189**, 419-426.

Ozacar, A. and S. L. Beck, (2004), The 2002 Denali Fault and 2001 Kunlun Fault earthquakes:
Complex rupture processes of two large strike-slip events, *Bull. Seismol. Soc. Am.*, **94**,
S278-S292.

Seismological Bureau of Qinghai Province and Institute of Crustal Dynamics, China Seismological Bureau (SBQP and ICD-CSB), (1999), *Eastern Kunlun Active Fault Zone*, Seismological
Press, Beijing, 186 pp. (in Chinese with English abstract).

Tapponnier, P., Z. Xu, F. Roger, B. Meyer, N. Arnaud, G. Wittinger, and J. Yang, (2001), Oblique
stepwise rise and growth of the Tibet plateau, *Science*, **294**, 1671-1677.

Tocheport, A., L. Rivera, and J. van der Woerd, (2006), A study of the Kokoxili, November 14,
2001, earthquake: History and geometry of the rupture from teleseismic data and field observations, *Bull. Seismol. Soc. Am.*, **96**, 1729-1741.

Unruh, J. R., W. R. Lettis, and J. M. Sowors, (1994), Kinematic interpretation of the 1992 Landers
earthquake, *Bull, Seismol. Soc. Am.*, **84**, 537-546.

van der Woerd, J., P. Tapponnier, J. Frederick, F. J. Ryerson, A. S. Meriaux, B. Meyer, Y. Gau-
demer, R. C. Finkel, M. W. Caffee, G. Zhao, and Z. Xu, (2002), Uniform postglacial slip-rate
along the central 600 km of the Kunlun Fault (Tibet), from $^{26}$Al, $^{10}$Be, and $^{14}$C dating of riser
offsets, and climatic origin of the regional morphology, *Geophys. J. Int.*, **148**, 356-388.

Xu, X., (2000), Seismic investigation on the Mani, northern Tibet earthquake. In: *China Earthquake Yearbook*, Seismological Press, Beijing, China, 327-329 (in Chinese).

Xu, X., W. Chen, W. Ma, G. Yu, and Y. Chen, (2002), Surface ruptures of the Kunlunshan earthquake (Ms 8.1), northern Tibetan Plateau, China, *Seismol. Res. Lett.*, **73**, 884-892.

Xu, X., G. Yu, Y. Klinger, P. Tapponnier, and J. van der Woerd, (2006), Reevaluation of surface
rupture parameters and faulting segmentation of the 2001 Kunlunshan earthquake (Mw7.8),
northern Tibetan Plateau, China, *J. Geophys. Res.*, **111**, B05316, doi:10.1029/20004JB003488.

Yeats, R. S., K. Sieh, and C. R. Allen, (1997), *The Geology of Earthquakes*, Oxford University
Press, New York, 568 pp.

# Aseismic-Seismic Transition and Fluid Regime along Subduction Plate Boundaries and a Fossil Example from the Northern Apennines of Italy

Paola Vannucchi
*Earth Science Department, University of Florence, Florence, Italy*

Francesca Remitti
*Earth Science Department, University of Modena and Reggio Emilia, Italy*

Jason Phipps-Morgan
*Earth Science Department, Cornell University, Ithaca, New York, U.S.A.*

Giuseppe Bettelli
*Earth Science Department, University of Modena and Reggio Emilia, Italy*

This chapter reviews observations and theories for the aseismic-seismic transition in the megathrust between the incoming and overriding plates at a subduction zone. The temperature of the aseismic-seismic transition appears to be quite similar at erosive and accretionary margins, despite large differences between them in the lithology of the seismogenic subduction channel that composes the "megathrust" plate interface. This fact, and the recent laboratory demonstration that both smectite and illite are velocity-strengthening in creep, suggests that the oft-postulated change in mechanical behavior of the megathrust due to a smectite-illite clay mineral transformation at ~150°C is not the cause of the onset in seismogenesis at these temperature conditions within the subduction channel. Field observations from fossil megathrust zones suggest that a temperature-dependent change in the availability of *in situ* fluid is likely to play a key role in the onset of seismogenesis. Perhaps the causal link is to the smectite-illite transformation and other metamorphic dewatering reactions that liberate water at ~150°C, under conditions where these reactions are an important local source of hydrous fluids. Field studies of fossil megathrusts support the hypothesis that fluids "control" seismogenesis, and indicate that there are large fluid pressure variations during the seismic cycle. In the fossil erosive megathrust system preserved in the

Apennines, two décollements are simultaneously active at the roof and base of the sub-duction channel. The uppermost (nonseismogenic) portion of the megathrust even appears to alternate between tensional and compressional modes of failure during the seismic cycle along the deeper portions of the megathrust.

## 1. INTRODUCTION

Investigation of earthquake sources is a highly active and rapidly developing field. Monitoring active faults both at the Earth's surface and at depth with boreholes seismometers gives the most information on earthquake generation mechanisms (*Ide et al.*, 1996; *Olsen et al.*, 1997; *Lee et al.*, 2002). The limi-tation of these techniques is the use of instruments located far from the seis-mic source, so that some fundamental aspects of earthquake mechanics, such as dynamic fault strength, the energy budget of an earthquake, and the heat production during seismic slip, remain almost unknown (*Kanamori and Heaton*, 2000). High-quality borehole data, though, can be integrated with real-time *in situ* measurements, as pore pressure, strain rate, stress, and with geodetic data such as GPS (Global Positioning System) and InSAR (Interfero-metric Synthetic Aperture Radar) at the surface. In addition, many experimen-tal studies on fault rock materials and friction instabilities have been conducted in the laboratory and have contributed much to the knowledge of the mechanics of seismic faulting (*Marone*, 1998; *Beeler et al.*, 2000; *Hirose and Shimamoto*, 2005).

Large earthquakes nucleate at ∼10 to 15 km depths in the Earth's crust (*Scholz*, 2002); depths still unreachable by drilling. Nonetheless, areas of smaller earthquake nucleation or shallower portions of seismically active faults have been drilled. Since the early 2000s, for example, active faults have been the target of several drilling projects, such as the high-angle reverse Nojima Fault responsible for the 1995 Kobe earthquake in Japan, which was penetrated at 426 m depth (*Boullier et al.*, 2001); the Chelungpu Fault, a low-angle reverse fault activated in 1999 during the Chichi earthquake in Taiwan, which was encountered at 1110 m depth (*Ma et al.*, 2006); and the right-lateral San Andreas Fault in the United States, which was drilled at ∼3000 m depth in an area of $M_w$ ∼2 earthquake nucleation (San Andreas Obser-vatory at Depth [SAFOD], www.earthscope.org/index.php/es_obs/safod_obs). Using a different approach, seismogenic faults have been the focus of the ongoing experiment in the Tau Tona deep gold mine in South Africa where semicontrolled earthquakes have been induced at 3600 m depth (*Heesakkers et al.*, 2005).

But the most ambitious of these projects are plans to drill the seismically active part of subduction megathrusts. Modern subduction zones are subma-rine, and drilling onshore is not feasible because the seismogenic zone below land is located too deep. To find a seismogenic zone as shallow as 5 to 7 km, drilling has to occur offshore, and, until now, only two locations meet this technological requirement: the Nankai Trough in southwest Japan, a classic accretionary margin (*Park et al.*, 2002), and the Central America Trench offshore southeast Costa Rica (Figure 1), an erosive end member margin (*Ranero et al.*, 2008). These two companion drilling projects have been designed for the Integrated Ocean Drilling Project (IODP), with the Nankai Trough Seismogenic Zone Experiment (NantroSEIZE), which started drilling operations in Fall 2007.

Many recent studies have concentrated on the geological characteristics of such faults. Of great interest to structural geologists is the fact that planned deep drilling will collect fault rocks from seismically active depths of the subduction plate boundary. Nonetheless, drilling has some big limitations, such as (1) deep drilling involves large costs, (2) the hypocenters of the most destructive earthquakes are located too deep to drill, and (3) drilling recovers a string of material and does not provide complete 3D information about the rupture network and fault rock distribution.

To overcome some of the above-mentioned limitations, geological studies have focused on former seismogenic faults from fossil subduction plate boundaries that are now exhumed at the Earth's surface (*Ikesawa et al.*, 2003; *Ujiie et al.*, 2007b). During the past few years, interactions between geologists, geophysicists, and rock mechanics specialists have been particularly frequent. In this chapter we have a dual goal: (1) to give an up-to-date overview of the most important information coming from experimental, seismological, and remote sensing studies conducted on subduction margins as useful background for a structural geologist working on exhumed subduction plate-boundary and (2) to describe some of the new results that have been learned from the study of a fossil seismogenic subduction zone. Of course,

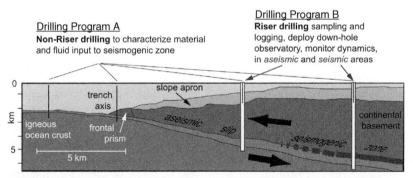

**FIGURE 1**    Example of the proposed drilling plan of the Costa Rica Seismogenesis Project (CRISP). Program A consists of shallow holes to be drilled with the conventional, nonriser drill technique. During Program A, data will be collected as rock composition and fluid regime, that will be used during Program B preparation. Program B plan is to drill the deep holes with riser technique, capable of controlling pressure changes. (**See Color Plate 9.**)

the choice of subjects and their organization is based on our experience and may differ from the experience of other scientists working in the field.

## 2. DEFORMATION AND SEISMOGENESIS AT ACCRETIONARY AND EROSIVE SUBDUCTION MARGINS

Different models have been proposed to explain the deformation processes related to subduction zones. Since the early 1970s, seismic-reflection profiles, deep-sea drilling data, and field studies on fossil subduction complexes led to the development of the accretionary prism model, where thrust sheets were imbricated at the frontal part of subduction margins (*Karig*, 1974; *Seely et al.*, 1974; *Karig and Sharman*, 1975) (Figure 2a). Later studies have shown how prisms were uplifted and thickened without shortening, suggesting that accretion is mostly realized through the underplating of duplexes (*Leggett et al.*, 1985; *Platt et al.*, 1985; *Silver et al.*, 1985; *Sample and Fisher*, 1986)

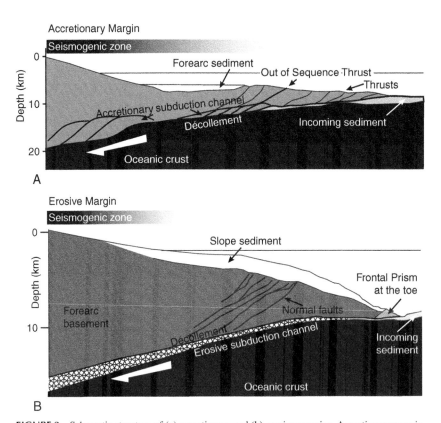

**FIGURE 2** Schematic structure of (a) accretionary and (b) erosive margins. Accretionary margin structure is based on seismic sections of Nankai Trough (*Park et al.*, 2002), and erosive margin is based on Central America (*Ranero and von Huene*, 2000).

(Figure 2a). Accretionary prisms developed by a normal thrust sequence can be thickened and internally deformed by out-of-sequence thrusts (OOST) cutting the interior of the wedge (*Morley*, 1988). OOST also maintain the critical taper of the wedge (*Chapple*, 1978; *Davis et al.*, 1983).

In margins dominated by accretion, the plate boundary thrust cuts through the trench sediment pile entering the subduction system (Figure 2a). These subduction thrust systems can be viewed as conveyor belts that transport trench sediments through an ever-increasing pressure-temperature-strain regime, and it is this material that represents the input into the seismogenic zone of the plate boundary (*Ruff*, 1989).

Further exploration at convergent margins revealed that uplift and contraction are not the only tectonic responses to subduction; instead, a large number of margins show deep scars on the trench slope, with progressive subsidence and upper plate thinning that suggests extensive removal of the base of the upper plate through a process called subduction erosion (*von Huene and Lallemand*, 1990; *Vannucchi et al.*, 2001; *von Huene et al.*, 2004a) (Figure 2b). Subduction erosion is a conspicuous process that accompanies plate convergence in ~60% of the total global active margin system (*Clift and Vannucchi*, 2004). Margins characterized by subduction erosion are fundamentally different from those characterized by accretion. Subduction erosion, in fact, involves the upward migration of the plate interface such that upper plate material is entrained in the subduction process, with subsequent movement of that material to deeper levels (Figure 2b).

Subduction accretion and erosion control the flux of material into the subduction zone along the plate boundary, and in this way they can influence the physical parameters affecting seismicity. At accretionary margins, a thick underthrusting sediment pile, which persists along the boundary down to depths characteristic of the seismogenic zone (>5 km), is systematically and progressively subject to thermally controlled diagenetic and metamorphic changes. These changes are thought to be associated with decreasing fluid overpressure at coseismic depths, which leads to increasing effective stress and ultimately controls seismic behavior (*Moore and Saffer*, 2001). The same authors, though, do not exclude the possibility that earthquakes could also be triggered by transient pulses of high fluid pressure, showing that the role of fluid pressure is still far from understood.

At erosional margins, in contrast, subduction megathrusts incorporate material from the upper plate, composed of either a fossil accretionary prism or crystalline/ophiolitic crust. It has been calculated that the Costa Rica erosive plate boundary offshore Nicoya Peninsula receives 20 times more material from the upper plate than from incoming sediments (*Vannucchi et al.*, 2003). The main physical problem of upward migration of the plate interface associated with seismogenesis is how the upper plate material weakens to promote the upward development of shear zones. To explain tectonic erosion, much of the scientific literature has inferred strong coupling and high friction

(*Jarrard*, 1986), but paradoxically, current investigations indicate low friction, at least in the first 10 to 15 km of the subduction zone (*von Huene and Ranero*, 2003; *von Huene et al.*, 2004b). It has been shown that fluids may play a key role in affecting the frictional behavior of the plate boundary (*Sibson*, 1974; *Segall and Rice*, 1995) and in weakening the upper plate (*von Huene et al.*, 2004a). High fluid pressure in erosive margins seems to characterize the frontal portion of the system to the up-dip limit of seismogenesis and it decreases at further depth along the plate boundary (*Ranero et al.*, 2008). These fluids may promote overpressuring and hydrofracturing, producing brecciation of the upper plate (*von Huene et al.*, 2004a). *Vannucchi and Leoni* (2007) reported that this kind of brecciation is associated with the shallow part of the erosive décollement drilled by ODP Leg 205.

The concept of a subduction-driven macroscopic flow along the plate boundary has been qualitatively described through the subduction channel model (*Cloos and Shreve*, 1988a, 1988b, 1996). The subduction channel model envisions that subduction-driven deformation is largely concentrated in a relatively thin layer of relatively rapid shearing, poorly consolidated sediment dragged down by the descending plate beneath the overriding plate (Figure 2). The flow of the material along the subduction channel can reverse and upwell from depths of ultra-high-pressure (UHP) rock formation (if metamorphism leads to a large enough decrease in viscosity and density). This model is particularly useful for field geologists because it is able not only to describe the process of prism accretion and subduction erosion, but also sediment subduction, mélange formation, and the potential for exhumation of UHP rocks.

## 3. SEISMOGENIC ZONE: DEFINITION

The stick-slip model for earthquake generation has become widely accepted since the late 1960s (*Brace and Byerlee*, 1966; *Dieterich*, 1978; *Scholz*, 1990). In this model, an earthquake is a dynamic instability created by a decrease in frictional resistance during slip along an existing fault surface. This accelerating slip behavior, also known as "velocity weakening," requires that the frictional resistance of the fault surface decreases faster during sliding than the force applied by the wall rocks. Velocity weakening contrasts with a slip event in which frictional resistance increases with sliding relative to the force applied by the wall rock—this latter behavior, known as "velocity strengthening," would result in creep. To allow the stress build up necessary for the stick-slip behavior, the fault needs to be essentially locked between earthquakes (*Scholz*, 1990).

The geological record shows evidence for both a large amount of fluid along the fault and fluid pressure often over hydrostatic conditions (*Parry*, 1998; *Miller et al.*, 2003; *Sibson and Rowland*, 2003). These observations imply that in earthquake mechanics, a key point is to understand the fluid pressure changes during the seismic cycle and its influence on both rupture nucleation and recurrence time through cyclical variations as suggested in the fault-valve model (*Sibson*, 1992; *Sibson and Rowland*, 2003).

Subduction megathrusts are characterized by creep and aseismic slip along their outermost shallow portions; only deeper does the thrust surface develop seismic behavior. The aseismic-seismic transition—known as the up-dip limit of seismogenesis—is commonly found at about 10 km depth, even though its depth can vary from ~5 to ~20 km depending, primarily, on the dip of the subducting slab and on the thermal regime (*Hyndman et al.*, 1997; *Oleskevich et al.*, 1999; *Pacheco et al.*, 1993). At least one of the main factors controlling the transition from stable slip to seismic slip along the plate boundary appears to be thermally controlled processes (*Moore and Saffer*, 2001; *Obana et al.*, 2003) that act in a system characterized by pervasive upward fluid flow (*Carson and Screaton*, 1998; *Moore and Vrolijk*, 1992; *Peacock*, 1990; *Ranero et al.*, 2008). Thermal modeling of modern subduction zones suggests that the up-dip limit of the seismogenic thrust interface corresponds to temperatures in the range of 100° to 150°C (*Hyndman et al.*, 1997; *Moore and Saffer*, 2001; *Obana et al.*, 2003; *Ranero et al.*, 2008). The seismogenic portion of the thrust interface generally extends down-dip to around $40 \pm 5$ km, ranging from 25 (Mexico) to 70 km (Java) (*Pacheco et al.*, 1993). The down-dip boundary of seismogenesis is characterized either by inferred temperatures reaching ~350°C or by the intersection of the subducting plate with serpentinized upper mantle below the Moho of the overriding plate (*Hyndman et al.*, 1997; *Oleskevich et al.*, 1999; *Moore and Saffer*, 2001). However, seismic observations in the Mariana Trench suggest that the seismogenic plate interface can sometimes extend well below the Moho—to depths of 50 km in places where the Moho is 20 km deep (*Wiens et al.*, 2007). The down-dip limit of modern seismogenic plate boundaries remains off limits to drilling, so our knowledge will have to continue to be based on geophysical data, laboratory modeling, and analyses of exhumed fossil faults. This latter opportunity has not yet received appropriate attention.

The up-dip limit of seismogenesis is an ongoing controversy, not only regarding the mechanisms for plate boundary locking, and possible interactions with fluid flow (*Ranero et al.*, 2008), but also because the actual meaning (let alone location!) of the limit can be differently interpreted according to the technique of investigation. Maybe the most intuitive way to define the up-dip limit of seismogenesis is by the actual presence of earthquakes (Figure 3). Thrust earthquakes have been recorded both teleseismically (*Bilek and Lay*, 1999) and, more accurately, with high-resolution ocean bottom seismometers (OBS) and hydrophones (OBH) able to record $M_w$ 2 earthquakes at 6000 m depths (*DeShon et al.*, 2003). Obana *et al.* (2003) were able to correlate the first occurrence of intraplate seismicity in Nankai Trough to inferred T ~150°C.

In accretionary margins, an important contribution to seismogenesis can be associated with OOST or megasplay faults branching from the subduction megathrust up to the forearc (*Wang and Hu*, 2006) (Figure 2a). In the Nankai

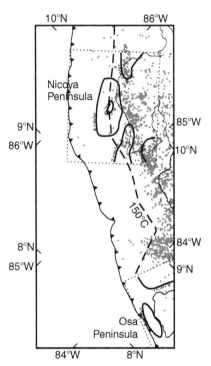

**FIGURE 3** Compilation map of the Central America trench projected along the strike of the trench offshore Nicoya Peninsula. The barbed line represents the trench. The 150°C temperature at the plate boundary is shown as isotherm (dashed line) calculated from heat flow probes and depth to the Bottom Simulating Reflector (*Ranero et al.*, 2008). Gray-filled circles are earthquakes from *Newman et al.* (2002), *Husen et al.* (2003), and *DeShon et al.* (2003). Within dotted areas, calculation of the plate boundary locking calculated in cm/yr (*Norabuena et al.*, 2004).

accretionary prism, tsunami and seismic waveform inversions indicate that the megasplay fault (imaged by *Park et al.*, 2002) is geographically associated with the up-dip termination of coseismic slip (~1 m) of the 1944 Tonankai event (*Tanioka and Satake*, 2001; *Baba et al.*, 2002; *Kikuchi et al.*, 2003). Moreover megasplay faults seem to be preferential pathways for fluid migration as inferred by submersible observations that indicate the presence of cold seeps around sea-floor scarps of the splay fault (*Ashi et al.*, 2001).

At Nankai, slip on the active megasplay fault appears to be an important mechanism that accommodates strain due to relative plate motion (*Park et al.*, 2002). Because of the small angle between the megasplay and the plate interface (~5°), once rupture begins to propagate on the megasplay, the stress around the rupture is released and it no longer drives rupture on the décollement. A mechanical approach seems to confirm that coseismic rupture would follow the megasplay and continue to the seafloor, assuming that the principal precompression applied to the plate interface is horizontal (*Kame et al.*, 2003).

Present studies focus on the partitioning of strain between the décollement and the megasplay system, and the nature and mechanisms of fault slip as a function of depth and time on the megasplay.

Nonetheless, earthquakes, especially great earthquakes, require a recharge interval between seismic slip events. During this time, the fault is thought to be mostly locked. In fact, some subduction zones, or portions of them, are characterized by a lack of microseismicity (*Obana et al.*, 2001), which is consistent with GPS measurements indicating coupling between the underthrusting and overthrusting plates (*Miyazaki and Heki*, 2001). Usually the portions of coupled plate boundary do not coincide with the areas of microseismic activity. As *Norabuena et al.* (2004) showed, for Central America, for example, the area of complete coupling offshore Costa Rica is trenchward of the microseismically active portion of the plate boundary (Figure 3).

Other problems when dealing with the seismogenic zone of subduction megathrusts is the patchiness of locked slip during the interseismic period (*Norabuena et al.*, 2004) and the distribution of shallow seismic events along the thrust interface (*Bilek and Lay*, 1999, 2002). Scientists have hypothesized that patchiness may be related to slip distribution during great earthquakes and to the factors controlling strain accumulation and rupture processes in small ($M < 6$) and large ($M > 8$) earthquakes (http://marginsreview.nsf-margins.org/SPSEIZE.html). Subduction earthquakes are characterized by a large range in the size of the rupture area reaching up to $\sim$1000 km along strike and $\sim$100 km wide—correspondent to $M > 9$. The region of larger slip during an earthquake is defined as an asperity (*Lay et al.*, 1982). The reason why earthquake magnitude can vary so much in subduction zone and why slip concentrates in particular areas (i.e., the physical nature of asperities) is still a matter of ongoing discussion within the scientific community. Asperities have been interpreted not only as areas of high friction able to accumulate large amounts of locked slip (*Cloos*, 1992) but also as areas of lower than average dynamic friction, therefore able to slip more during an earthquake (*Cochard and Madariaga*, 1994). Although different earthquakes recurring on the same fault can rupture in different areas (*Shimazaki and Nakata*, 1980; *Ruff*, 1992), it has been observed that often earthquakes tend to reuse the same asperity (*Vidale et al.*, 1994; *Nadeau and McEvilly*, 1997; *Waldhauser and Ellsworth*, 2002; *Igarashi et al.*, 2003). In this latter case, they should be controlled by specific fault properties, and asperities should have a structural meaning.

Within the stick-slip conceptual model, the onset of fault locking is a fundamental component of the aseismic to seismic transition and in general of the possibility to generate an earthquake reusing a preexisting surface: in order to be able to accumulate strain to be released in the next slip event, the slip event must be followed by a healing or restrengthening of the fault zone. In the particular case of subduction zones, they are characterized by the continuous increase of PT (pressure–temperature) conditions at progressively greater depths, which drives progressive dehydration and metamorphism along the plate boundary. PT-increasing conditions trigger incremental processes such as lithification, mineralogical phase transformations, or permeability controlled changes in fluid pressure; some of these changes have the potential to lead to locking at a specific depth along subduction thrusts (*Moore and Saffer*, 2001). Therefore, an important set of studies has concentrated on the physical properties and the deformation state of the material along the plate boundary. Somewhat surprisingly, in spite of the obvious apparent importance of material behavior and of the different material input into the seismogenic zone of accretionary and erosive margins, both types of margins have not only the same abundance and recurrence rate of great earthquakes (*Uyeda and Kanamori*, 1979), but in both the aseismic-seismic transition seems to happen close to the 150°C temperature (*Ranero et al.*, 2008). These seismogenic similarities suggest that in both margin types the processes leading to the seismic behavior of the plate boundary are not very different, and the role of the material input, although important, is not the primary controlling factor.

The geomechanical definition of seismogenesis is difficult for the following reasons:

- The great variation in seismic style exhibited by thrust interfaces: some appear to be strongly coupled, remaining locked between successive great earthquakes, whereas others are weakly coupled with varying proportions of aseismic slip, microseismicity, and moderate to large events that never occupy the full down-dip width of the interface.
- The difficulty that lab experiments can only examine smaller spatial scales and faster strain rates than typical of natural systems.
- The natural tendency to extrapolate simple rheological models of fault zones developed in quartzo-feldspatic crust to subduction settings: since the greater lithological heterogeneity, the complex thermal structure (partly stress dependent) and the massive fluid-release associated with subduction may lead to critical controlling factors that differ from those for slip in a simple quartzo-feldspatic system.

However, structural geology can make a fundamental contribution in identifying seismic structures and the seismic deformation environment. In this context, two main criteria have been adopted to geologically define a seismogenic portion of a subduction megathrust: (1) the identification of seismically produced structural features and (2) the definition of the seismogenic portion of the fossil plate boundary through its PT conditions during deformation, to verify if they were between 150° to 350°C and 5 to 40 km depth.

## 4. SLOW SLIP EVENTS AND SEISMIC TREMORS

Since the late 1990s, high-quality geodetic networks monitoring subduction margins revealed the complexity of the earthquake cycle through the detection of strain-release processes like slow slip events (SSEs), deep episodic tremors, low-frequency earthquakes (LFE), very-low-frequency earthquakes, and silent earthquakes. These events are characterized by longer durations and lower seismic energy radiation than earthquakes.

These slow events are thought to include the strain release processes necessary to account for the entire slip budget accompanying plate tectonic displacements at different timescales. SSEs, for example, propagate at timescales typical of creep from $10^{-7}$ (*Shibazaki and Shimamoto*, 2007) to close to plate motion velocities (*Schwartz and Rokosky*, 2007) in contrast with typical earthquake slip rates of m/s (*Scholz*, 2002).

The stimulating discovery is that processes like SSEs and tremors, often together, seem to closely correlate with the subduction process within several subduction margins such as Cascadia (*Rogers and Dragert*, 2003), Japan (*Hirose et al.*, 1999; *Obara*, 2002; *Obara et al.*, 2004; *Ito and Obara*, 2006; *Obara and Hirose*, 2006), Costa Rica (*Protti et al.*, 2004), and Mexico (*Lowry et al.*, 2001). Moreover, these events, mostly associated with fluid processes either involving shear failure or hydrofracturing or dehydration reactions (see *Schwartz and Rokosky*, 2007, for a full review), are causing a great interest both in the rock mechanics community and among structural geologists.

SSEs were first discovered by continuous GPS monitoring along the northern portion of the Cascadia margin (*Dragert et al.*, 2001). SSEs are not detected seismically; events last for hours to days. They differ from slow earthquakes that still radiate seismic energy although characterized by slow source processes. Typically the GPS signal reveals a contractive movement with strain accumulation suddenly reversed for hours to weeks. Modeling suggests rupture along the plate interface (*Dragert et al.*, 2001) in a patchy slip distribution consistent with a fault plane containing asperities (*Melbourne et al.*, 2005). More recently, SSEs have been found to follow and precede large earthquakes, as well as being present during interseismic periods. A major limitation to their detection is adequate coverage of the GPS network and the onland location of the instruments. SSEs in subduction zones, in fact, have been primarily reported from the down-dip edge of the seismogenic zone, between 25 and 45 km (*Schwartz and Rokosky*, 2007, and references therein), usually a portion that corresponds to onshore areas at the surface.

In the Cascadia margin and beneath Shikoku, Japan, SSEs were temporally and spatially correlated with deep episodic tremors (*Rogers and Dragert*, 2003; *Shelly et al.*, 2007). Seismic tremor, first described in association with volcanic activity (*Sassa*, 1935), is a long-lasting vibration with dominant frequencies lower than local seismicity and without a clear body wave arrival (*Schwartz and Rokosky*, 2007). In their original volcanic setting, tremors were linked to underground movements of fluids. The source of tremors has been recognized at the deeper portion of the slip area of megathrust earthquakes between 30 and 40 km depth (*Obara*, 2002; *Sagiya and Thatcher*, 1999), which is the transition region from seismic to aseismic slip with source depths focused near the Moho discontinuity (*Obara*, 2002). Their location suggests slab dehydration as a possible source mechanism. Other possibilities, suggested by their lack of harmonic character, include hydrofracturing during fluid migration (*Katsumata and Kamaya*, 2003) and shear failure on the plate interface (*Obara and Hirose*, 2006; *Shelly et al.*, 2006; *Shelly et al.*, 2007).

The association of SSEs and tremors in subduction zones, although often complicated, has favored new interpretations of slow slip associated with the action of fluids. In particular, fluid migration is suggested by the propagation of tremors with slow slip (*Obara*, 2002; *Rogers and Dragert*, 2003). Another possibility suggested by numerical modeling and SSE source location at the down-dip limit of seismogenic zone (i.e., the transition from unstable to stable sliding) is the heterogeneous stress distribution caused by frictional properties variations (*Kuroki et al.*, 2004; *Liu and Rice*, 2005; *Yoshida and Kato*, 2003).

The shallow portions of subduction zones are usually submarine and more difficult to instrument. Nevertheless, in the last years SSEs have been reported, for example, on the shallow subduction zone following the 1995 Northern Chile earthquake (*Pritchard and Simons*, 2006) or both up-dip and down-dip of the main shock asperity that break the 2005 Nias earthquake (*Kreemer et al.*, 2006). Shallow SSEs were suggested as possible processes responsible for correlated fluid flow and shallow seismic tremors recorded at the seafloor of the Costa Rica subduction margin (*Brown et al.*, 2005) and for transient fluid pressure recorded in boreholes at the Nankai Trough (*Davis et al.*, 2006). Low-frequency earthquakes and pressure transients have been detected in the Nankai accretionary prism at the up-dip portion of the Nankai subduction zone and interpreted as connected with SSEs (*Ito and Obara*, 2006). Fluids responsible for fault zone weakening seem to be a key to interpreting these events, characterizing both accretionary and erosive forearcs.

## 5. SEISMICALLY PRODUCED STRUCTURES

The relationship between fault fabric and seismic slip has been the subject of much ongoing research. The most common path has been to find field evidence that can be related to dynamic weakening processes. Melt lubrication (*Di Toro et al.*, 2006; *Sibson*, 1975; *Ujiie et al.*, 2007b), thermal pressurization (*Sibson*, 1973; *Wibberley and Shimamoto*, 2005; *Rempel and Rice*, 2006; *Rice*, 2006), acoustic fluidization (*Melosh*, 1996), normal interface vibrations generated by dynamic changes in normal stress (*Brune et al.*, 1993) that could result in the fluidization of the granular material (*Chester and Chester*, 1998; *Monzawa and Otsuki*, 2003; *Otsuki et al.*, 2003), hydrodynamic effects developed during rapid slip on saturated gouges (*Brodsky and Kanamori*, 2001; *Ma et al.*, 2003; *Ujiie*, 2005), flash melting (*Rempel*, 2006) and Joule-Thompson effects (*O'Hara*, 2005) at the asperity contacts, and silica gel lubrication (*Goldsby and Tullis*, 2002;

*Di Toro et al.*, 2004) have been, so far, the most promising mechanisms by which to link specific processes to structural features.

Tectonic pseudotachylytes (i.e., glassy veins produced by rapid quenched friction-induced melts) have been recognized for a long time as the product of seismic slip (*Sibson*, 1975). However, many questions still remain open, such as (1) their limited distribution when compared to the seismic activity on Earth (*Sibson and Toy*, 2006), (2) the effect of rock-melting on the dynamic friction (*Fialko and Khazan*, 2005), (3) the relationship between pseudotachylyte and fluids (*John and Schenk*, 2006), and (4) the uncommon occurrence of frictional melting along large faults (*Chester et al.*, 1993).

Pseudotachylytes have been found along thrusts and out-of-sequence-thrusts in the Cretaceous-Tertiary Shimanto Belt in Japan, a fossil accretionary prism inland of the modern Nankai Trough (*Ikesawa et al.*, 2003; *Okamoto et al.*, 2006; *Ujiie et al.*, 2007b), and in the Kodiak accretionary complex in Alaska (*Rowe et al.*, 2005). The Shimanto Belt pseudotachylytes, less than a few mm in thickness, developed within clay-rich sediments of eastern (Mugi area) and western (Okitsu area) Shikoku (*Ikesawa et al.*, 2003; *Ujiie et al.*, 2007b) and on Kyushu (Nobeoka area) (*Okamoto et al.*, 2006). The fossil fault zone assemblage in the Mugi area is believed to have been a principal thrust zone. Here the ~1.5-m-thick fault zone is composed of cataclasite with laumontite and quartz mineralized veins. Several generations of cross-cutting pseudotachylytes suggest repeated seismic slip. The host rock is a sedimentary mélange with sandstone and claystone and the approximate depth of pseudotachylyte generation was 3.2 to 4 km at a temperature of 170° to 190°C. In the Nobeoka area, an out-of-sequence-thrust displays a subsidiary fault in the hanging wall with a fault assemblage formed by pseudotachylytes and carbonate-cemented implosion breccias (Figure 4). The thrust zones of the Shimanto Belt have been interpreted to be characterized by mineralized veins and cementation associated with fault zone fluid-saturation

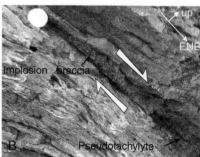

**FIGURE 4**    (a) Photograph of the 20-cm-thick Nobeoka Thrust fault core and pseudotachylyte-bearing subsidiary fault. (b) Close-up of Nobeoka Thrust subsidiary fault zone characterized by a dilational jog with implosion breccias and pseudotachylyte. (Pictures taken during a field trip led by G. Kimura in 2006.) (**See Color Plate 10.**)

during the repeated deformation (possibly seismic cycle) recorded in the pseudotachylyte-bearing fault (*Ujiie et al.*, 2007b) and even during the seismic event (*Okamoto et al.*, 2006). Although high fluid flow is typical of subduction zones, a "wet" seismogenic zone has been long thought to inhibit frictional melting and pseudotachylyte formation (*Sibson*, 1975; *Sibson and Toy*, 2006) and to trigger other dynamic weakening processes such as thermal pressurization. This idea has been challenged by both theoretical studies and observations. Theoretical studies suggested that frictional melting could occur even in the fluid-infiltrated slip zone if the thickness of the slip zone is thin ($\sim$1 mm) (*Fialko*, 2004; *Bizzarri and Cocco*, 2006) and has high permeability and compressibility (*Lachenbruch*, 1980; *Mase and Smith*, 1987). The pseudotachylytes in the Shimanto accretionary complex are generally less than a few millimeters thick and show a high $H_2O$ content possibly representing the formation of a hydrous melt layer during earthquakes (*Ujiie et al.*, 2007b). Okamoto *et al.* (2006) pointed out that in the Nobeoka Thrust zone (Figure 4) frictional melting, hydrofracturing, and mineral precipitation (i.e., implosion brecciation) occurred during a single seismic event. In this latter case, hydrofracturing, mineralization, and frictional melting should have occurred during the very short (1 to 10 s) slip duration (the rise time) of a moderate to large earthquake. In any case, the important observation is that the presence of a mineralized vein along the slip zone and implosion breccias cannot be explained by using the evidence of past thermal pressurization. The implosion brecciation was not derived from fluid pressurization by frictional heating but resulted from the depressurization at a dilational jog during faulting.

Another stimulating data set is the detection of coseismic frictional heating on active faults. The Taiwan Chelungpu-fault Drilling Project detected frictional heating during the 1999 Chi-Chi earthquake using temperature logging (*Kano et al.*, 2006) and magnetic susceptibility (*Mishima et al.*, 2006; *Tanikawa et al.*, 2007). In addition to pseudotachylytes, vitrinite reflectance (*O'Hara*, 2004) and electron spin resonance (*Fukuchi et al.*, 2005) may also be useful in identifying past frictional heating on the fault. Ujiie *et al.* (2008) demonstrated that the stretching of fluid inclusions in calcite could be an indicator of frictional heating on faults.

Implosion breccias, formed by a sudden drop in fluid pressure, are usually found in dilation jogs where angular clasts are irregularly distributed within a mineralized cement. A possible example of implosion brecciation within a dilational jog a few tens of meters long was reported from a fossil subduction thrust exhumed from the seismogenic depths (*Ujiie et al.*, 2007b). The collapse of the fracture wall is thought to be produced by rapid slip transfer during rupture propagation leading to abrupt localized reduction in fluid pressure triggering brecciation by hydraulic implosion (*Sibson*, 1986).

Generally a large amount of syntectonic fault veins suggests a valving action, where rupturing of overpressured crust leads to postfailure fluid discharge due to the enhanced fault-permeability (*Sibson*, 1992; *Robert et al.*, 1995). Overpressure, then, has to be restored before a new event (*Sibson*, 1992), and the widespread mineralization leads to significant strength variations that in turn affect rupture nucleation and recurrence (*Parry*, 1998).

Acoustic fluidization has been linked to the structural characteristics of fault zones observed in the field. Fluidization has been associated with a typical particle size distribution within cataclasites as reported by Monzawa and Otsuki (2003) and Otsuki *et al.* (2003). Fluidized cataclasites agree well with the transient nature of the fluid overpressure developed during seismogenesis and also fit in with a fluid-rich subduction environment. Fluidization from paleosubduction thrusts has been reported from the Kodiak accretionary complex (*Rowe et al.*, 2005) and the Shimanto Belt (*Ujiie et al.*, 2007a). Fluidized material in the paleosubduction thrust in the Shimanto accretionary complex suggests fluidization induced by thermal-pressurization (*Ujiie et al.*, 2007a).

Injections and pulverized rocks, the latter characterized by the preservation of original fabric and crystal boundaries (*Dor et al.*, 2006), have been linked to pressurization-depressurization processes associated with transient seismic slip, but the association of fault rock assemblages with seismic processes remains unclear. Both pulverization and fluidization could contribute to the reduction in the normal stress on faults during earthquakes.

Lab experiments have demonstrated that velocity weakening mechanisms lead to more localized shear zones. Specifically, the transition from distributed deformation to shear localization occurs with the onset of velocity weakening in controlled experimental conditions (*Beeler et al.*, 1996; *Mair and Marone*, 1999). Shear localization is recognized by "Riedel" and "boundary" shears directly cutting and bounding shear zones several mm thick. A period of fault healing and strengthening explicitly follows the microseismic event and is required for the repetition of slip events. Healing is known to occur at low temperatures by compaction and the increase in contact area in gouges (*Marone*, 1998). Angevine (1982) argued for the role of pressure solution and cementation in healing and locking the fault surface at higher temperatures and pressures during the interseismic interval. Observational evidence shows that cementation and fracturing are recurrent along crustal fault zones such as those of the San Andreas System (e.g., *Chester et al.*, 1993). Pressure-solution is a slower process, and although triggered by postseismic stress drop, it is more typical of the interseismic period. However, the actual rates of natural pressure-solution processes are a major uncertainty in the tectonic interpretation of fossil seismic zones.

Experimental studies performed at slip velocities higher than 1 m/s demonstrated that the mixture of phyllosilicates and halite exhibits the velocity weakening behavior (*Niemeijer and Spiers*, 2006, 2007). The generated microstructures can be compared to those observed in the fault rocks of an ancient accretionary complex exhumed from seismogenic depths.

## 6. THE UP-DIP LIMIT OF SEISMOGENESIS IN A FOSSIL EROSIVE SUBDUCTION CHANNEL

So far fossil erosive subduction margins have never been described, and all the data related to dynamic weakening processes came from indirect monitoring of active margins. Here we describe a fossil erosive margin preserved in the Northern Apennines of Italy. At this margin, an erosive subduction channel was formed during the Tertiary transition from oceanic subduction to continental collision (*Coward and Dietrich*, 1989) and the latter caused deactivation, uplift, and fossilization of the plate boundary (*Vannucchi et al.*, 2008).

The subduction channel is presently sandwiched between the late Cretaceous-early Eocene accretionary prism, which was formed at the front of the active European margin (*Vannucchi and Bettelli*, 2002) and the Oligo-Miocene foredeep turbidites of the subduction Adria plate (*Cibin et al.*, 2001) (Figure 5).

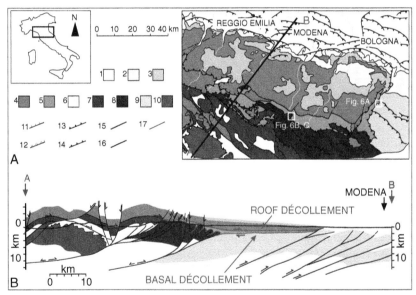

**FIGURE 5**   (a) Schematic geological map of the Northern Apennines with its geographical location shown in the inset. In the map, there is the location of outcrops photographed in Figures 6a, 6b, and 6c. Key: 1. Quaternary deposits; 2. late Miocene-Pleistocene marine deposits; 3. forearc slope deposits; 4. oceanic units of the late Cretaceous-early Eocene accretionary prism, European plate; 5. Sestola-Vidiciatico tectonic unit-subduction channel; 6. Mesozoic carbonate units of the Adria plate; 7. late Oligocene-early Miocene (Aquitanian) trench turbidites of the Adria plate; 8. early Miocene (Aquitanian-Burdigalian) foredeep turbidites of the Adria plate; 9. middle Miocene-late Miocene (Langhian-Messinian) foredeep turbidites of the Adria plate; 10. metamorphic continental units of the Adria plate; 11. normal faults; 12. normal faults (subsurface); 13. thrust faults and overthrusts; 14. thrust faults (subsurface); 15. strike-slip faults; 16. high-angle faults of unknown displacement (subsurface); 17. lithological boundaries. (b) Geological cross section of the Northern Apennines as marked on map (A-B). The thickness of the SVU is about 500 m, slightly decreasing toward the NE. (modified from *Vannucchi et al.*, 2008) **(See Color Plate 11.)**

The subduction channel appears as a mélange formed by tectonically and gravitationally reworked blocks of (1) the previous, late Cretaceous-early Eocene accretionary prism, (2) debris flow deposits of the frontal prism, and (3) late Eocene-early Miocene slope sediments deposited on top of the frontal prism (*Remitti et al.*, 2007). In general, the mélange is composed of a strongly deformed matrix of shales that encloses lesser-deformed blocks of shales, marls, limestones, and sandstones of different age and size. Based on the age of the slope sediment within the channel, we can infer the time interval of activity from late Eocene ($\sim$35 Ma) to at least middle Miocene ($\sim$15 Ma), which is also the age of the channel. The Apennine subduction channel is about 500 m thick with an along strike development of around 100 km, and it is formally known among Apennine geologists as the Sestola-Vidiciatico Unit (SVU).

In the deepest portion of the erosive subduction channel clay mineral assemblages, fluid inclusions, reflectance of organic matter (*Reutter et al.*, 1992), fission tracks (*Zattin et al.*, 2000), indicate that the sediment reached $\sim$150°C (*Reutter et al.*, 1992; *Zattin et al.*, 2000) corresponding to a depth of $\sim$5km. This portion, although shallow, should show the key threshold structural changes in the portion recognized in a modern subduction zone as marking the up-dip limit of seismogenesis (*Moore and Saffer*, 2001; *Obana et al.*, 2003).

Here we will concentrate on two features of the subduction channel: the subduction channel/shear zone architecture and the internal fabric of the subduction channel.

## 6.1. Subduction Channel Architecture

Differently from what has been proposed so far from the subduction erosion mechanism (*von Huene et al.*, 2004a), the subduction channel is characterized by a roof and a basal décollement (Figure 5) contemporaneously active (*Vannucchi et al.*, 2008). The upper tectonic contact of the SVU with the overlying fossil accretionary prism maintains a planar geometry at the regional scale, indicating that it has never been involved in the collisional fold-and-thrust development even after the fossilization of the subduction channel. This subhorizontal thrust is then maintaining a geometry typical of subduction. On a smaller scale, the geometry of the upper tectonic contact varies from flat to wiggly.

The well-exposed lower tectonic contact of the SVU on the foredeep sequences (Figure 6a) has a ramp and flat geometry where the flats are nearly parallel to the bedding of the foredeep rocks in the footwall. Unlike the upper tectonic contact, the deeper part of the lower décollement is cut by a series of map-scale thrusts and folds, indicating that this deeper portion was deactivated while the shallower portion of this boundary was still active and responsible for the migration of the upper plate on the foredeep turbidites together with the upper tectonic contact as demonstrated by the age of the underthrust deposits (*Vannucchi et al.*, 2008) (Figure 5).

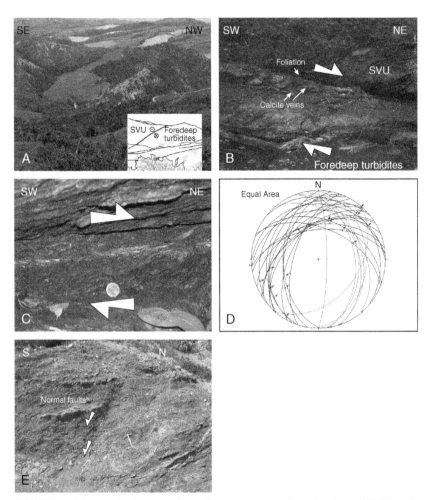

**FIGURE 6**   (a) Panoramic view of the basal décollement of the subduction channel (SVU) on the foredeep turbidites. (b) Photograph of the basal décollement of the Apennine subduction channel. (c) Detail of the basal décollement of the Apennine subduction channel showing a top-to-the-NE movement. (d) Stereographic projection (equal area lower hemisphere) of the structures characterizing the décollement of Figure 6b and 6c. Faults, black circles, with kinematic indication of normal or reverse movement; foliation, red circles. All the structures show a consistent movement direction toward NE. (e) Normal faults cutting blocks of the late Cretaceous-early Eocene accretionary prism forming the upper plate at the time of the channel activity and incorporated in the channel. The block of material from the accretionary prism shows the block-in matrix fabric acquired during accretion. Fault kinematic is indicated by the calcite fibers forming the shear veins. **(See Color Plate 12, part d only.)**

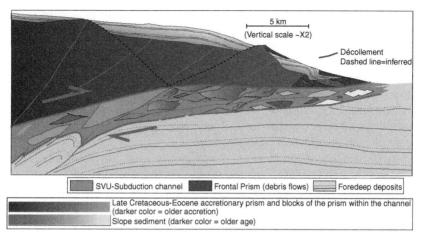

**FIGURE 7**   Scheme of the erosive subduction channel. **(See Color Plate 13.)**

The presence of a roof and a basal décollement simultaneously cutting through the margin toe implies that they were able to incorporate intact slices of the frontal prism through a process of frontal tectonic erosion (Figure 7).

The subduction channel presents vertical compositional differences with younger blocks laying closer to the basal décollement, whereas the older are located closer to the top (*Bettelli and Panini*, 1992) (Figure 7). Although at present there are no recognized differences in the strain evolution between the material at the top and that at the bottom of the subduction channel, this vertical tectonic stratigraphy is indicating a velocity gradient. The subduction channel, then, can be described as a shear zone characterized by multiple décollements. The observations that the lower part of the subduction channel is cut by thrust-and-folds and, as described in the following chapter, by normal faults indicate that the basal décollement and those in the lower part of the channel were deactivated while the topmost part was still active.

The model of the subduction channel requires material flow along the plate boundary (*Cloos and Shreve*, 1988a). The velocity gradient within the channel supports the interpretation of a viscous rheology of the material within the channel at the 10 to 100 m scale. There are several physical processes that can cause this velocity gradient between the layers, but those are beyond the scope of this chapter.

## 6.2. Subduction Channel Internal Structure: A Low-Friction Plate Boundary

The fault zones in the deepest part of the subduction channel, where the defined PT conditions are in agreement with the up-dip limit of the seismogenic zone, are characterized by a strong foliation, they are typically 10 cm to 1 m thick, and they are permeated by a dense array of calcite veins (Figures 6b, 6c, and 6d). The foliation in the fault zones is mainly defined by reorientation of phyllosilicates. Cataclasite and fractures are also present along the fault zones. The fault zones cut through material where the fabric is formed by mm-spaced foliation parallel to subparallel to the fault zones. This foliation is particularly developed in shale and marl. In contrast with the fault zones, this fabric lacks evidence for brittle deformation. Clay mineral assemblage indicates partial to total smectite-to-illite transformation in the material of the subduction channel.

Inside the subduction channel, planar calcite veins are commonly shear veins either parallel to the décollement (i.e., subhorizontal to gently dipping toward NE) or at a high angle to the décollement with fibers indicating normal sense of shear and cutting through all the components of the mélange (Figure 6e). Both vein types are characterized by crack-seal, and they show evidence of pressure solution following mineral crystallization (Figures 8a and 8b) (*Vannucchi et al.*, 2008). The planar veins border and die into dilational jogs (Figures 8c and 8d). Dilational jogs characterize these fault zones as asymmetric extensional cracks with a length of tens of cm parallel to the slip direction of the fault zone. Some of the jogs present repeated events of

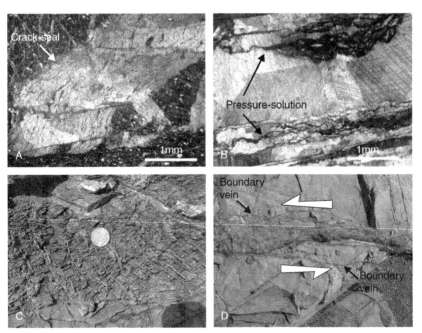

**FIGURE 8**   (a) Crack-and-seal texture in calcite of shear vein associated with the deformation of the Apennine subduction channel. (b) Pressure-solution seams alternating with calcite precipitation within the veins. (c) Mesoscopic repetition of extensional calcite veins. (d) Dilational jog.

extensional vein opening, giving a texture similar to the microscopic crack-and-seal, visible also at the mesoscopic scale (Figure 8c). But commonly jogs are formed by calcite surrounding clasts of material from the wall rock and from previously precipitated veins (Figure 8d). These clasts have angular shapes, they show neither preferential orientation nor gradation, and in some cases they show that the jog formed by a single stage. In both cases, parallel extensional veins or breccias, these structures are always inside dilational jogs and are always bordered by planar veins, indicating that they occur concurrently and that their formation is associated with the subduction channel deformation.

The internal structure of the dilational jogs suggests fluid implosion with fracturing occurring during a fast slip event. This occurrence suggests that during each single slip event with several tens of cm displacement, fracturing was concentrated at the jog. On the contrary, each vein records cyclical fluid pressure rises and drops that were followed by mineral precipitation and then a new loading phase.

Implosion breccias indicate a sudden drop in fluid pressure. The collapse of the fracture walls, in fact, agrees with the sudden depressurization during a rupture propagation in a dilational jog along the fault surface (*Sibson*, 1986). The geometry of the system, with extensional veins associated with the jog, implies a high fluid pressure regime, but it does not allow for characterizing the fluid pressure level.

Thrust fault zones structures are cut by ~10-m-spaced normal faults. Faults are 10 cm to 1 m thick, and, as already described, they are permeated by dense arrays of extensional calcite veins. Each fault accommodates centimeter-scale displacement that is reached after a great number of repeated smaller events. The cross-cutting relationship indicates that slip along normal faults was contemporaneous with the subduction channel activity. Although normal faults could be cut by successive slip along the thrusts, they commonly represent deactivation of the thrusts and transfer of slip to a structurally higher fault zone. Normal faults also cut the basal décollement. These structures intersect all the SVU components indicating that by this stage the unit had a homogeneous mechanical behavior (Figure 6d).

Extensional shear strains also characterize the channel through the development of sets of extensional shear fractures cutting the mélange (*Vannucchi et al.*, 2008), implying that the maximum principal stress is at a high angle to the décollements according to their weak nature and low strength. Moreover, this extensional strain regime recorded inside the subduction channel defines the fluid pressure level not above lithostatic pressure.

In the deeper portion of the outcropping subduction channel, the basal décollement is also overprinted by contractional structures. The involvement of the basal décollement in the fold-and-thrust deformation indicates locking, although fluid circulation was still active as shown by vein development. These contractional locking structures develop while the structures of the intermediate portions are still actively forming, as indicated by cross-cutting relationships. The structures in the field show cyclic behavior and the localization of shear along the décollement progressively activated from the bottom to the top of the channel.

## 7. DISCUSSION AND COMPARISON BETWEEN EROSIVE AND ACCRETIONARY SEISMOGENIC ZONES

The analysis in the Apennines reveals that strain concentration within the erosive subduction channel developed in a regime associated with a fluid pressure cycle (*Sibson*, 1990, 1992). Two competing processes accommodate this deformation: (1) calcite veins and implosion breccias indicating relatively fast events and stress drop and (2) slower pressure-solution. Here transient phases of stress accumulation are all worked by fluids as a key deformation component within this convergent plate boundary.

Also the work done in the Shimanto Belt points toward a key role of fluids (*Okamoto et al.*, 2006) rather than major influence by the progressive change of physical properties of material along the plate boundary. Indeed both erosive and accretionary subduction systems are characterized by a disproportionate amount of fluid flow (*Moore and Vrolijk*, 1992; *Ranero et al.*, 2008) and subduction thrusts systems are conduits of focused fluid flow (*Gieskes et al.*, 1990; *Kastner et al.*, 1991; *Silver et al.*, 2000).

Measurements and related models of fluid circulation in modern subduction zones cover a whole spectrum from continuous to transient and heterogeneous fluid events (*Carson and Screaton*, 1998; *Ranero et al.*, 2008). Transient fluid flow is also in agreement with pumping mechanisms solicited by the seismic cycle (*Sibson*, 1992; *Wang and Hu*, 2006). Repeated and transient fluid events are likely to cause overpressure and hydrofracturing. Such a mechanism would be responsible for the progressive migration of slip from the bottom part to the top of the subduction channel as observed in the Northern Apennines. Fluid migration can also take into account the asymmetric development of the plate interface as inferred by *von Huene et al.* (2004a).

It has been a common trend to infer that the aseismic/seismic transition would occur where fluids are less abundant and the fluid pressure can no longer reduce the effective normal stress (*Moore and Saffer*, 2001; *Ranero et al.*, 2008). However, this picture is not what has been observed in the erosive subduction channel of the Northern Apennines, where not only the pervasive presence of veins indicating high fluid content but the occurrence of dilational jogs indicate that fracturing occurs as result of transient stress concentration around the propagating fault due to hydrofracturing. We can infer a large fluid pressure gradient between the fluid in the jog (i.e., preceding cracks) and the pore fluids in the fault zone.

In the fossil erosive subduction channel of the Northern Apennines, migrating fluids change the frictional nature of the fault zone (especially the time-dependent friction), and they also heal and strengthen the fault zone during interseismic intervals. Differently from what was observed in the Shimanto Belt, where the abrupt extensional fracturing and the sudden fluid pressure drop led to the recovery of the effective shear strength and shear heating could take over producing frictional melting at the slip plane (i.e., pseudotachylyte) (*Okamoto et al.*, 2006), in the Northern Apennines and for the observed PT conditions, effective normal stress did not reach the level of frictional melting.

## 8. CONCLUSIONS AND FUTURE PERSPECTIVE

Perhaps the most basic seismogenic observation at erosive and accretionary subduction margins is that the up-dip onset of seismogenesis along the megathrust occurs at roughly the same *in situ* temperature, corresponding to ~150°C. This is also the temperature of the smectite-illite clay mineral transformation. However, the fact that the onset of seismogenesis appears to be independent of material input into the margin (e.g., at erosive margins there can be little smectite along the subduction thrust unless there are conditions where the eroded base of the upper plate is very clay-mineral rich) and the fact that the lab experiments have now demonstrated that illite does not have velocity-weakening properties in slip (*Saffer and Marone*, 2003) both provide compelling evidence that material property changes associated with the smectite-illite transition are not the cause of the onset of seismogenesis. Instead, the onset of seismogenesis appears to be linked to the necessary conditions for fluid overpressure during a seismic event and fluid undersaturation during the interseismic phase of stress buildup. The smectite-illite transformation and other ~150°C metamorphic dewatering transformations may be linked to these conditions. It is clear in structural studies of accretionary and erosive fossil margins that fluid flow is even more pervasive at depths shallower than the depth of seismogenesis than at the onset of seismogenesis (*Moore and Vrolijk*, 1992; *Vannucchi et al.*, 2008). Pseudotachylytes have been found in fossil accretionary seismogenic zones but not in the erosive Apennine seismogenic zone—we think because the Apennines' fossil zone has not been exhumed to the structural depths where pseudotachylytes would occur.

The down-dip limit of seismogenesis is even more poorly understood at present. It is well known that seismogenesis within subducting slabs can persist until lower mantle depths (670 km). Until recently it was believed that seismogenesis along the megathrust interface stopped at the crust-mantle transition in the overriding plate—at first contact with a serpentinized mantle wedge. However, a microseismic study of the Marianas megathrust appears to find megathrust seismogenesis to depths of ~50 km, significantly below the Moho of the overriding plate and within a region characterized by abundant serpentine diapirism in the Marianas forearc (*Wiens et al.*, 2007).

We have now recognized several good field examples of fossil accretionary and erosive seismogenic zones. At accretionary zones (Shimanto Belt,

Japan) there are clear examples of pseudotachylyte formation associated with seismogenic slip, which is turn is evidence of significant thermal pressurization during a seismic event (*Okamoto et al.*, 2006; *Ujiie et al.*, 2007b). In the erosive Apennines fossil seismogenic zone, there is evidence that the subduction channel had simultaneously active basal and roof faults and that the process of subduction erosion itself may be linked to progressive "up-stepping" of the subduction channel into the overriding plate. All fossil seismogenic zones show evidence of pervasive fluid flow and of a fluid pressure cycle linked to seismogenesis, with earthquakes themselves linked to sudden depressurization along the subduction channel (*Ranero et al.*, 2008).

At present there are several ongoing deep-drilling initiatives to drill through active fault zones, and there is a continuing rapid increase in observations of active and fossil seismogenic faults. We think it likely that after the deep-drilling projects are completed there will be a much better understanding of seismogenesis, and we hope that the links between seismogenesis and fluid flow highlighted in this chapter will not prove to be a complete red herring!

## REFERENCES

Angevine, C. L., D. L. Turcotte, and M. D. Furnish, (1982), Pressure solution lithification as a mechanism for stick-slip behavior of faults, *Tectonics*, **1**(2), 151-160.

Ashi, J., S. Kuramoto, S. Morita, U. Tsunogai, and K. Kameo, (2001), Structure and cold seep of the Nankai accretionary prism off Kumano, *Eos Trans. AGU*, **82**(47), Fall Meet. Suppl. Abstract, T41A-0859.

Baba, T., Y. Tanioka, P. R. Cummins, and K. Uhira, (2002), The slip distribution of the 1946 Nankai earthquake estimated from tsunami data using a new model for the Philippine Sea Plate. *Phys. Earth Planet. Interi.*, **132**(1-3), 59-73.

Beeler, N. M., R. W. Simpson, S. H. Hickman, and D. A. Lockner, (2000), Pore fluid pressure, apparent friction, and Coulomb failure, *J. Geophys. Res.*, **105**(B11), 25533-25542.

Beeler, N. M., T. E. Tullis, M. L. Blanpied, and J. D. Weeks, (1996), Frictional behavior of large displacement experimental faults, *J. Geophys. Res.*, **101**(B4), 8697-8715.

Bettelli, G. and F. Panini, (1992), Liguridi, mélanges e tettoniti nel complesso caotico lungo la "linea del Sillaro" (Appennino Settentrionale, Provincie di Firenze e Bologna), *Memorie Descrittive della Carta Geologica d'Italia*, **39**, 91-125.

Bilek, S. L. and T. Lay, (1999), Rigidity variations with depth along interplate megathrust faults in subduction zones, *Nature*, **400**, 443-446.

Bilek, S. L. and T. Lay, (2002), Tsunami earthquakes possibly widespread manifestations of frictional conditional stability, *Geophys. Res. Lett.*, **29**(14), 1673, doi:10.1029/2002GL015215.

Bizzarri, A. and M. Cocco, (2006), A thermal pressurization model for the spontaneous dynamic rupture propagation on a three-dimensional fault: 1. Methodological approach, *J. Geophys. Res.*, **111**, B05303, doi:10.1029/2005JB003862.

Boullier, A.-M., T. Ohtani, K. Fujimoto, H. Ito, and M. Dubois, (2001), Fluid inclusions in pseudotachylytes from the Nojima fault, Japan, *J. Geophys. Res.*, **106**(B10), 21965-21977.

Brace, W. F. and J. D. Byerlee, (1966), Stick slip as mechanism for earthquakes, *Science*, **153**, 990-992.

Brodsky, E. E. and H. Kanamori, (2001), Elastohydrodynamic lubrication of faults, *J. Geophys. Res.*, **106**(B8), 16357-16374.

Brown, K., M. D. Tryon, H. R. DeShon, L. M. Dorman, and S. Schwartz, (2005), Correlated transient fluid pulsing and seismic tremor in the Costa Rica subduction zone, *Earth Planet. Sci. Lett.*, **238**(1-2), 189-203.

Brune, J. N., S. Brown, and P. A. Johnson, (1993), Rupture mechanism and interface separation in foam rubber models of earthquakes: A possible solution to the heat flow paradox and the paradox of large overthrusts, *Tectonophys.*, **218**(1-3), 59-67.

Carson, B. and E. J. Screaton, (1998), Fluid flow in accretionary prisms: Evidence for focused, time-variable discharge, *Rev. Geophys.*, **36**(3), 329-351.

Chapple, W. M., (1978), Mechanics of thin-skinned fold-and-thrust belts, *Geol. Soc. Am. Bull.*, **89**(8), 1189-1198.

Chester, F. M. and J. S. Chester, (1998), Ultracataclasite structure and friction processes of the Punchbowl fault, San Andreas system, California, *Tectonophys.*, **295**(1-2), 199-221.

Chester, F. M., J. P. Evans, and R. L. Biegel, (1993), Internal structure and weakening mechanisms of the San-Andreas fault, *J. Geophys. Res.*, **98**(B1), 771-786.

Cibin, U., E. Spadafora, G. G. Zuffa, and A. Castellarin, (2001), Continental collision history from arenites of episutural basins in the Northern Apennines, Italy, *Geol. Soc. Am. Bull.*, **113**(1), 4-19.

Clift, P. and P. Vannucchi, (2004), Controls on tectonic accretion versus erosion in subduction zones: Implications for the origin and recycling of the continental crust, *Rev. Geophys.*, **42**, RG2001, doi:10.1029/2003RG000127.

Cloos, M., (1992), Thrust-type subduction-zone earthquakes and seamount asperities—A physical model for seismic rupture, *Geology*, **20**(7), 601-604.

Cloos, M. and R. L. Shreve, (1988a), Subduction-channel model of prism accretion, mélange formation, sediment subduction, and subduction erosion at convergent plate margins: 1. Background and description, *Pure Appl. Geophys.*, **128**(3-4), 455-500.

Cloos, M. and R. L. Shreve, (1988b), Subduction-channel model of prism accretion, mélange formation, sediment subduction, and subduction erosion at convergent plate margins: 2. Implications and discussion, *Pure Appl. Geophys.*, **128**(3-4), 501-545.

Cloos, M. and R. L. Shreve, (1996), Shear-zone thickness and the seismicity of Chilean- and Marianas-type subduction zones, *Geology*, **24**(2), 107-110.

Cochard, A. and R. Madariaga, (1994), Dynamic faulting under rate-dependent friction, *Pure Appl. Geophys.*, **142**(3-4), 419-445.

Coward, M. P. and D. Dietrich, (1989), Alpine tectonics—An overview, *Special Publication, London, Geological Society*, **45**, 1-29.

Davis, D., J. Suppe, and F. A. Dahlen, (1983), Mechanics of fold-and-thrust belts and accretionary wedges, *J. Geophys. Res.*, **88**(B2), 1153-1172.

Davis, E. E., K. Becker, K. Wang, K. Obara, Y. Ito, and M. Kinoshita, (2006), A discrete episode of seismic and aseismic deformation of the Nankai Trough subduction zone accretionary prism and incoming Philippine Sea plate, *Earth Planet. Sci. Lett.*, **242**(1-2), 73-84.

DeShon, H. R., S. Y. Schwartz, S. L. Bilek, L. M. Dorman, V. Gonzalez, J. M. Protti, E. R. Flueh, and T. H. Dixon, (2003), Seismogenic zone structure of the southern Middle America Trench, Costa Rica, *J. Geophys. Res.*, **108**(B10), 2491, doi:10.1029/2002JB002294.

Di Toro, G., D. L. Goldsby, and T. E. Tullis, (2004), Friction falls towards zero in quartz rock as slip velocity approaches seismic rates, *Nature*, **427**, 436-439.

Di Toro, G., T. Hirose, S. Nielsen, G. Pennacchioni, and T. Shimamoto, (2006), Natural and experimental evidence of melt lubrication of faults during earthquakes, *Science*, **311**, 647-649.

Dieterich, J. H., (1978), Time dependent friction and the mechanics of stick slip, *Pure Appl. Geophys.*, **116**(4-5), 790-806.

Dor, O., Y. Ben-Zion, T. K. Rockwell, and J. N. Brune, (2006), Pulverized rocks in the Mojave section of the San Andreas Fault Zone, *Earth Planet. Sci. Lett.*, **245**(3-4), 642-654.

Dragert, H., K. Wang, and T. S. James, (2001), A silent slip event on the deeper Cascadia subduction interface, *Science*, **292**, 1525-1528.

Fialko, Y., (2004), Temperature fields generated by the elastodynamic propagation of shear cracks in the Earth, *J. Geophys. Res.*, **109**, B01303, doi:10.1029/2003JB002497.

Fialko, Y. and Y. Khazan, (2005), Fusion by earthquake fault friction: Stick or slip? *J. Geophys. Res.*, **110**, B12407, doi:10.1029/2005JB003869.

Fukuchi, T., K. Mizoguchi, and T. Shimamoto, (2005), Ferrimagnetic resonance signal produced by frictional heating: A new indicator of paleoseismicity, *J. Geophys. Res.*, **110**, B12404, doi:10.1029/2004JB003485.

Gieskes, J. M., P. Vrolijk, and G. Blanc, (1990), Hydrogeochemistry of the northern Barbados accretionary complex transect—Ocean Drilling Project Leg 110, *J. Geophys. Res.*, **95**(B6), 8809-8818.

Goldsby, D. L. and T. E. Tullis, (2002), Low frictional strength of quartz rocks at subseismic slip rates, *Geophys. Res. Lett.*, **29**(17), 1844, doi:10.1029/2002GL01240.

Heesakkers, V., S. K. Murphy, G. van Aswegen, R. Domoney, S. Addams, T. Dewers, M. Zechmeister, and Z. Reches, (2005), The rupture zone of the M=2.2 earthquake that reactivated the ancient Pretorius Fault in TauTona Mine, South Africa, *Eos Trans. AGU*, **86**(52), Fall Meet. Suppl. Abstract, S31B-04.

Hirose, H., K. Hirahara, F. Kimata, N. Fujii, and S. Miyazaki, (1999), A slow thrust slip event following the two 1996 Hyuganada earthquakes beneath the Bungo Channel, southwest Japan, *Geophys. Res. Lett.*, **26**(21), 3237-3240.

Hirose, T. and T. Shimamoto, (2005), Slip-weakening distance of faults during frictional melting as inferred from experimental and natural pseudotachylytes, *Bull. Seismol. Soc. Am.*, **95**, 1666-1673.

Husen, S., R. Quintero, E. Kissling, and B. Hacker, (2003), Subduction-zone structure and magmatic processes beneath Costa Rica constrained by local earthquake tomography and petrological modelling, *Geophys. J. Int.*, **155**(1), 11-32.

Hyndman, R. D., M. Yamano, and D. A. Oleskevich, (1997), The seismogenic zone of subduction thrust faults, *Island Arc*, **6**(3), 244-260.

Ide, S., M. Takeo, and Y. Yoshida, (1996), Source process of the 1995 Kobe earthquake: Determination of spatio-temporal slip distribution by Bayesian modeling, *Bull. Seismol. Soc. Am.*, **86**(3), 547-566.

Igarashi, T., T. Matsuzawa, and A. Hasegawa, (2003), Repeating earthquakes and interplate aseismic slip in the northeastern Japan subduction zone, *J. Geophys. Res.*, **108**(B5), 2249, doi:10.1029/2002JB001920.

Ikesawa, E., A. Sakaguchi, and G. Kimura, (2003), Pseudotachylyte from an ancient accretionary complex: Evidence for melt generation during seismic slip along a master décollement? *Geology*, **31**(7), 637-640.

Ito, Y. and K. Obara, (2006), Dynamic deformation of the accretionary prism excites very low frequency earthquakes, *Geophys. Res. Lett.*, **33**, L02311, doi:10.1029/2005GL025270.

Jarrard, R. D., (1986), Relations among subduction parameters, *Rev. Geophys.*, **24**(2), 217-284.

John, T. and V. Schenk, (2006), Interrelations between intermediate-depth earthquakes and fluid flow within subducting oceanic plates: Constraints from eclogite facies pseudotachylytes, *Geology*, **34**(7), 557-560.

Kame, N., J. R. Rice, and R. Dmowska, (2003), Effects of pre-stress state and rupture velocity on dynamic fault branching, *J. Geophys. Res.*, **108**(B5), 2265, doi:10.1029/2002JB002189.

Kanamori, H. and T. H. Heaton, (2000), Microscopic and macroscopic physics of earthquakes, In: *Geocomplexity and the Physics of Earthquakes*, edited by Rundle, J., D. L. Turcotte, and W. Klein, *Geophysical Monograph Series*, **120**, 147-163, American Geophysical Union, Washington, D.C.

Kano, Y., J. Mori, R. Fujio, H. Ito, T. Yanagidani, S. Nakao, and K.-F. Ma, (2006), Heat signature on the Chelungpu fault associated with the 1999 Chi-Chi, Taiwan earthquake, *Geophys. Res. Lett.*, **33**, L14306, doi:10.1029/2006GL026733.

Karig, D. E., (1974), Evolution of arc systems in the Western Pacific, *Ann. Rev. Earth Planet. Sci.*, **21**, 51-75.

Karig, D. E. and G. F. Sharman, (1975), Subduction and accretion in trenches, *Geol. Soc. Am. Bull.*, **86**(3), 377-389.

Kastner, M., H. Elderfield, and J. B. Martin, (1991), Fluids in convergent margins—What do we know about their composition, origin, role in diagenesis and importance for oceanic chemical fluxes? *Phil. Trans. Roy. Soc.*, Ser. A, **335**, 243-259.

Katsumata, A. and N. Kamaya, (2003), Low-frequency continuous tremor around the Moho discontinuity away from volcanoes in the southwest Japan, *Geophys. Res. Lett.*, **30**(1), 1020, doi:10.1029/2002GL015981.

Kikuchi, M., M. Nakamura, and K. Yoshikawa, (2003), Source rupture processes of the 1944 Tonankai earthquake and the 1945 Mikawa earthquake derived from low-gain seismograms, *Earth, Planets Space*, **55**(4), 159-172.

Kreemer, C., G. Blewitt, and F. Maerten, (2006), Co- and postseismic deformation of the 28 March 2005 Nias $M_w$ 8.7 earthquake from continuous GPS data, *Geophys. Res. Lett.*, **33**, L07307, doi:10.1029/2005GL025566.

Kuroki, H., H. M. Ito, H. Takayama, and A. Yoshida, (2004), 3-D Simulation of the occurrence of slow slip events in the Tokai Region with a rate- and state-dependent friction law, *Bull. Seismol. Soc. Am.*, **94**(6), 2037-2050.

Lachenbruch, A. H., (1980), Frictional heating, fluid pressure, and the resistance to fault motion, *J. Geophys. Res.*, **85**(B11), 6097-6112.

Lay, T., H. Kanamori, and L. Ruff, (1982), The asperity model and the nature of large subduction zone earthquakes, *Earthquake Prediction Research*, **1**, 3-71.

Lee, W. H. K., H. Kanamori, P. Jennings, and C. Kisslinger, (2002), *International Handbook of Earthquake & Engineering Seismology*, Part A, Academic Press, Amsterdam, 1200 pp.

Leggett, J., Y. Aoki, and T. Toba, (1985), Transition from frontal accretion to underplating in a part of the Nankai Trough Accretionary Complex off Shikoku (SW Japan) and extensional features of the lower slope, *Marine Petrol. Geol.*, **2**(2), 131-141.

Liu, Y. and J. R. Rice, (2005), Aseismic slip transient emerge spontaneously in three-dimensional rate and state modeling of subduction earthquake sequences, *J. Geophys. Res.*, **110**, B08307, doi:10.1029/2004JB003424.

Lowry, A. R., K. M. Larson, V. Kostoglodov, and R. Bilham, (2001), Transient fault slip in Guerrero, southern Mexico, *Geophys. Res. Lett.*, **28**(19), 3753-3756.

Ma, K.-F., H. Tanaka, S.-R. Song, C.-Y. Wang, J.-H. Hung, Y.-B. Tsai, J. Mori, Y.-F. Song, E.-C. Yeh, W. Soh, H. Sone, L.-W. Kuo, and H.-Y. Wu, (2006), Slip zone and energetics of a large earthquake from the Taiwan Chelungpu-fault Drilling Project, *Nature*, **444**, 473-476.

Ma, K.-F., E. E. Brodsky, J. Mori, C. Ji, T.-R. A. Song, and H. Kanamori, (2003), Evidence for fault lubrication during the 1999 Chi-Chi, Taiwan, earthquake (Mw7.6), *Geophys. Res. Lett.*, **30**(5), 1244, doi:10.1029/2002GL015380.

Mair, K. and C. Marone, (1999), Friction of simulated fault gouge for a wide range of velocities and normal stresses, *J. Geophys. Res.*, **104**(B12), 28899-28914.

Marone, C., (1998), Laboratory-derived friction laws and their application to seismic faulting, *Ann. Rev. Earth Planet. Sci.*, **26**, 643-696.

Mase, C. W. and L. Smith, (1987), Effects of frictional heating on the thermal, hydrologic, and mechanical response of a fault, *J. Geophys. Res.*, **92**(B7), 6249-6272.

Melbourne, T. I., W. M. Szeliga, M. M. Miller, and V. M. Santillan, (2005), Extent and duration of the 2003 Cascadia slow earthquake, *Geophys. Res. Lett.*, **32**, L04301, doi:10.1029/2004GL021790.

Melosh, H. J., (1996), Dynamical weakening of faults by acoustic fluidization, *Nature*, **379**, 601-606.

Miller, S. A., W. van der Zee, D. L. Olgaard, and J. A. D. Connolly, (2003), A fluid-pressure feedback model of dehydration reactions: Experiments, modelling, and application to subduction zones, *Tectonophys.*, **370**(1-4), 241-251.

Mishima, T., T. Hirono, W. Soh, and S.-R. Song, (2006), Thermal history estimation of the Taiwan Chelungpu fault using rock-magnetic methods, *Geophys. Res. Lett.*, **33**, L23311, doi:10.1029/2006GL028088.

Miyazaki, S. and K. Heki, (2001), Crustal velocity field of southwest Japan: Subduction and arc-arc collision, *J. Geophys. Res.*, **106**(B3), 4305-4326.

Monzawa, N. and K. Otsuki, (2003), Comminution and fluidization of granular fault materials: Implications for fault slip behavior, *Tectonophys.*, **367**(1-2), 127-143.

Moore, J. C. and D. Saffer, (2001), Updip limit of the seismogenic zone beneath the accretionary prism of southwest Japan: An effect of diagenetic to low-grade metamorphic processes and increasing effective stress, *Geology*, **29**(2), 183-186.

Moore, J. C. and P. Vrolijk, (1992), Fluids in accretionary prisms, *Rev. Geophys.*, **30**(2), 113-135.

Morley, C. K., (1988), Out of sequence thrusts, *Tectonics*, **7**(3), 539-561.

Nadeau, R. M. and T. V. McEvilly, (1997), Seismological studies at Parkfield V: Characteristic microearthquake sequences as fault-zone drilling targets, *Bull. Seismol. Soc. Am.*, **87**(6), 1463-1472.

Newman, A. V., S. Y. Schwartz, V. Gonzalez, H. R. DeShon, J. M. Protti, and L. M. Dorman, (2002), Along-strike variability in the seismogenic zone below Nicoya Peninsula, Costa Rica, *Geophys. Res. Lett.*, **29**(20), 1977, doi:10.1029/2002GL015409.

Niemeijer, A. R. and C. J. Spiers, (2006), Velocity dependence of strength and healing behaviour in simulated phyllosilicate-bearing fault gouge, *Tectonophys.*, **427**(1-4), 231-253.

Niemeijer, A. R. and C. J. Spiers, (2007), A microphysical model for strong velocity weakening in phyllosilicate-bearing fault gouges, *J. Geophys. Res.*, **112**, B10405, doi:10.1029/2007JB005008.

Norabuena, E., T. H. Dixon, S. Schwartz, H. DeShon, A. Newman, M. Protti, V. Gonzalez, L. Dorman, E. R. Flueh, P. Lundgren, F. Pollitz, and D. Sampson, (2004), Geodetic and seismic constraints on some seismogenic zone processes in Costa Rica, *J. Geophys. Res.*, **109**, B11403, doi:10.1029/2003JB002931.

O'Hara, K., (2004), Paleo-stress estimates on ancient seismogenic faults based on frictional heating of coal, *Geophys Res. Lett.*, **31**, L03601, doi:10.1029/2003GL018890.

O'Hara, K., (2005), Evaluation of asperity-scale temperature effects during seismic slip, *J. Struct. Geol.*, **27**(10), 1892-1898.

Obana, K., S. Kodaira, Y. Kaneda, K. Mochizuki, M. Shinohara, and K. Suyehiro, (2003), Micro-seismicity at the seaward updip limit of the western Nankai Trough seismogenic zone, *J. Geophys. Res.*, **108**(B10), 2459, doi:10.1029/2002JB002370.

Obana, K., S. Kodaira, K. Mochizuki, and M. Shinohara, (2001), Micro-seismicity around the seaward updip limit of the 1946 Nankai earthquake dislocation area, *Geophys. Res. Lett.*, **28**(12), 2333-2336.

Obara, K., (2002), Nonvolcanic deep tremor associated with subduction in southwest Japan, *Science*, **296**, 1679-1681.

Obara, K., and H. Hirose, (2006), Non-volcanic deep low-frequency tremors accompanying slow slips in the southwest Japan subduction zone, *Tectonophys.*, **417**(1-2), 33-51.

Obara, K., H. Hirose, F. Yamamizu, and K. Kasahara, (2004), Episodic slow slip events accompanied by non-volcanic tremors in southwest Japan subduction zone, *Geophys. Res. Lett.*, **31**, L23602, doi:10.1029/2004GL020848.

Okamoto, S., G. Kimura, and H. Yamaguchi, (2006), Earthquake fault rock including a coupled lubrication mechanism, *eEarth Discussions*, **1**(2), 135-149.

Oleskevich, D. A., R. D. Hyndman, and K. Wang, (1999), The updip and downdip limits to great subduction earthquakes: Thermal and structural models of Cascadia, south Alaska, SW Japan, and Chile, *J. Geophys. Res.*, **104**(B7), 14965-14991.

Olsen, K. B., R. Madariaga, and R. J. Archuleta, (1997), Three-dimensional dynamic simulation of the 1992 Landers earthquake, *Science*, **278**, 834-838.

Otsuki, K., N. Monzawa, and T. Nagase, (2003), Fluidization and melting of fault gouge during seismic slip: Identification in the Nojima fault zone and implications for focal earthquake mechanisms, *J. Geophys. Res.*, **108**(B4), 2192, doi:10.1029/2001JB001711.

Pacheco, J. F., L. R. Sykes, and C. H. Scholz, (1993), Nature of seismic coupling along simple plate boundaries of the subduction type, *J. Geophys. Res.*, **98**(B8), 14133-14159.

Park, J. O., T. Tsuru, S. Kodaira, P. R. Cummins, and Y. Kaneda, (2002), Splay fault branching along the Nankai subduction zone, *Science*, **297**, 1157-1160.

Parry, W. T., (1998), Fault-fluid compositions from fluid-inclusion observations and solubilities of fracture sealing minerals, *Tectonophys.*, **290**(1-2), 1-26.

Peacock, S. M., (1990), Fluid processes in subduction zones, *Science*, **248**, 329-337.

Platt, J. P., J. K. Leggett, J. Young, H. Raza, and S. Alam, (1985), Large-scale sediment under-plating in the Makran accretionary prism, Southwest Pakistan, *Geology*, **13**(7), 507-511.

Pritchard, M. E. and M. Simons, (2006), An aseismic slip pulse in Northern Chile and along-strike variations in seismogenic behavior, *J. Geophys. Res.*, **111**, B08405, doi:10.1029/2006JB004258.

Protti, M., V. Gonzales, T. Kato, T. Iinuma, S. Miyasaki, K. Obana, Y. Kaneda, P. LaFemina, T. Dixon, and S. Schwartz, (2004), A creep event on the shallow interface of the Nicoya Peninsula, Costa Rica seismogenic zone, *Eos Trans. AGU*, **85**(47), Fall Meet. Suppl. Abstract, S41D-07.

Ranero, C. R., I. Grevemeyer, U. Sahling, U. Barckhausen, C. Hensen, K. Wallmann, W. Weinrebe, P. Vannucchi, R. von Huene, and K. McIntosh, (2008), Hydrogeological system of erosional convergent margins and its influence on tectonics and interplate seismogenesis, *Geochem. Geophys. Geosyst.*, **9**, Q03S04, doi:10.1029/2007GC001679.

Ranero, C. R. and R. von Huene, (2000), Subduction erosion along the Middle America convergent margin, *Nature*, **404**, 748-752.

Remitti, F., G. Bettelli, and P. Vannucchi, (2007), Internal structure and tectonic evolution of an underthrust tectonic mélange: The Sestola-Vidiciatico tectonic unit of the Northern Apennines, Italy, *Geodinamica Acta*, **20**(1-2), 37-51.

Rempel, A. W., (2006), The effects of flash-weakening and damage on the evolution of fault strength and temperature, In: *Radiated Energy and the Physics of Faulting*, edited by Abercrombie, R., A. McGarr, G. Di Toro, and H. Kanamori, *Geophysical Monograph Series*, **170**, 263-270, American Geophysical Union, Washington D.C.

Rempel, A. W. and J. R. Rice, (2006), Thermal pressurization and onset of melting in fault zones, *J. Geophys. Res.*, **111**, B09314, doi:10.1029/2006JB004314.

Reutter, K. J., I. Heinitz, and R. Eusslin, (1992), Structural and geothermal evolution of the Modino-Cervarola Unit, *Memorie Carta Geologica d'Italia*, **46**, 257-266.

Rice, J. R., (2006), Heating and weakening of faults during earthquake slip, *J. Geophys. Res.*, **111**, B05311, doi:10.1029/2005JB004006.

Robert, F., A.-M. Boullier, and K. Firdaous, (1995), Gold-quartz veins in metamorphic terranes and their bearing on the role of fluids in faulting, *J. Geophys. Res.*, **100**(B7), 12861-12879.

Rogers, G. and H. Dragert, (2003), Episodic tremor and slip on the Cascadia subduction zone: The chatter of silent slip, *Science*, **300**, 1942-1943.

Rowe, C. D., J. C. Moore, F. Meneghini, and A. W. McKiernan, (2005), Large-scale pseudotachylytes and fluidized cataclasites from an ancient subduction thrust fault, *Geology*, **33**(12), 937-940.

Ruff, L. J., (1989), Do trench sediments affect great earthquake occurrence in subduction zones?, *Pure Appl. Geophys.*, **129**(1-2), 263-282.

Ruff, L. J., (1992), Asperity distributions and large earthquake occurrence in subduction zones, *Tectonophys.*, **211**(1-4), 61-83.

Saffer, D. M. and C. Marone, (2003), Comparison of smectite- and illite-rich gouge frictional properties: Application to the updip limit of the seismogenic zone along subduction megathrusts, *Earth Planet. Sci. Lett.*, **215**(1-2), 219-235.

Sagiya, T. and W. Thatcher, (1999), Coseismic slip resolution along a plate boundary megathrust: The Nankai Trough, southwest Japan, *J. Geophys. Res.*, **104**(B1), 1111-1129.

Sample, J. C. and D. M. Fisher, (1986), Duplexes and underplating in an ancient accretionary complex, Kodiak islands, Alaska, *Geology*, **14**(2), 160-163.

Sassa, K., (1935), Geophysical studies on the volcano Aso. Part 1: Volcanic micro-tremors and eruptive earthquakes, *Mem. Coll. Sci., Univ. Kyoto*, **18**, 255-293.

Scholz, C. H., (1990), Geophysics—Earthquakes as chaos, *Nature*, **348**, 197-198.

Scholz, C. H., (2002), *The Mechanics of Earthquakes and Faulting, 2^{nd} ed.*, Cambridge University Press, Cambridge, 471 pp.

Schwartz, S. Y. and J. M. Rokosky, (2007), Slow slip events and seismic tremor at circum-pacific subduction zones, *Rev. Geophys.*, **45**, RG3004, doi:10.1029/2006RG000208.

Seely, D. L., P. R. Vail, and G. G. Walton, (1974), Trench slope model, In: *Geology of the Continental Margins*, eds. Burk, C.A., and Drake, C.L., Springer-Verlag, New York, 249-260.

Segall, P. and J. R. Rice, (1995), Dilatancy, compaction, and slip instability of a fluid-infiltrated fault, *J. Geophys. Res.*, **100**(B11), 22155-22171.

Shelly, D. R., G. C. Beroza, and S. Ide, (2007), Non-volcanic tremor and low-frequency earthquake swarms, *Nature*, **446**, 305-307.

Shelly, D. R., G. C. Beroza, S. Ide, and S. Nakamura, (2006), Low-frequency earthquakes in Shikoku, Japan, and their relationship to episodic tremor and slip, *Nature*, **442**, 188-191.

Shibazaki, B. and T. Shimamoto, (2007), Modelling of short-interval silent slip events in deeper subduction interfaces considering the frictional properties at the unstable-stable transition regime, *Geophys. J. Int.*, **171**(1), 191-205.

Shimazaki, K. and T. Nakata, (1980), Time-predictable recurrence model for large earthquakes, *Geophys. Res. Lett.*, **7**(4), 279-282.

Sibson, R. H., (1973), Interactions between temperature and pore fluid pressure during earthquake faulting—A mechanism for partial or total stress relief, *Nature*, **243**, 66-68.

Sibson, R. H., (1974), Frictional constraints on thrust, wrench and normal faults, *Nature*, **249**, 542-544.

Sibson, R. H., (1975), Generation of pseudotachylyte by ancient seismic faulting, *Geophys. J. Roy. Astr. Soc.*, **43**(3), 775-794.

Sibson, R. H., (1986), Brecciation processes in fault zones—Inferences from earthquake rupturing, *Pure Appl. Geophys.*, **124**(1-2), 159-175.

Sibson, R. H., (1990), Conditions for fault-valve behaviour, *Geological Society, London, Special Publications*, **54**, 15-28.

Sibson, R. H., (1992), Implications of fault-valve behavior for rupture nucleation and recurrence, *Tectonophys.*, **211**(1-4), 283-293.

Sibson, R. H. and J. V. Rowland, (2003), Stress, fluid pressure and structural permeability in seismogenic crust, North Island, New Zealand, *Geophys. J. Int.*, **154**(2), 584-594.

Sibson, R. H. and V. Toy, (2006), The habitat of fault-generated pseudotachylyte: Presence vs. absence of friction-melt, In: *Radiated Energy and the Physics of Faulting*, edited by Abercrombie, R., A. McGarr, G. Di Toro, and H. Kanamori, *Geophysical Monograph Series*, **170**, 153-166, American Geophysical Union, Washington, D.C.

Silver, E., M. Kastner, A. Fisher, J. Morris, K. McIntosh, and D. Saffer, (2000), Fluid flow paths in the Middle America Trench and Costa Rica margin, *Geology*, **28**(8), 679-682.

Silver, E. A., M. J. Ellis, N. A. Breen, and T. H. Shipley, (1985), Comments on the growth of accretionary wedges, *Geology*, **13**(13), 6-9.

Tanikawa W., T. Mishima, T. Hirono, W. Lin, T. Shimamoto, W. Soh, and S.-R. Song, (2007), High magnetic susceptibility produced in high-velocity frictional tests on core samples from the Chelungpu fault in Taiwan, *Geophys. Res. Lett.*, **34**, L15304, doi:10.1029/2007GL030783.

Tanioka, Y. and K. Satake, (2001), Detailed coseismic slip distribution of the 1944 Tonankai earthquake estimated from tsunami waveforms, *Geophys. Res. Lett.*, **28**(6), 1075-1078.

Ujiie, K., (2005), Fault rock analysis of the northern part of the Chelungpu Fault and its relation to earthquake faulting of the 1999 Chi-Chi earthquake, Taiwan, *Island Arc*, **14**(1), 2-11.

Ujiie, K., A. Yamaguchi, G. Kimura, and S. Toh, (2007a), Fluidization of granular material in a subduction thrust at seismogenic depths, *Earth Planet. Sci. Lett.*, **259**(3-4), 307-318.

Ujiie, K., H. Yamaguchi, A. Sakaguchi, and S. Toh, (2007b), Pseudotachylytes in an ancient accretionary complex and implications for melt lubrication during subduction zone earthquakes, *J. Struct. Geol.*, **29**(4), 599-613.

Ujiie, K., A. Yamaguchi, and S. Taguchi, (2008), Stretching of fluid inclusions in calcite as an indicator of frictional heating on faults, *Geology*, **36**(2), 111-114.

Uyeda, S. and H. Kanamori, (1979), Back-arc opening and the mode of subduction, *J. Geophys. Res.*, **84**(B3), 1049-1061.

Vannucchi, P. and G. Bettelli, (2002), Mechanisms of subduction accretion as implied from the broken formations in the Apennines, Italy, *Geology*, **30**(9), 835-838.

Vannucchi, P. and L. Leoni, (2007), Structural characterization of the Costa Rica decollement: Evidence for seismically-induced fluid pulsing, *Earth Planet. Sci. Lett.*, **262**(3-4), 413-428.

Vannucchi, P., C. R. Ranero, S. Galeotti, S. M. Straub, D. W. Scholl, and K. McDougall-Ried, (2003), Fast rates of subduction erosion along the Costa Rica Pacific margin: Implications for nonsteady rates of crustal recycling at subduction zones, *J. Geophys. Res.*, **108**(B11), 2511, doi:10.1029/2002JB002207.

Vannucchi, P., F. Remitti, and G. Bettelli, (2008), Geologic record of fluid flow and seismogenesis along an erosive subducting plate boundary, *Nature*, **451**, 699-703.

Vannucchi, P., D. W. Scholl, M. Meschede, and K. McDougall-Reid, (2001), Tectonic erosion and consequent collapse of the Pacific margin of Costa Rica: Combined implications from ODP Leg 170, seismic offshore data, and regional geology of the Nicoya Peninsula, *Tectonics*, **20**(5), 649-668.

Vidale, J. E., W. L. Ellsworth, A. Cole, and C. Marone, (1994), Variations in rupture process with recurrence interval in a repeated small earthquake, *Nature*, **368**, 624-626.

von Huene, R. and S. Lallemand, (1990), Tectonic erosion along the Japan and Peru convergent margins, *Geol. Soc. Am. Bull.*, **102**(6), 704-720.

von Huene, R. and C. R. Ranero, (2003), Subduction erosion and basal friction along the sediment-starved convergent margin off Antofagasta, Chile, *J. Geophys. Res.*, **108**(B2), 2079, doi:10.1029/2001JB001569.

von Huene, R., C. R. Ranero, and P. Vannucchi, (2004a), Generic model of subduction erosion, *Geology*, **32**(10), 913-916.

von Huene, R., C. R. Ranero, and P. Watts, (2004b), Tsunamigenic slope failure along the Middle America Trench in two tectonic settings, *Marine Geology*, **203**(3-4), 303-317.

Waldhauser, F. and W. L. Ellsworth, (2002), Fault structure and mechanics of the Hayward Fault, California, from double-difference earthquake locations, *J. Geophys. Res.*, **107**(B3), 2054, doi:10.1029/2000JB000084.

Wang, K. L. and Y. Hu, (2006), Accretionary prisms in subduction earthquake cycles: The theory of dynamic Coulomb wedge, *J. Geophys. Res.*, **111**, B06410, doi:10.1029/2005JB004094.

Wibberley, C. A. J. and T. Shimamoto, (2005), Earthquake slip weakening and asperities explained by thermal pressurization, *Nature*, **436**, 689-692.

Wiens, D. A., S. Pozgay, J. Conder, E. Emry, M. Barklage, H. Shiobara, and H. Sugioka, (2007), Seismogenic characteristics and seismic structure of the Mariana Arc System: Comparison with Central America, In: *Workshop to Integrate Subduction Factory and Seismogenic Zone Studies in Central America*, edited by Silver, E., T. Plank, K. Hoernle, M. Protti, G. Alvarado, and V. González, Heredia, Costa Rica, 121.

Yoshida, S. and N. Kato, (2003), Episodic aseismic slip in a two-degree-of-freedom block-spring model, *Geophys. Res. Lett.*, **30**(13), 1681, doi:10.1029/2003GL017439.

Zattin, M., A. Landuzzi, V. Picotti, and G. G. Zuffa, (2000), Discriminating between tectonic and sedimentary burial in a foredeep succession, Northern Apennines, *J. Geol. Soc. London*, **157**(3), 629-633.

# Fault Zone Structure and Deformation Processes along an Exhumed Low-Angle Normal Fault

## Implications for Seismic Behavior

**Cristiano Collettini**
*Geologia Strutturale e Geofisica, Dipartimento di Scienze della Terra, Università degli Studi di Perugia, Perugia, Italy*

**Robert E. Holdsworth**
*Reactivation Research Group, Department of Earth Sciences, University of Durham, Durham, United Kingdom*

**Steven A. F. Smith**
*Reactivation Research Group, Department of Earth Sciences, University of Durham, United Kingdom*

Significant controversy exists as to whether low-angle normal faults (i.e., normal faults dipping less than 30°) can be seismically active in the brittle upper crust. In this chapter, we present a comprehensive review of our field studies on the exhumed Zuccale low-angle normal fault (central Italy) aimed at characterizing the fault zone structure, and we use geological evidence to make inferences about the possible seismic behavior of the detachment. The key component of the fault zone is a pervasively foliated fault core, crosscut by numerous carbonate mineral veins interpreted to be hydrofractures, which is sandwiched between hanging wall and footwall blocks where the deformation is entirely brittle. To explain this, we suggest that an initial phase of frictional brecciation and cataclasis increased fault zone permeability and facilitated the pervasive grain-scale influx of $CO_2$-rich hydrous fluids into the fault zone. The fluids reacted with the fine-grained cataclasites and triggered low-grade alteration and the onset of stress-induced dissolution and precipitation processes (i.e., pressure solution). The resulting switch from frictional to pressure-solution accommodated deformation led to shear localization and the formation of a foliated fault core rich in phyllosilicates. We predict a fault slip-behavior in which aseismic creep occurs by pressure-solution-accommodated "frictional-viscous" slip along the phyllosilicate-rich foliate at low-friction coefficients (0.2 to 0.3), with transient small ruptures induced by local buildups of fluid overpressure trapped below the low-permeability fault core.

---

## 1. INTRODUCTION

Field geologists interested in fault mechanics study exhumed fault zones to characterize the fault zone structure (geometry of the fault core and damage zone, fault kinematics, fault rock distribution, and microstructural evolution) and infer the deformation processes operating at depth. The results of this type of investigation, although predominantly qualitative, are essential for understanding the mechanical, hydraulic, and rheological properties of the fault zones (e.g., *Holdsworth et al.*, 2001; *Handy et al.*, 2007). However, except for the small percentage of faults containing pseudotachylytes, for which there is a consensus about their earthquake-induced nature (e.g., *Sibson*, 1975; *Spray*, 1992; *Di Toro and Pennacchioni*, 2005), fault structures or fault rocks that unambiguously represent a seismic rupture have not been identified (see *Cowan*, 1999, for a comprehensive review).

One way to bridge the gap between the study of exhumed faults and earthquakes is to compare ancient exhumed faults with active faults possessing similar tectonic characteristics. For example, the structural and mechanical characteristics of the Carboneras fault, a crustal-scale strike-slip fault possessing a 1-km-wide fault core made of several strands of phyllosilicate-rich fault gouge (*Faulkner et al.*, 2003), have been interpreted as an exhumed analog of the creeping section of the San Andreas Fault near Parkfield (*Faulkner et al.*, 2003). The narrow and low-permeability central slip zone well documented along the Median Tectonic Line in Japan (*Wibberley and Shimamoto*, 2003; *Jefferies et al.*, 2006) has been interpreted as the most likely site for earthquake propagation induced by thermal pressurization processes (*Wibberley and Shimamoto*, 2005). A detailed study of the Punchbowl Fault, an exhumed strand of the San Andreas system, has defined a ∼30-cm-thick fault core consisting of two lithologically distinct ultracataclasite layers separated by a principal slip zone less than 2 cm thick, across which at least 2 km of strike-slip displacement is accommodated (*Chester and Chester*, 1998). By analogy with the seismic activity found along portions of the San Andreas Fault, the 2-cm-thick principal slip zone is inferred to have developed by seismogenic processes (*Chester and Chester*, 1998).

In this chapter, we contribute to the field-based characterization of fault zone structure by studying an exhumed low-angle normal fault cropping out on the isle of Elba (in central Italy). We integrate information from our previously published papers (*Collettini and Holdsworth*, 2004; *Collettini et al.*, 2006b; *Smith et al.*, 2007) and present a comprehensive analysis of fault zone structure and deformation processes operating at depth. Finally we compare our field data with both laboratory experiments and seismicity found along an active structure of the same tectonic system to infer the character of slip along low-angle normal faults at depths greater than 3 km in the upper crust.

## 2. REGIONAL SETTING

The Northern Apennines consist of a northeast verging thrust-fold belt formed as the result of the collision between the European continental margin (Sardinia-Corsica block) and the Adriatic microplate (e.g., *Alvarez*, 1972; *Reutter et al.*, 1980). Northeastward migration of compression is contemporaneous with hinterland extension. Extension is largely accommodated along shallowly (~15°) east-dipping normal faults (*Barchi et al.*, 1998; *Decandia et al.*, 1998; *Collettini et al.*, 2006b), including the microseismically active Alto Tiberina Fault located in the presently extending Umbria region (*Barchi et al.*, 1998; *Boncio et al.*, 2000; *Chiaraluce et al.*, 2007), and the older exhumed Zuccale Fault (ZF) on Elba (*Keller and Pialli*, 1990; *Collettini and Holdsworth*, 2004).

The superposition of extension following compression, which is typical of the Northern Apennines, is well documented on Elba. Geologically, the structure of the island (Figure 1) consists of five thrust sheets stacked up during late Cretaceous to early Miocene compression (*Trevisan*, 1950; *Keller and Pialli*, 1990). These thrust sheets are crosscut by later mid-Miocene to early Pliocene extensional faults (*Trevisan*, 1950; *Keller and Pialli*, 1990). Extension was broadly associated with intrusion of granite porphyry sheets (~8 Ma) (*Rocchi et al.*, 2002), the Monte Capanne granodiorite (~6.9 Ma) (*Saupé et al.*, 1982; *Dini et al.*, 2002; *Gagnevin et al.*, 2004), and the Porto Azzurro monzogranite and associated dykes (~5.9 Ma) (*Saupé et al.*, 1982; *Maineri et al.*, 2003). The Zuccale Fault is the most prominent extensional structure on the island and cuts down-section across the thrust sheet stack formed during Cretaceous to early Miocene compression. This produces a repetition of the five thrust "complexes" in central and eastern Elba (complexes I-V; Figure 1). Stratigraphic separations across the Zuccale Fault suggest a normal fault offset of 6 to 8 km (*Keller and Coward*, 1996; *Collettini and Holdsworth*, 2004). Exhumation of the Zuccale Fault (i.e., the depths at which the extension-related fault rocks formed) is not well constrained. However, the juxtaposition of Cretaceous flysch from the uppermost part of the thrust sheet stack (complex V) against rocks of the lowermost thrust sheet (complex I) suggests an exhumation of at least 3 to 6 km, the original thickness of the thrust sequence before exhumation (*Collettini and Holdsworth*, 2004).

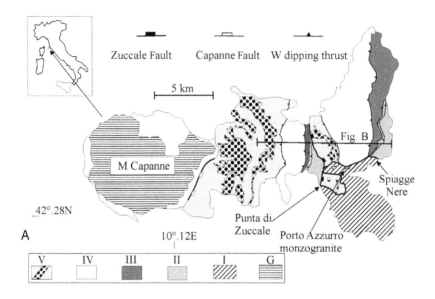

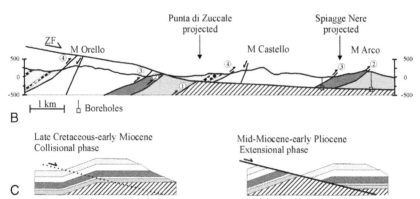

**FIGURE 1** (a) Schematic geological and structural map of the isle of Elba: complex V, late Cretaceous flysch intruded by granite porphyry (~8 Ma); complex IV, Ligurian ophiolites; complex III, Tuscan carbonate sequence; complex II, Tuscan metamorphic sequence; complex I, basement schists; G, Mt. Capanne granodiorite (~6.9 Ma) and Porto Azzurro monzogranite (~5.9 Ma). (b) Geological cross section through central and eastern Elba. The section has been constructed along section 1 of the geological map of Elba (*Trevisan et al.*, 1967). The Zuccale Fault position at depth has been located by using borehole data (*Bortolotti et al.*, 2001). (c) Schematic diagram reconstructing the nappe structure at the end of the collisional phase (early Miocene) which was crosscut by the Zuccale Fault (middle Miocene-early Pliocene). The repetition of the complexes in central and eastern Elba is due to the activity of the Zuccale Fault that cuts down-section and displaces the thrust sheets eastward (modified from *Trevisan*, 1950; *Keller and Coward*, 1996).

## 3. FAULT ZONE ARCHITECTURE

### 3.1. Geometry and Kinematics

The geometry and kinematics of the Zuccale Fault have been studied in spec-
tacular, almost completely exposed coastal sections at Punta di Zuccale and
Spiagge Nere (Figure 1). At Punta di Zuccale (Figure 2), the Zuccale Fault
separates a hanging wall sequence of upper Cretaceous flysch (complex V)
from a footwall of Palaeozoic basement schists and Triassic metasediments
of the Verrucano formation (complex I). At Spiagge Nere, the Zuccale Fault
detaches a hanging wall sequence of various rock types belonging to complexes II and III from basement schists of complex I. At both localities, the
fault dips up to 15° east but has an apparently undulating geometry, with a
local region of subhorizontal to westward dips. West-dipping segments of the
Zuccale Fault are interpreted to have been back-rotated by small-displacement
(up to few meters) listric normal faults that are well exposed in the footwall
(*Collettini and Holdsworth*, 2004; *Smith et al.*, 2007). Many footwall structures
link directly into the lower parts of the Zuccale Fault core and cause the
core to increase in thickness locally from ~3 to ~8m (*Smith et al.*, 2007).
At Punta di Zuccale, the fault zone is also crosscut by a complex system of
carbonate veins (Figure 3), many of which are hosted within, or are linked to,

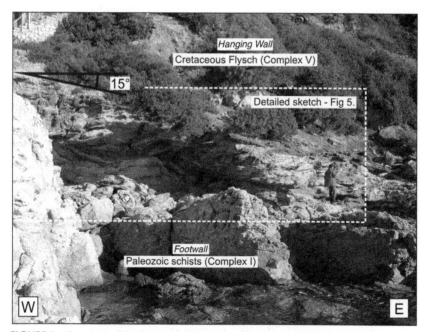

**FIGURE 2**   Geometry of the Zuccale Fault: outcrop photographs of the fault, which separates a
hanging wall sequence of upper Cretaceous flysch from a Palaeozoic basement. Person for scale.

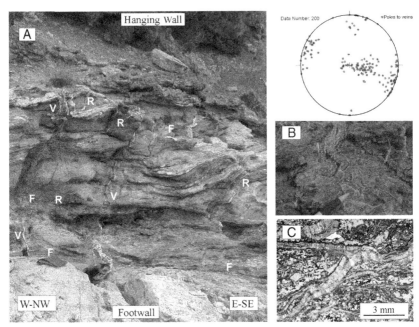

**FIGURE 3** Zuccale Fault zone hydrofractures. (a) Subvertical N-S trending (labeled V), rotated west-dipping (labeled R), and subhorizontal foliation parallel (labeled F) hydrofractures. The vertical hydrofracture located in the middle of the picture is linked to a footwall fault. (b) Details of crack-seal texture and zoned mineralization in a vertical N-S trending hydrofracture. (c) Rotated N-S trending subvertical vein. Stereoplot (equal area projection lower hemisphere) showing poles to hydrofractures.

footwall faults. These are interpreted to be hydrofractures that are arranged into a system of three orthogonal vein sets (stereoplot in Figure 3) with crosscutting relationships that suggest a complex history of development (*Collettini et al.*, 2006a). Two vertical sets trend approximately north-south and approximately east-west, and one ranges from vertical north-south to horizontal. Many of the veins show a zoned mineralization and crack-and-seal textures (Figure 3b), suggesting repeated hydrofracturing events followed by mineral precipitation processes (e.g., *Secor*, 1965; *Ramsay*, 1980). Earlier veins are locally synthetically rotated, folded, or sheared into parallelism with the main foliation (Figure 3c). The direct linkage of veins into footwall faults, together with the local rotation of veins, and development of crack-seal textures suggests that the veins formed syntectonically. Additionally, some of the veins are truncated by discrete, low-angle brittle slip surfaces within the core of the Zuccale Fault.

Kinematically, the internal part of the Zuccale Fault zone (the "fault core") is characterized by shallowly plunging east-west lineations and slickenlines (Figure 4a and stereoplot 1) with abundant C structures and C'-type shear

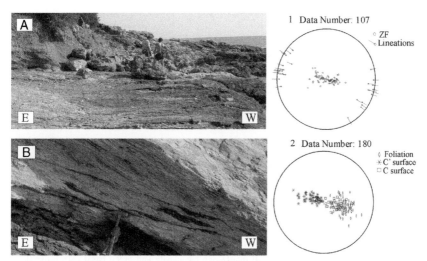

**FIGURE 4**  Geometry and kinematics of the Zuccale Fault. (a) Subhorizontal to gently west-dipping detachment within the fault zone with east-west slickenlines (coastal outcrop immediately north of Punta di Zuccale). (b) Top-to-the-east C-type shear bands in the phyllosilicate-rich foliated fault core. Stereoplot 1 (equal area projection lower hemisphere) shows poles to fault planes and slickenline lineations, whereas stereoplot 2 (equal area projection lower hemisphere) shows poles to fault core foliation, poles to C surfaces, and poles to C' surfaces.

bands bounding S-foliated domains (Figure 4b and stereoplot 2). Shear sense is consistently top-to-the-east (*Collettini et al.*, 2006b).

## 3.2. Fault Rock Distribution and Microstructures

To investigate the distribution of fault rocks, samples were collected along a series of subvertical traverses across the fault zone. Microstructural studies were supplemented by SEM (scanning electron microscopy) and XRD (X-ray diffraction) analyses to help in the identification of specific fine-grained mineral phases.

In the low-angle fault core and its immediate footwall, up to five typical fault rock units may be present, although these are not always preserved everywhere (Figure 5) (*Collettini and Holdsworth*, 2004; *Smith et al.*, 2007). From bottom to top these are L1, footwall frictional breccias, and cataclasite (Figure 5b). In some exposures, the top few centimeters of this fault rock unit are represented by a basal green cataclasite; L2, foliated phyllonites with C-type and C'-type shear bands (Figure 5c); L3, highly foliated unit of green fault rocks derived from an ultramafic protolith (Figure 5b); L4, foliated cataclasite and carbonate vein-rich domain incorporating sigmoidal pods and lenses of carbonate, calc-schists, and ultramafic material (Figure 5b); and L5, foliated fault breccia and gouge (Figure 5d).

Individual fault rock units vary considerably in thickness in the Zuccale Fault. In particular, the phyllonites (L2) are found only as isolated lenses

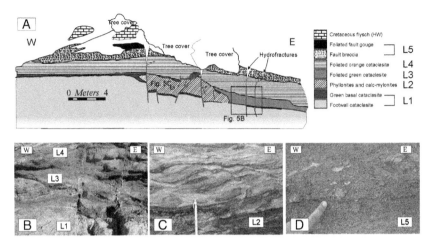

**FIGURE 5**    Zuccale Fault rock zonation (after *Smith et al.*, 2007). (a) Field sketch of the Zuccale Fault at Punta di Zuccale. (b) Footwall quartz-rich cataclasite and green basal cataclasite (L1), foliated green tremolite-rich fault rocks (L3), and foliated cataclasite (L4). (c) Foliated phyllonites preserved in the hanging wall block of a footwall structure with C-type and C'-type shear bands (see text for explanation). (d) Upper foliated fault gouge (L5) showing kinematic indicators pointing to a top-to-the-east shear.

of material in the immediate downthrown hanging wall regions of small-displacement, steeply dipping footwall faults where they locally thicken the fault core (*Smith et al.*, 2007). For example, at Punta di Zuccale, along the east-west coastal section, the phyllonites are found as an isolated, fault-bounded lens ∼8 m long and ∼3 m high (Figure 5), whereas in the north-south section they are interlayered with the green tremolite-rich fault rocks (L3) and occur in a fault-bounded lens ∼20 m long and ∼5 m high. Small outcrops of phyllonites are also preserved in the immediate hanging wall of footwall faults at Spiagge Nere (*Smith et al.*, 2007). In areas where the fault core of the Zuccale Fault is not down-faulted (e.g., Figure 5b), the fault rock sequence is dominated by the foliated cataclasites (L4). In the field, the base of L4 is marked by a discrete brittle detachment surface, which crosscuts and terminates all high-angle footwall faults. This suggest that during the development of the Zuccale Fault, extensional movements along the low-angle detachments and small-displacement, high-angle footwall structures have alternated (for details, see *Smith et al.*, 2007). The alternating high-angle and low-angle scenario provides a compelling explanation for the current preservation or excision of the early formed phyllonites in the core of the Zuccale Fault.

The fault rock zonation is clearly preserved at the microscale at Punta di Zuccale (Figure 6). In L1 (Figure 6a), just below the base of the low-angle foliated fault core, the fault rocks are represented by a breccias and cataclasites with quartz clasts derived from the Verrucano formation (complex I), set in a carbonate-chlorite-quartz matrix. On approaching the fault core, the

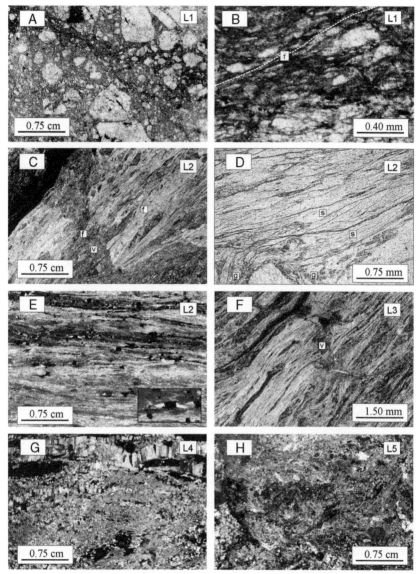

**FIGURE 6**   Zuccale Fault rocks. (a) Cataclasite of the footwall block with randomly oriented quartz-rich clasts set in a carbonate-chlorite-quartz matrix. (b) At the footwall-fault core boundary, the cataclasite starts to be smeared out into foliation planes (f). (c) Foliated chlorite and talc-rich phyllonites with crosscutting relationships between a calcite vein (v) and phyllosilicate-rich layers (f). In the middle of the vein some phyllosilicate layers cut across the vein (f). (d) Dark dissolution seam (s) and fibrous overgrowth (g) in the foliated fault core. (e) Iron-rich layers with iron oxides and chlorite fringes. (f) Tremolite talc and chlorite foliation within the green fault rocks. Partially folded calcite vein (v). (g) Calcite-rich veins within the foliated cataclasite. (h) Clay-rich fault gouge at the top of the fault zone. Images are taken across polarized light and are oriented east-west (east to the right). L1 to L5 represent the position in the tectonic stratigraphy (compare with Figure 5).

cataclasite becomes more fine-grained, and the small clasts seem to float in a matrix rich in chlorite and calcite. At the same time, the cataclastic texture becomes increasingly "smeared out" into a new, overprinting foliation that develops subparallel to the low-angle detachment faults (Figure 6b). L2 is a strongly foliated unit rich in phyllonites, with a significant proportion of talc present (e.g., Figure 6c). The mutual crosscutting relationships observed between phyllosilicate-rich layers and carbonate veins suggest their contemporaneous development during fault activity (Figure 6c). In L2, a substantial component of the foliation and lineation derives from the development of numerous solutions seams and fibrous overgrowths (Figure 6d). During fault activity, many foliation planes in L2 were pervaded by iron-rich fluids forming reddish horizons rich in iron oxides with the associated development of fibrous chlorite overgrowths in strain fringes (Figure 6e). In the green tremolite-rich fault rocks, L3, the foliation is defined by fine-grained aligned aggregates of tremolite, talc, and chlorite, and, once again, the development of hydrous minerals along the foliation planes is contemporaneous with the development of calcite-rich veins. Hence, some veins crosscut while others are folded and are progressively transposed into the foliation (Figure 6f). L4 is a foliated cataclasite full of calcite-rich veins, the most recent of which show a crack-seal texture, whereas the others have been folded and reworked during the fault activity (Figure 6g). The uppermost unit of the fault zone, L5, comprises a mixture of coarse foliated breccias and clay-rich fault gouges (Figure 6h).

## 4. DISCUSSION

### 4.1. Fault Rock Evolution

The key feature within the Zuccale Fault zone is the preservation of a highly strained, foliated fault core, sandwiched between a footwall and the hanging wall in which top-to-the-east extensional fault-related deformation is exclusively brittle.

Most of the foliated fault rocks found within the core of the Zuccale Fault preserve typical cataclastic textures that are overprinted by a foliation (Figures 6a to 6f). The development of foliations can occur during wholly cataclastic faulting processes (e.g., *Chester et al.*, 1985), but there is an abundance of evidence to suggest that this is not the case for the Zuccale Fault. The widespread preservation of cataclastic textures certainly suggests that grain-scale deformation processes were initially frictional and brittle throughout most of the fault zone. This is consistent with a total exhumation in the range of 3 to 6 km. To judge from the intensity of later alteration in the fault core, it appears that the initial development of fracture networks and grain-scale brittle dilatancy led to an increase in permeability that promoted fluid influx into the fault core (e.g., Figure 6b). Fluids reacted with the fine-grained, precrushed cataclasites, triggering the widespread growth of fine-grained aggregates of weak hydrous phases such as chlorite, talc, and other clay minerals (Figures 6c-6f). This led to a general framework collapse and compaction in the fault core generating a strong alignment of the fine-grained phyllosilicates subparallel to the low-angle fault zone. A substantial component of the foliation within the fault core results from the development of numerous solution seams and fibrous overgrowths (Figures 6c to 6e), fluid-filled veins and fractures (Figures 6b, 6e, 6f, and 6g), and foliation-parallel mineralized horizons (Figure 6e), structures that are consistent with the widespread onset and operation of fluid-assisted diffusive mass transfer mechanisms such as pressure solution. The asymmetric character of these features, consistently showing a top-to-the-east shear sense, strongly suggests that operation of dissolution-precipitation deformation mechanism was synchronous with activity along the Zuccale Fault. At the same time, the development of a highly foliated phyllosilicate-rich fault core enhanced the sealing capacity of the Zuccale Fault and seems to have promoted the attainment of local fluid overpressures that led to transient embrittlement episodes in the fault zone indicated by the development of syntectonic vein systems characterized by crack-seal textures (Figure 3).

## 4.2. The Mechanical Paradox of Low-Angle Normal Faults

The possibility that low-angle normal faults like the Zuccale Fault (i.e., normal faults dipping less than 30°) can accommodate significant extension in the brittle upper crust is still a strongly debated topic in the literature (discussed later). In extensional settings, Anderson-Byerlee frictional fault mechanics, assuming a state of stress characterized by vertical $\sigma1$ and faults possessing static friction coefficients, $\mu s$, in the range $0.6 < \mu s < 0.85$, predicts no slip on normal faults dipping less than 30°. This is because, under these conditions, it is mechanically easier to form a new fault instead of reactivating an existing fault dipping less than 30° (*Sibson*, 1985). This mechanical approach is consistent with global seismological data where no moderate-to-large normal-slip earthquakes occur in the brittle intracontinental crust on normal faults dipping less than 30° (*Jackson and White*, 1989; *Collettini and Sibson*, 2001). Potential examples of seismic activity on low-angle normal faults have been proposed for a few moderate and large earthquakes (*Abers*, 1991; *Wernicke*, 1995; *Axen*, 1999), but the low-angle interpretation of the ruptures in these isolated examples remains controversial. By contrast, geological evidence for active low-angle normal faulting is widespread and documented mainly by field-based structural studies (*Lister and Davis*, 1989; *Axen*, 1999; *Sorel*, 2000; *Hayman et al.*, 2003; *Collettini and Holdsworth*, 2004; *Smith et al.*, 2007) and from the interpretation of seismic reflection profiles (*Roy and Kenneth*, 1992; *Barchi et al.*, 1998; *Laigle et al.*, 2000; *Floyd et al.*, 2001).

To explain movements on low-angle normal faults, we need to look beyond Anderson-Byerlee frictional theory. We therefore propose a slip model for the low-angle normal faults (LANFs) of the Northern Apennines based on field and microstructural observations along the Zuccale Fault and striking similarities that exist with the results of analog experiments set up to investigate weakening mechanisms in phyllosilicate-rich fault rocks.

## 4.3. A Slip Model for Low-Angle Normal Faults (Evidences That ZF Was Active as LANF)

There is compelling evidence to suggest that only minor passive reorientation of the Zuccale Fault has occurred on a regional scale and that it was active as a low-angle (dip <20°), east-dipping structure within an extensional stress field with a vertical $\sigma 1$ axis. In particular,

1. Moderate-to-large offsets of the Zuccale detachment along later, more steeply dipping normal faults are not observed.
2. Preexisting thrusts in both the hanging wall and footwall of the detachment are west-dipping and show typical Andersonian dips (e.g., 30°) consistent with the earlier phase of east-northeast-verging Apenninic compression (see Figure 1b and *Collettini and Holdsworth*, 2004, for details).
3. Within the fault zone, the latest hydrofracture veins are subvertical tensile fractures (Figure 3) consistent with a vertical $\sigma 1$ axis (*Collettini et al.*, 2006b), whereas earlier veins are sheared synthetically into parallelism with the foliation. East-dipping vein sets, as would be required in a passive rotation model of the detachment, are not observed.
4. The synchronous or alternating activity of the small-offset high-angle footwall structures and the larger offset low-angle detachments in the Zuccale Fault core suggest that the detachment has not experienced a decrease in dip of more than ~5° (*Smith et al.*, 2007).
5. Offshore seismic reflection profiles show that the Zuccale Fault dips gently to the east over a depth range from the surface—12 km. High-angle, west-dipping antithetic structures are present in its hanging wall (*Keller and Coward*, 1996).

In the presently extending area of the Northern Apennines, a dense seismic network highlighted the presence of a microseismically active low-angle normal fault (*Chiaraluce et al.*, 2007). The low-angle, east-dipping plane defined by microseismicity coincides with the geometry of the known Alto Tiberina normal fault (ATF) retrieved from geological observations and the interpretation of depth-converted seismic profiles (*Barchi et al.*, 1998; *Boncio et al.*, 2000). Inversion of focal mechanisms in the area shows that the microseismically active detachment is moving within a brittle crust characterized by a vertical $\sigma 1$ axis (*Boncio et al.*, 2000; *Chiaraluce et al.*, 2007).

The fault rock characteristics and evolution preserved along the Zuccale Fault suggest an evolution with time from frictional deformation that produced breccias and cataclasites to a regime increasingly dominated by dissolution-precipitation accommodated slip that is associated with the development of

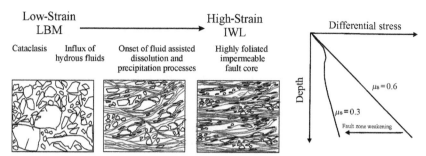

**FIGURE 7**   Long-term fault zone weakening of the Zuccale Fault. The switch from a cataclastic (i.e., a load-bearing microstructure [LBM]) to a dissolution-precipitation accommodated deformation (i.e., an interconnected weak layer [IWL]) due to fluid-rock interaction produced a chemical weakening of the fault zone—that is, a decrease in friction from Byerlee's values to values of 0.3 or less.

the foliated fault core (Figure 7). The localization of strain into the foliated phyllosilicate-rich fault core suggests that the fault became weak, and the widespread mineral alteration and onset of pressure solution suggest significant chemical fluid-rock interaction. An investigation into the origins of the fluids responsible for weakening will be an important next step in our understanding of this fault system and is currently under way.

Laboratory experiments designed to investigate the effects of pressure solution in phyllosilicate-rich fault rocks record a remarkable and almost identical deformation history, from initial cataclastic deformation to pressure-solution-accommodated slip in the presence of a saturated pore fluid (*Bos and Spiers*, 2001, 2002). In the laboratory, pressure-solution-accommodated slip ("frictional-viscous creep") produces highly organized foliated microstructures with abundant C-type and C'-type shear bands similar to those preserved along the Zuccale Fault. In the experimental analogs, the switch to stress-induced dissolution-precipitation is associated with a decrease in friction coefficients from Byerlee's values of ~0.6 to 0.85 to values of 0.3 or less.

The integration of field and laboratory data suggests that a low-angle normal fault like the Zuccale Fault or the microseismically active Alto Tiberina Fault (*Chiaraluce et al.*, 2007) can move in a stress field characterized by a vertical $\sigma 1$ axis because the development of a phyllosilicate-rich fault core lowers the static friction coefficient and the fault is no longer severely misoriented for reactivation. At the same time, the most likely behavior of the detachment would be aseismic creep because at low slip-rates, the pressure-solution-accommodated deformation is a velocity-strengthening process (*Niemeijer and Spiers*, 2005). The presence of talc makes this all the more likely because experimental results show that this weak mineral consistently exhibits velocity-strengthening behavior during deformation across a broad range of crustal temperatures (*Moore and Lockner*, 2007). We suggest that the local short-lived attainment of fluid overpressures may trigger frictional instabilities producing hydrofractures, as documented by carbonate veining along the Zuccale Fault, possibly explaining the microseismicity observed along the active Alto Tiberina Fault (*Collettini and Barchi*, 2002; *Collettini and Holdsworth*, 2004; *Chiaraluce et al.*, 2007). It is worth noting that (1) the observed microseismicity along the Alto Tiberina Fault is not enough to achieve the observed long-term slip rate of 1 mm/y and (2) no moderate-to-large historical earthquakes have been recorded for this sector of the Northern Apennines in the past 2000 years (*Collettini and Holdsworth*, 2004; *Chiaraluce* et al., 2007). These two observations suggest that the fault is incapable of sustaining significant stress levels and that a large amount of its displacement may be accommodated aseismically, perhaps by dissolution-precipitation mechanisms like those responsible for producing the foliated fault rocks preserved along the Zuccale Fault.

## 5. CONCLUSIONS

In the Northern Apennines of Italy, a large amount of extension is accommodated along shallowly (~15°) east-dipping normal faults. In this chapter we have reviewed our field studies on the exhumed Zuccale low-angle normal fault cropping out in the isle of Elba.

The fault dips up to 15° east but has an undulating geometry, with local areas of subhorizontal to westward dips. The west-dipping segments seem to be associated with back-rotated arrays of small-displacement (up to a few meters) listric normal faults. Some footwall structures cut directly into the base of the Zuccale Fault and cause the fault core to increase in thickness from ~3 to ~8m. They can be used as a reliable indication that the Zuccale Fault has not experienced significant passive reorientation. Kinematically, the internal part of the Zuccale Fault is characterized by shallowly plunging east-west lineations and slickenlines with abundant C structures and C'-type shear bands that display consistent top-to-the-east shear sense.

A characteristic and striking feature of the Zuccale Fault zone is the preservation of a highly strained, foliated fault core, sandwiched between a footwall and hanging wall in which top-to-the-east extensional fault-related deformation is exclusively brittle. The foliated fault rocks present are all derived from initially cataclastically deformed protoliths and include chlorite and talc-rich phyllonites, green tremolite-rich fault rocks, and coarse foliated fault breccia and gouge. Carbonate veins, interpreted as hydrofractures, are preserved widely both as early features pervasively sheared into concordance with the foliation and as later crosscutting features.

Field-based and microstructural studies of the Zuccale Fault suggest that an initial phase of frictional cataclasis increased fault zone permeability favoring the influx of hydrous fluids. These fluids reacted with the fine-grained cataclasite and triggered framework collapse resulting in the formation of a foliated fault core dominated by weak hydrous phases such as chlorite, talc, and clay minerals, which helped facilitate pressure-solution-accommodated slip processes. The foliated fault core also acted as a barrier to fluids migrating within the footwall, leading to the localized buildup of fluid overpressure and transient embrittlement of the fault as testified in the field by the syntectonic vein system.

We predict a slip model for the Zuccale Fault in which aseismic dissolution-precipitation creep occurs on a weak, slow-moving (slip rate ∼1 mm/a), low-angle normal fault interspersed with small seismic ruptures caused by local short-lived buildups of fluid overpressure. This model is consistent with (1) the deformation behavior of fault rock analogs and minerals like talc studied in the laboratory (*Bos and Spiers*, 2000, 2001; *Moore and Lockner*, 2007; *Moore and Rymer*, 2007; *Niemeijer and Spiers*, 2005), (2) the seismic behavior of the Alto Tiberina Fault located in the presently extending area of the Northern Apennines, and (3) the absence of moderate-to-large extensional ruptures on low-angle normal fault planes worldwide.

## REFERENCES

Abers, G. A., (1991), Possible seismogenic shallow-dipping normal faults in the Woodlark-D'Entrecasteaux extensional province, Papua New Guinea, *Geology*, **19**, 1205-1208.

Alvarez, W., (1972), Rotation of the Corsica-Sardinia microplate, *Nature*, **248**, 309-314.

Axen, G. J., (1999), Low-angle normal fault earthquakes and triggering, *Geophys. Res. Let.*, **26**, 3693-3696.

Barchi, M. R., G., Minelli, and G. Pialli, (1998), The crop 03 profile: A synthesis of results on deep structures of the Northern Apennines, *Mem. Soc. Geol. It.*, **52**, 383-400.

Boncio, P., F. Brozzetti, and G. Lavecchia, (2000), Architecture and seismotectonics of a regional low-angle normal fault zone in central Italy, *Tectonics*, **19**, 1038-1055.

Bortolotti, V., M. Fazzuoli, E. Pandeli, G. Principi, A. Babbini, and S. Corti, (2001), Geology of central and Eastern Elba Island, Italy, *Ofioliti*, **26**, 97-150.

Bos, B. and C. J. Spiers, (2001), Experimental investigation into the microstructural and mechanical evolution of phyllosilicate-bearing fault rock under conditions favouring pressure solution, *J. Struct. Geol.*, **23**, 1187-1202.

Bos, B. and C. J. Spiers, (2002), Frictional-viscous flow of phyllosilicate-bearing fault rock: Microphysical model and implications for crustal strength profiles, *J. Geophys. Res.*, **107** (B2), 2028, doi:10.1029/2001JB000301.

Chester, F. M. and J. S. Chester, (1998), Ultracataclasite structure and friction processes of the Punchbowl fault, San Andreas system, California, *Tectonophys.*, **295**(1-2), 199-221.

Chester, F. M., M. Friedman, and J. M. Logan, (1985), Foliated cataclasites, *Tectonophys.*, **11**, 139-146.

Chiaraluce L., C. Chiarabba, C. Collettini, D. Piccinini, and M. Cocco, (2007), Architecture and mechanics of an active low-angle normal fault: Alto Tiberina Fault, northern Apennines, Italy, *J. Geophys. Res.*, **112**, B10310, doi:10.1029/2007JB005015.

Collettini, C. and R. H. Sibson, (2001), Normal faults, normal friction?, *Geology*, **29**, 927-930.

Collettini, C. and M. R. Barchi, (2002), A low angle normal fault in the Umbria region (Central Italy): A mechanical model for the related microseismicity, *Tectonophys.*, **359**, 97-115.

Collettini, C. and R. E. Holdsworth, (2004), Fault zone weakening processes along low-angle normal faults: Insights from the Zuccale Fault, Isle of Elba, Italy, *J. Geol. Soc.*, **161**, 1039-1051.

Collettini, C., N. De Paola, and N. R. Goulty, (2006a), Switches in the minimum compressive stress direction induced by overpressure beneath a low-permeability fault zone, *Terra Nova*, **18**, 224-231.

Collettini, C., N. De Paola, R. E. Holdsworth, and M. R. Barchi, (2006b), The development and behaviour of low-angle normal faults during Cenozoic asymmetric extension in the Northern Apennines, Italy, *J. Struct. Geol.*, **28**, 333-352.

Cowan, D. S., (1999), Do faults preserve a record of seismic slip? A field geologist's opinion, *J. Struct. Geol.*, **21**(8-9), 995-1001.

Decandia, F. A., A. Lazzarotto, D. Liotta, L. Cernobori, and R. Nicolich, (1998), The CROP 03 traverse: Insights on post-collisional evolution of the Northern Apennines, *Mem. Soc. Geol. It.*, **52**, 413-425.

Dini, A., F. Innocenti, S. Rocchi, S. Tonarini, and D. S. Westerman, (2002), The magmatic evolution of the late Miocene laccolith-pluton-dyke granitic complex of Elba Island, Italy, *Geol. Mag.*, **139**, 257-279.

Di Toro, G. and G. Pennacchioni, (2005), Fault plane processes and mesoscopic structure of a strong-type seismogenic fault in tonalities (Adamello batholith, Southern Alps), *Tectonophys.*, **402**, 55-80.

Faulkner, D. R., A. C. Lewis, and E. H. Rutter, (2003), On the internal structure and mechanics of large strike-slip fault zones: Field observations on the Carboneras fault in southern Spain, *Tectonophys.*, **367**, 235-251.

Floyd, J. S., J. C. Mutter, A. M. Goodliffe, and B. Taylor, (2001), Evidence for fault weakness and fluid flow within active low-angle normal fault, *Nature*, **411**, 779-783.

Gagnevin, D., J. S. Daly, and G. Poli, (2004), Petrographic, geochemical and isotopic constraints on magma dynamics and mixing in the Miocene Monte Capanne monzogranite (Elba Island, Italy), *Lithos*, **78**, 157-195.

Handy, M. R., G. Hirth, and N. Hovius, (2007), Tectonic faults: Agents of change on a dynamic Earth, In: *The Dynamics of Fault Zones*, edited by Handy, M. R., G. Hirth, and N. Hovius, MIT Press, Cambridge, Massachusetts, 1-8.

Hayman, N. W., J. R. Knott, D. S. Cowan, E. Nemser, and A. M. Sarna-Wojcicki, (2003), Quaternary low-angle slip on detachment faults in Death Valley, California, *Geology*, **31**, 343-346.

Holdsworth, R. E., M. Stewart, J. Imber, and R. A. Strachan, (2001), The structure and rheological evolution of reactivated continental fault zones: A review and case study, In: *Continental Reactivation and Reworking*, edited by Miller, J. A., R. E. Holdsworth, I. S. Buick, and M. Hand, *Geological Society, London, Special Publications*, **184**, 115-137.

Jackson, J. A. and N. J. White, (1989), Normal faulting in the upper continental crust: Observation from regions of active extension, *J. Struct. Geol.*, **11**, 15-36.

Keller, J. V. A. and M. P. Coward, (1996), The structure and evolution of the Northern Tyrrhenian Sea, *Geol. Mag.*, **133**, 1-16.

Keller, J. V. A. and G. Pialli, (1990), Tectonics of the island of Elba: A reappraisal, *Boll. Soc. Geol. It.*, **109**, 413-425.

Laigle, M., A. Hirn, M. Sachpazi, and N. Roussos, (2000), North Aegean crustal deformation: An active fault imaged to 10 km depth by reflection seismic data, *Geology*, **28**, 71-74.

Lister, G. S. and G. A. Davis, (1989), The origin of metamorphic core complexes and detachment faults formed during Tertiary continental extension in the northern Colorado River region, USA, *J. Struct. Geol.*, **11**, 65-93.

Maineri, C., M. Benvenuti, P. Costaglioli, A. Dini, P. Lattanzi, G. Ruggieri, and I. M. Villa, (2003), Sericitic alteration at the La Crocceta deposit (Elba Island, Italy): Interplay between magmatism, tectonics and hydrothermal activity, *Mineralium Deposita*, **38**, 67-86.

Moore, D. E. and D. A. Lockner, (2007), Comparative deformation behaviour of minerals in serpentinized ultramafic rock: Application to the slab-mantle interface in subduction zones, *International Geology Review*, **49**, 401-415.

Moore, D. E. and M. J. Rymer, (2007), Talc-bearing serpentinite and the creeping section of the San Andreas Fault, *Nature*, **448**, 795-797.

Niemeijer, A. R. and C. J. Spiers, (2005), Influence of phyllosilicates on fault strength in the brittle–ductile transition: Insights from rock analogue experiments, In: *Microstructural Evolution and Physical Properties in High Strain Zones*, edited by Bruhn, D. and L. Burlini, *Geological Society, London, Special Publications*, **245**, 303-327.

Ramsay, J. G., (1980), The crack-seal mechanism of rock deformation, *Nature*, **284**, 135-139.

Reutter, K. J., P. Giese, and H. Closs, (1980), Lithospheric split in the descending plate: Observations from the Northern Apennines, *Tectonophys.*, **64**, T1-T9.

Rocchi, R., S. D. Westerman, A. Dini, F. Innocenti, and S. Tonarini, (2002), Two-stage growth of laccoliths at Elba Island, Italy, *Geology*, **30**, 983-986.

Roy, A. J. and L. L. Kenneth, (1992), Seismic reflection evidence for seismogenic low-angle faulting in south-eastern Arizona, *Geology*, **20**, 597-600.

Saupé, F., C. Marignac, B. Moine, J. Sonet, and J. L. Zimmerman, (1982), Datation par les méthodes K/Ar et Rb/Sr de quelques roches de la partie orientale de l'ile d'Elbe Province de Livourne, Italie, *Bulletin de Minéralogie*, **105**, 236-245.

Secor, D. T., (1965), Role of fluid pressure in jointing, *Am. J. Science*, **263**, 633-646.

Sibson, R. H., (1985), A note on fault reactivation, *J. Struct. Geol.*, **7**, 751-754.

Sibson, R. H., (1975), Generation of pseudotachylyte by ancient seismic faulting, *Geophys. J. Roy. Astr. Soc.*, **43**(3), 775-794.

Smith, S. A. F., R. E. Holdsworth, C. Collettini, and J. Imber, (2007), Using footwall structures to constrain the evolution of low-angle normal faults, *J. Geol. Soc. London*, **164**, 1187-1192.

Sorel, D., (2000), A Pleistocene and still-active detachment fault and the origin of the Corinth-Patras rift, Greece, *Geology*, **28**, 83-86.

Spray, J. G., (1992), A physical basis for the frictional melting of some rock-forming minerals, *Tectonophys.*, **204**(3-4), 205-221.

Trevisan, L., (1950), L'Elba orientale e la sua tettonica di scivolamento per gravità, *Memorie dell'Istituto Geologico dell'Università di Padova*, **16**, 1-30.

Trevisan, L., G. Marinelli, F. Barberi, G. Giglia, F. Innocenti, G. Raggi, P. Squarci, L. Taffi, and C. A. Ricci, (1967), *Carta Geologica dell'Isola d'Elba*. Scala 1:25.000, Consiglio Nazionale delle Ricerche, Gruppo di Ricerca per la Geologia dell'Appennino centro-settentrionale e della Toscana, Pisa.

Wernicke, B., (1995), Low-angle normal faults and seismicity: A review, *J. Geophys. Res.*, **100**, 20159-20174.

Wibberley, C. A. J. and T. Shimamoto, (2003), Internal structure and permeability of major strike-slip fault zones: The Median Tectonic Line in W. Mie Prefecture, S. W. Japan, *J. Struct. Geol.*, **25**, 59-78.

Wibberley, C. A. J. and T. Shimamoto, (2005), Earthquake slip weakening and asperities explained by thermal pressurization, *Nature*, **436**, 689-692.

## Chapter 5

# Pseudotachylytes and Earthquake Source Mechanics

**Giulio Di Toro**
*Dipartimento di Geoscienze, Università di Padova, Italy*
*Istituto Nazionale di Geofisica e Vulcanologia, Rome, Italy*

**Giorgio Pennacchioni**
*Dipartimento di Geoscienze, Università di Padova, Italy*
*Istituto Nazionale di Geofisica e Vulcanologia, Rome, Italy*

**Stefan Nielsen**
*Istituto Nazionale di Geofisica e Vulcanologia, Rome, Italy*

Destructive earthquakes nucleate at depth (10 to 15 km), therefore monitoring active faults at the Earth's surface, or interpreting seismic waves, yields only limited information on earthquake mechanics. Tectonic pseudotachylytes (solidified friction-induced melts) decorate some exhumed ancient faults and remain, up to now, the only fault rocks recognized as the unambiguous signature of seismic slip. It follows that pseudotachylyte-bearing fault networks might retain a wealth of information on seismic faulting and earthquake mechanics. In this chapter, we will show that in the case of large exposures of pseudotachylyte-bearing faults, as the glacier-polished outcrops in the Adamello massif (Southern Alps, Italy), we might constrain several earthquake source parameters by linking field studies with microstructural observations, high-velocity rock friction experiments, modeling of the shear heating and melt flow, and dynamic rupture models. In particular, it is possible to estimate the rupture directivity and the fault dynamic shear resistance. We conclude that the structural analysis of exhumed pseudotachylyte-bearing faults is a powerful tool for the reconstruction of the earthquake source mechanics, complementary to seismological investigations.

## 1. INTRODUCTION

Large earthquakes critical for human activities nucleate at ∼7 to 15 km depth (*Scholz*, 2002). The sources of these earthquakes and the process of rupture propagation can be investigated by the geophysical monitoring of active faults from the Earth's surface or by the interpretation of seismic waves; most

**Fault-Zone Properties and Earthquake Rupture Dynamics**

information on earthquake mechanics is retrieved from seismology (*Lee et al.*, 2002). However, these indirect techniques yield incomplete information on fundamental issues of earthquake mechanics (e.g., the dynamic fault strength and the energy budget of an earthquake during seismic slip remain unconstrained; *Kanamori and Brodsky*, 2004) and on the physical and chemical processes active during the seismic cycle.

To gain direct information on seismogenic sources, fault-drilling projects have been undertaken in several active faults, such as the Nojima Fault in Japan (*Ohtani et al.*, 2000; *Boullier et al.*, 2001), the Chelungpu Fault in Taiwan (*Ma et al.*, 2006), and the San Andreas Fault in the United States (*Hickman et al.*, 2004). Fault drilling allows integration of real-time *in situ* measurements (strain rate, pore pressure, etc.) and sampling with high-quality seismological data, collected by seismometers located at depth, and geodetic data at the surface (GPS, InSAR, etc.). However, fault drilling has several limitations: (1) to date, drilling is confined to shallow depths (<3 km); (2) the investigated fault volume is too small to provide representative 3D information on fracture networks and fault rock distribution (i.e., large earthquakes rupture faults with areas >100 km$^2$); and (3) the costs are high.

An alternative and complementary approach to gain direct information about earthquakes is the investigation of exhumed faults showing evidence of ancient seismic ruptures (a direct approach to the earthquake engine). However, the use of exhumed faults to constrain the mechanics of earthquakes also has limitations: (1) alteration during exhumation and weathering may erase the pristine coseismic features produced at depth; (2) reactivation of a fault zone by repeated seismic slip events may render it difficult or impossible to distinguish the contribution of individual ruptures; (3) single faults may record seismic and aseismic slip and there might be the need to distinguish between microstructures produced during the different stages of the seismic cycle (coseismic, postseismic, interseismic, etc.); (4) the microstructural proxies used to recognize the coseismic nature of a fault rock have not yet been identified with certainty except in some cases; and (5) the ambient (e.g., pressure, temperature) conditions and the stress tensor coeval with seismic faulting are often difficult to estimate with precision.

Therefore, the use of exhumed faults to retrieve information on earthquakes rely on (1) the recognition of faults rocks produced during seismic slip that have escaped significant structural overprinting and alteration until exhumation to the Earth's surface and (2) the presence of tight geological constraints that allow the determination of ambient conditions during seismic faulting. To date, the only fault rock recognized as a signature of an ancient earthquake is pseudotachylyte (*Cowan*, 1999). Pseudotachylyte is the result of solidification of friction-induced melt produced during seismic slip (*McKenzie and Brune*, 1972; *Sibson*, 1975; *Spray*, 1987, 1995). This chapter presents the study of an exceptional exposure of pseudotachylyte-bearing faults where many of the above-listed limitations are overcome. It will be shown that a multidisciplinary approach, which includes field and laboratory study of the natural pseudotachylytes integrated with theoretical and rock friction experiments, may yield fundamental information on earthquake mechanics and are complementary to seismological investigations. This new approach is inspired by the pioneering work of *Sibson* (1975). First we will briefly review the literature about pseudotachylytes, whose main geochemical, microstructural, and mesostructural features are summarized in detail by *Lin* (2007).

## 2. PSEUDOTACHYLYTES

The term *pseudotachylyte* was introduced by *Shand* (1916) to describe a dark, aphanitic, glassy-looking rock similar to basaltic glasses (or *tachylytes*: Shand in his 1916 paper used the wrong spelling, *tachylyte*, for the basaltic glass, or *tachilyte*. This resulted in the use of both the words pseudo*tachylyte* and pseudo*tachilyte* in the literature. In this chapter, we will use the word introduced by Shand: *pseudotachylyte*) and filling networks and veins in the Old Granite of the Parijs region of the Vredefort Dome in South Africa. Pseudotachylytes have been found in numerous localities and different genetic environments within silicate-built rocks as impact structures (*Shand*, 1916; *Reimold*, 1998), "superfaults" (or large displacement faults related to the collapse of large structures as impact craters and calderas; *Spray*, 1997), rock landslides (*Scott and Drever*, 1953; *Masch et al.*, 1985; *Lin et al.*, 2001), pyroclastic flows (*Grunewald et al.*, 2000), and faults (*Sibson*, 1975). The latter, referred to as tectonic pseudotachylytes, are the most common form, though considered rare between fault rocks by some authors (*Snoke et al.*, 1998; *Blenkinsop*, 2000; *Sibson and Toy*, 2006). This chapter deals with tectonic pseudotachylytes.

Despite a long-lasting debate about the origin of tectonic pseudotachylytes (e.g., *Philpotts*, 1964; *Francis*, 1972; *Wenk*, 1978; *Spray*, 1995), they are now recognized as the product of comminution and friction-induced melting along a fault surface during seismic slip (i.e., at slip rates of 1 to 10 m s$^{-1}$). In fact, by definition pseudotachylyte is a fault rock that shows evidence of melting (*Magloughlin and Spray*, 1992). Although evidence for a quenching origin of pseudotachylytes has been occasionally reported in the literature since the beginning of the 20th century (*Holland*, 1900), *Scott and Drever* (1953) described a vesicular glassy rock in a Himalayan Thrust (later recognized as the product of a large landslide; *Masch et al.*, 1985), only in 1975 did *Sibson* show unambiguously from field evidence that frictional melting was possible during seismic faulting. From a theoretical point of view, *Jeffreys* (1942) demonstrated that friction-induced melting could occur along fault surfaces during coseismic slip. *McKenzie and Brune* (1972) investigated in detail the process of frictional melting and proposed that, given the stress conditions and the elevated strain rates achieved during seismic slip, frictional melting should be widespread in nature. *Wenk* (1978) later questioned the nature of pseudotachylytes as quenched melts. He emphasized that few pseudotachylytes contain glass and suggested ultracomminution as the main mechanism responsible for the development of most fault rocks referred to as pseudotachylytes. Despite the fact that ultracomminution remains a valid alternative to frictional melting to explain the origin of some pseudotachylyte-looking fault rocks, a wealth of data has provided evidence that most of these fault rocks have indeed been through a melt phase (e.g., *Maddock*, 1983; *Shimamoto and Nagahama*, 1992; *Lin*, 1994). These data include field and microstructural observations as well as mineralogical, petrographical, and geochemical data (see Sections 2.1 and 2.2). In the lab, *Spray* (1987, 1988, 1995) clearly showed the origin of pseudotachylyte by frictional melting by using a frictional welding apparatus. In 1995, *Spray* demonstrated experimentally that grain size reduction during frictional sliding is a precursor of melting during coseismic slip: comminution and frictional melting are two related processes. In further experiments, *Tsutsumi and Shimamoto* (1997a) measured the evolution of the friction coefficient during sliding at high slip rates, showing that local melting and welding of the asperity contacts occurred before the bulk melting of the sliding surface. Thus, the comminution model proposed by *Wenk* (1978) was rejected. The lack of glass (often replaced by a cryptocrystalline matrix) in tectonic pseudotachylyte is explained by its instability in fault zones.

## 2.1. Mesoscale Geometry of Pseudotachylyte

Pseudotachylytes commonly occur in the field as sharply bounded veins associated with faults. The typical vein thickness is in the range of a few millimeters to several centimeters, though meter-thick veins are reported in some major faults (e.g., Outer Hebrides Thrust, *Sibson*, 1975; the Woodroffe Thrust, *Camacho et al.*, 1995). The veins lay parallel to, and decorate discontinuously, the fault surface (fault veins, *Sibson*, 1975) or intrude the host rocks branching off the slip surface (injection veins, *Sibson*, 1975) (Figures 1a and 1b). Fault veins are interpreted to decorate the generation surfaces where frictional melt is produced during seismic slip and from where most of the melt is extruded to form injection veins. The latter are assumed to result from fracturing induced by fault pressurization due to thermal expansion during generation of the frictional melt (*Sibson*, 1975; *Swanson*, 1989, 1992). *Di Toro et al.* (2005a) proposed a complementary model for the production of injection veins, associated to the dynamics of propagation of an earthquake rupture.

Pseudotachylyte may occur in complex geometric arrangements including pseudotachylyte-cemented breccias, paired shears, duplexes, and side-wall rip-outs (*Grocott*, 1981; *Swanson*, 1988, 1989, 2006). Contrasting geometries of the pseudotachylyte networks apparently develop as a function of the degree of host rock anisotropy (*Swanson*, 2006). Because seismic slip commonly occurs along preexisting planes of weakness, the geometry of precursor structures plays a major role in determining the architecture of pseudotachylyte networks.

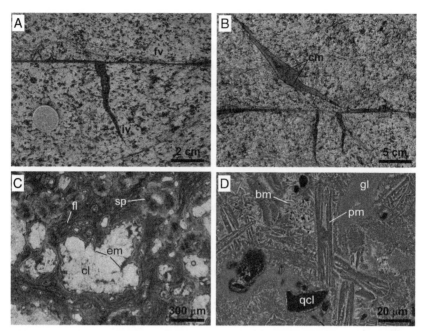

**FIGURE 1**   Pseudotachylyte from the Gole Larghe Fault Zone (Adamello, Italy). (a) Field image of a pseudotachylyte fault vein (fv) and an injection vein (iv) within the Adamello tonalites. North is pointing downward. (b) Pseudotachylyte-bearing faults showing injection veins filling both fractures formed during coseismic slip (veins approximately orthogonal to the fault vein and intruding the block in the lower part of the photos) and fractures predating pseudotachylyte generation and associated with cataclastic faulting (vein oblique to the fault vein in the upper part of the photo) (see Section 4). Note the zoning (cm = chilled margin) in the latter vein (for microstructural description of chilled margins in these pseudotachylyte veins from the Gole Larghe Fault; see *Di Toro and Pennacchioni*, 2004). North is pointing downward. (c) Optical microstructures (plane-polarized light) of a pseudotachylyte. The pseudotachylyte consists of a brown cryptocrystalline matrix with flow structures (fl) and spherulites (sp) and including several clasts (cl). The matrix intrudes the clasts forming deep embayments (em). All these features, as discussed in Section 2.2, suggest that pseudotachylytes are solidified melts. (d) Back scatter scanning electron microscope image of the pseudotachylyte. Quartz clasts (qcl; black in color) are immersed in a fine matrix made of biotite microlites (bm; white), plagioclase microlites (pm; dark gray), and devitrified glass (gl; light gray).

## 2.2. Microstructures and Geochemistry in Pseudotachylytes

Pseudotachylytes consist of a microlitic to cryptocrystalline or glassy (more rare) matrix embedding survivor clasts of the host rock (Figures 1c and 1d). Several microstructures of pseudotachylyte matrix are identical or closely resemble microstructures of volcanic rocks (e.g., *Maddock*, 1983; *McPhie et al.*, 1993) indicating rapid cooling of a melt. In fact, cooling to solidus temperature is in the range of a few seconds to minutes at seismogenic depths

(10–15 km) for typical pseudotachylytes (1 to 20 mm thick) (*Boullier et al.*, 2001; *Di Toro and Pennacchioni*, 2004). These microstructures include (1) microlites and spherulites with a wide variety of shapes (*Philpotts*, 1964; *Maddock*, 1983; *Magloughlin*, 1992; *Lin*, 1994, 2007; *Shimada et al.*, 2001; *Di Toro and Pennacchioni*, 2004) often arranged to define a symmetric zoning of the pseudotachylyte vein (chilled margin); (2) quenched sulfide droplets related to immiscibility in the silicate melt (*Magloughlin*, 1992, 2005); (3) vesicles and amygdales (*Scott and Drever*, 1953; *Maddock*, 1986; *Magloughlin*, 1992), and (4) flow structures (e.g., *Lin*, 1994) (Figures 1c and 1d). Glass has been reported rarely in pseudotachylytes (*Toyoshima*, 1990; *Lin*, 1994; *Obata and Karato*, 1995), which reflects its high instability under geological conditions, but pseudotachylytes may show secondary (devitrification) microstructures (*Maddock*, 1983; *Lin*, 1994).

Clasts within the pseudotachylyte include both single-mineral or lithic clasts from the host rock. Host rock minerals are not equally represented between the clasts mainly due to selective nonequilibrium melting of the different mineral species and, therefore, to preferential consumption of the low-melting point minerals (*Shand*, 1916; *Spray*, 1992). In many pseudotachylytes within granitoids, phyllosilicates and amphiboles are rare between clasts, whereas plagioclase and, especially, quartz tend to survive in the melt (*Di Toro and Pennacchioni*, 2004). The type of survivor clasts also depends on other physical properties of the mineral, such as fracture toughness and thermal conductivity (*Spray*, 1992). Clasts are angular to rounded and show embayment in the case of melting (*Magloughlin*, 1989). They commonly act as nuclei for the growth of radially arranged microlites of the same mineral species to develop spherulitic microstructures (*Shimada et al.*, 2001; *Di Toro and Pennacchioni*, 2004; ). The clast size distribution within pseudotachylytes is fractal in the clast size range of 10 to 2000 µm, with a fractal dimension D of about 2.5 (*Shimamoto and Nagahama*, 1992; *Di Toro and Pennacchioni*, 2004). The clast size distributions typically have a kink at grain size of about 5 µm and show "fractal" values D < 2.5 for the smaller grain fraction that is interpreted as due to the preferential assimilation of the finer-grained grains in the melt (*Shimamoto and Nagahama*, 1992; *Ray*, 1999; *Tsutsumi*, 1999). This observation is a further support for the origin of pseudotachylytes from a melt because comminution alone cannot explain the decrease in smaller grains (*Shimamoto and Nagahama*, 1992). Clasts smaller than 1 µm in size are uncommon in the pseudotachylyte matrix, probably in relation to "the critical limit at which the energy for melting becomes smaller than that for comminution" (*Wenk et al.*, 2000, p. 271; for a discussion, see also *Pittarello et al.*, 2008).

Because of the preferential melting of mafic minerals, the pseudotachylyte melts are commonly more mafic in composition than the host rock. This is the strongest evidence that disequilibrium (i.e., single mineral) melting, rather than eutectic melting, occurs during pseudotachylyte generation. Because of the presence of clasts within the matrix, geochemical studies on the matrix are usually done using microprobe analysis with a defocused beam (e.g., *Ermanovics et al.*, 1972; *Spray*, 1988) or by subtracting the clast content from the bulk XRF (X-ray fluorescence) composition (*Sibson*, 1975; *Di Toro and Pennacchioni*, 2004). Composition of the matrix has been compared with the composition of the associated fault rock (i.e., cataclasites; *Magloughlin*, 1992; *Di Toro and Pennacchioni*, 2004) or of the bounding rocks (e.g., *Maddock*, 1992). Several authors have described pseudotachylyte compositions related to disequilibrium partial melts (*Sibson*, 1975; *Allen*, 1979; *Maddock*, 1986, 1992; *Magloughlin*, 1989; *Bossière*, 1991; *Spray*, 1992, 1993; *Camacho et al.*, 1995; ). *Philpotts* (1964) and later *Ermanovics et al.* (1972) showed near total melting of the host rock with the exception of quartz. *O'Hara and Sharp* (2001) used isotope composition to show the large contribution of biotite and K-feldspar and minor quartz in the production of frictional melt. Total melting of the fault rock assembly was proposed for pseudotachylyte associated with ultramafics (*Obata and Karato*, 1995) and mylonites (*Toyoshima*, 1990).

In summary, when a fault rock records several types of evidence from outcrop scale to microscale that suggest a melt origin, it can be called a pseudotachylyte (*Magloughlin and Spray*, 1992; *Passchier and Trouw*, 1996; *Reimold*, 1998). Among the structural evidence for a melt origin of pseudotachylytes, the foremost is the intrusive habit of injection veins and the presence of flow structures (*Sibson*, 1975). Second, pseudotachylyte veins often exhibit microlitic and spherulitic textures, chilled margins, or glass, indicating rapid chilling of a melt (*Maddock*, 1983; *Lin*, 1994). However, the presence of glass is not a necessary feature of these rocks, because the environmental conditions under which pseudotachylytes form (usually between 3 to 15 km in depth) are unfavorable for glass preservation (*Maddock*, 1986; *Lin*, 1994). Third, fractal analysis of size distribution of clasts in pseudotachylytes shows that the number of small grains (<5 micron) is very small compared to the number of larger grains. This is indicative of preferential melting of the finest clasts (Shimamoto and Nagahama, 1992). Lastly, frictional melting is a nonequilibrium process, with the consequence that pseudotachylytes usually contain survivor clasts of quartz and feldspar and the matrix is enriched in Fe, Mg, Al, Ca, and $H_2O$ compared to the host rock or to the cataclastic precursor (*Sibson*, 1975).

## 2.3. Temperature Estimate of Frictional Melts

The value of peak temperature of the friction-induced melt is an important parameter to estimate the energy budget during seismic slip as well as on the lubricating effect of the melt (see Section 5.1). In natural pseudotachylytes, melt temperatures were deduced by different methods, including the following:

1. $SiO_2$ glass composition ($T_{melt}$ 1450°C; *Lin*, 1994).
2. Microlite mineralogy ($T_{melt} = 890$ to 1100°C: two pyroxene geothermometer; *Toyoshima*, 1990; $T_{melt} = 790°$ to 820°C: omphacite-garnet

geothermometer; *Austrheim and Boundy*, 1994; $T_{melt} = 1000°C$: pigeonite crystallization; *Camacho et al.*, 1995; $T_{melt} = 1200°C$, mullite crystallization, *Moecher and Brearley*, 2004).

3. Mineralogy of survivor clasts ($T_{melt} = 1000°C$; *Maddock*, 1983).
4. Numerical modeling by matching melt cooling curves with the presence of microlitic versus spherulitic zoning in thick veins assuming that the different microstructures are the result of contrasting cooling rates at the center and periphery of the pseudotachylyte vein ($T_{melt} = 1450°C$; *Di Toro and Pennacchioni*, 2004).
5. Volume ratio between lithic clasts and matrix (*O'Hara*, 2001).

In high-velocity rock friction experiments, melt temperatures of $1000°$ to $1550°C$ were measured by means of a radiation thermometer and thermocouples (*Tsutsumi and Shimamoto*, 1997b; *Lin and Shimamoto*, 1998; *Hirose and Shimamoto*, 2005a; *Spray*, 2005; *Del Gaudio et al.*, 2006). In summary, the estimates and measures of the temperature of melts in natural and experimental pseudotachylytes are in the range of $750°$ to $1550°C$. Given the fact that pseudotachylyte is often hosted in granitoids or in rock with a silica-rich composition, these temperature estimates support the idea that friction-induced melts are produced by nonequilibrium melting (*Spray*, 1992) and that they are superheated (*Di Toro and Pennacchioni*, 2004). In fact, the equilibrium melting temperature for granitoid systems ranges between $700°$ and $850°C$ (*Philpotts*, 1990). However, most temperature estimates and measures probably underestimate the peak temperature achieved by the melt. For instance, microlite mineralogy yields a cooling temperature. In experiments, the radiation thermometer measures the temperature over a spot size of $400 \mu m$ in diameter, whereas the slipping zone thickness measured at the end of the experiment is $\sim 180 \mu m$ thick only (*Del Gaudio et al.*, 2006). It follows that the measured temperature is an average between that of the wall rocks and the slipping zone. Thermocouples, inserted in the specimen, measure the temperature inside the sample and not in the slipping zone, and the temperature of the melt is estimated through numerical modeling refinement (*Tsutsumi and Shimamoto*, 1997b). Because some pseudotachylytes record melting microstructures of quartz and apatite (e.g., embayed clasts), temperatures as high as $1700°C$ can be locally achieved if disequilibrium melting and $H_2O$-free conditions (in the case of quartz melting) are assumed. Together with frictional melting, a possible mechanism for melt superheating is viscous shear heating in the melt layer (*Nielsen et al.*, 2008).

## 2.4. Distribution of Tectonic Pseudotachylytes

Most pseudotachylytes are clearly linked to brittle (elastico-frictional, *Sibson*, 1977) and are associated with cataclasites (e.g., *Magloughlin*, 1992; *Fabbri et al.*, 2000; *Di Toro and Pennacchioni*, 2004) or fluidized gouge (*Otsuki et al.*, 2003; *Rowe et al.*, 2005). Other pseudotachylytes, instead, appear closely linked to ductile (quasi-plastic, *Sibson*, 1977) regimes because they overprint and are in turn reworked by ductile shear zones (*Sibson*, 1980; *Passchier*, 1982; *White*, 1996). These latter pseudotachylytes are found associated with greenschist facies mylonites (*Passchier*, 1982; *Takagi et al.*, 2000), amphibolite facies mylonites (*Passchier*, 1982; *White*, 1996; *Pennacchioni and Cesare*, 1997), granulitic facies mylonites (*Clarke and Norman*, 1993), and spinel-lherzolite facies mylonites (*Ueda et al.*, 2008). Thus, pseudotachylytes are produced both at shallow (2 to 10 km) and midcrustal levels (10 to 20 km). Pseudotachylytes can be produced during intermediate and deep earthquakes also, but probably by processes other than frictional melting (we will not describe these mechanisms here). Eclogitic facies pseudotachylytes (> 60 km in depth) were found in the Bergen Arc of Western Norway (*Austrheim and Boundy*, 1994). The production of frictional melts (and plasmas) was invoked for the deep focus (637 km in depth) 1994 Bolivian mantle earthquake ($M_w$ 8.3) (*Kanamori et al.*, 1998).

## 2.5. Production of Pseudotachylytes

In this chapter, we focus on the mechanism of pseudotachylyte production in the elastico-frictional continental crust (<12 to 15 km depth for a geothermal gradient of 25°C km$^{-1}$, *Sibson*, 1977). During rupture propagation and coseismic slip (e.g., Figure 2a), the elastic energy stored in the wall rock is released (*Reid*, 1910). Part of the released elastic energy is dissipated in frictional work $W_f$ on the fault. We may assume that $W_f$ is partitioned in (*Kostrov and Das*, 1988)

$$W_f = Q + U_s \qquad (1)$$

where $Q$ is heat and $U_s$ is surface energy for gouge and fracture formation (see Table 1 for a list of symbols used in this chapter). Because $U_s$ is considered negligible (*Lockner and Okubo*, 1983), most (>95%) of the work done in faulting is converted to heat (*Scholz*, 2002, p. 155), especially in the lower part of the elastico-frictional crust (e.g., 10 km depth, *Pittarello et al.*, 2008). It follows that the amount of heat generated during seismic slip for unit area of the fault is (*Price and Cosgrove*, 1990)

$$W_f = \mu(\sigma - p_p)d \approx Q \qquad (2)$$

where $\mu$ is the friction coefficient, $\sigma$ is the normal stress, $p_p$ is the pore fluid pressure, and $d$ is the coseismic slip. The heat flux generated per unit area of the fault is (*McKenzie and Brune*, 1972)

$$q \approx \mu(\sigma - p_p)V \qquad (3)$$

where $V$ is the slip rate.

The thermal penetration distance $x$ in the bounding host rock as a function of slip duration $t$ is

$$x \sim (\kappa\, t)^{0.5} \qquad (4)$$

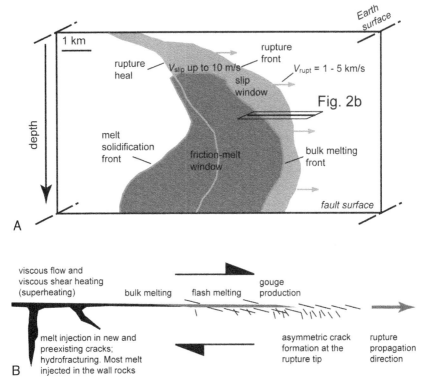

**FIGURE 2** Schematic models of pseudotachylytes production (modified from *Swanson*, 1992). (a) Model of propagation of a seismic self-healing pulse mode II crack. The fracture propagates toward the right side (as indicated by the gray arrows) at a speed $V_{rupt}$ in the range between 1 and 5 km/s. Behind the rupture front, slip is restricted to a band (light gray area, in large part overlapped by the dark gray area of presence of friction melts). The slipping area is a few km wide. As the slipping area moves behind the fracture front, friction melt is produced after some refinement along the fault plane and may survive for some time (friction-melt window) after the cessation of slip. (b) Enlargement of the rupture along the fault section shown in Figure 2a. The rupture propagates to the right and the sense of shear is dextral. The stress perturbation induced by the propagation of the crack produces fractures in the lower block under tensional transient stress (see Section 4). Gouge production and comminution at the rupture tip are followed by flash melting (i.e., melting at asperity contacts) and bulk melting. Wall rocks interaction and viscous shear heating in the melt layer allow the achievement of extremely high temperatures (superheated frictional melts; see Section 5). Most melt is injected in the wall rocks.

where $\kappa$ is thermal diffusivity of the host rock. The heat produced is large even for small slips (Equation 3) (*Sibson and Toy*, 2006) given (1) the high effective stress in the seismic source area (hundreds of MPa at depths in the range of 2 to 15 km) and (2) the high particle velocities (1 to 10 m s$^{-1}$) localized in thin slipping zones (few mm at most; *Sibson*, 2003). *McKenzie and Brune* (1972) estimated that melting could occur for fault slips as small as 1 to 3 mm for driving stresses of 100 MPa, and production of seismic melts

**TABLE 1** Latin and Greek Symbols

**Latin Symbols**

| | |
|---|---|
| $A$ | Fault surface area ($m^2$) |
| $c_p$ | Solid specific heat at constant pressure ($J\ K^{-1}\ kg^{-1}$) |
| $c_{pm}$ | Melt specific heat at constant pressure ($J\ K^{-1}\ kg^{-1}$) |
| $d$ | Displacement (m) |
| $E$ | Heat (J) |
| $E_m$ | Total energy input for unit mass ($J\ kg^{-1}$) |
| $H$ | Latent heat of fusion ($J\ kg^{-1}$) |
| $L$ | Effective latent heat of fusion ($J\ kg^{-1}$) |
| $L_e$ | Dimension related to the escaping distance of the frictional melt from the sliding surface |
| $N_f$ | Normalizing factor ($Pa^{0.75}$) |
| $p_p$ | Pore pressure (MPa) |
| $q$ | Heat flux per unit area ($J\ m^{-2}\ s^{-1}$) |
| $Q$ | Heat density ($J\ m^{-2}$) |
| $r$ | Specimen radius (m) |
| $r_1$ | Specimen inner radius (m) |
| $r_2$ | Specimen outer radius (m) |
| $R$ | Revolution rate of the motor ($s^{-1}$) |
| $t$ | Time (s) |
| $T_c$ | Characteristic temperature (K) |
| $T_m$ | Rock melting temperature (K) |
| $T_M$ | Maximum temperature achieved by the melt (K) |
| $T_{hr}$ | Host rock temperature (K) |
| $U_s$ | Surface energy for gouge and surface formation ($J\ m^{-2}$) |
| $V$ | Slip rate ($m\ s^{-1}$) |
| $V_e$ | Equivalent slip rate ($m\ s^{-1}$) |
| $V_{Ray}$ | Rayleigh rupture propagation velocity, about 90% of $V_{shear}$ ($m\ s^{-1}$) |
| $V_{rupt}$ | Rupture propagation velocity ($m\ s^{-1}$) |
| $V_{shear}$ | Shear wave velocity ($m\ s^{-1}$) |

*Continued*

**TABLE 1**    Latin and Greek Symbols – (Cont'd)

Latin Symbols

| | |
|---|---|
| $w$ | Melt half thickness (m) |
| $w_{av}$ | Slipping zone average thickness (m) |
| $W$ | Factor with velocity dimensions (m s$^{-1}$) |
| $W_f$ | Frictional work density done in faulting (J m$^{-2}$) |
| $x$ | Thermal penetration distance (m) |

Greek Symbols

| | |
|---|---|
| $\dot{\varepsilon}$ | Shear rate (s$^{-1}$) |
| $\eta_c$ | Characteristic viscosity (Pa s) |
| $\eta_e$ | Equivalent viscosity (Pa s) |
| $\kappa$ | Solid thermal diffusivity (m$^2$ s$^{-1}$) |
| $\kappa_m$ | Melt thermal diffusivity (m$^2$ s$^{-1}$) |
| $\mu$ | Coefficient of friction |
| $v$ | Shortening rate (m s$^{-1}$) |
| $\rho$ | Solid density (kg m$^{-3}$) |
| $\rho_m$ | Melt density (kg m$^{-3}$) |
| $\sigma$ | Normal stress (MPa) |
| $\tau$ | Shear stress (MPa) |
| $\tau_p$ | Peak shear stress (MPa) |
| $\tau_{ss}$ | Steady-state shear stress (MPa) |
| $\phi$ | Ratio between the volume of lithic clasts and the total volume of pseudotachylyte |
| $\omega$ | Rotary speed (s$^{-1}$) |

for such small slips is confirmed by field evidence (*Griffith et al.*, 2008). Because of the low value of $\kappa$ in crustal rocks ($10^{-6}$ m$^2$ s$^{-1}$) and the short duration of seismic slip at a point of a fault (few seconds at most), heat remains *in situ* and the process is adiabatic at seismogenic depths (see *Di Toro et al.*, 2006b, for a detailed discussion). Given these deformation conditions, fault rocks as well as the host rocks immediately adjacent to the slipping surface are heated and eventually melt (e.g., *Fialko*, 2004; *Bizzarri and Cocco*, 2006; *Nielsen et al.*, 2008).

A possible scenario for frictional melt production in the elastico-frictional crust is shown in Figure 2 (modified from *Swanson*, 1992). Unstable sliding occurs due to the velocity weakening behavior of most of the rock materials (i.e., friction coefficient decreases with increasing sliding speed; e.g., *Tullis*, 1988) and the rupture propagates as a mode II self-healing pulse (*Heaton*, 1990) (Figure 2a). At the rupture tip (Figure 2b), the stress perturbation during crack propagation induces fracturing in the wall rock under tension (the southern block for a rupture propagating toward the east along a dextral strike-slip fault; see Section 4). The opposite sliding surfaces are then uncoupled and a cushion of gouge develops in between during sliding. Rupture propagation is followed by surface refinement through crushing, clast rotation, and flash heating and melting (*Archard*, 1958; *Rempel*, 2006; *Rice*, 2006) of the initiation gouge. Bulk melting occurs where the gouge is highly comminuted and strain rate is higher, though it mainly involves the minerals with the lowest melting point (see Sections 2.2 and 2.3) (Figure 2b). With increasing slip, the opposite surfaces are separated by a thin layer of melt, which is further heated due to viscous shear heating (superheated melts). Melting occurs at the wall rocks for rock-rock interaction and for phase transition at the melt-wall rock and melt-survivor clast boundaries (*Nielsen et al.*, 2008). The melt produced in the slipping zone (i.e., the fault vein) is mostly injected in the wall rocks, along (1) prerupture fractures (if the rupture is propagating along a preexisting fault) and (2) new fractures produced under the dynamic transient stress field at the rupture tip during propagation and due to the volume increase related to the melting of the rock. The melt is largely dragged and injected in the wall rocks, but a small percentage of melt still remains along the slipping zone. Under these conditions, the fault is lubricated by friction melts (*Nielsen et al.*, 2008). Once the elastic strain energy stored in the wall rocks that drives the propagation of the rupture is released, the slip rate drops down, and the viscous strength of the melt layer increases instantaneously (see Section 5), leading to the healing of the rupture. This process may occur in 1 to 10 s at most, consistent with rise times typical of earthquakes. Instead, the melt injected into the wall rocks or pounding in dilational jogs along the fault vein (i.e., reservoirs), cools slowly, from seconds to minutes, depending on the peak temperature of the melt, the temperature of the host rock, and the thickness of the melt layer: melt might be still present after the healing of the rupture (friction melt window in Figure 2a). The solidification of the melt produces the pseudotachylyte.

### 2.5.1. The Role of Water

The role and the origin of water during frictional melting is still debated (*Sibson*, 1973, 1975; *Allen*, 1979; *Magloughlin*, 1992; *Magloughlin and Spray*, 1992; *O'Hara and Sharp*, 2001). The presence of fluid inclusions, vesicles, and amygdales in some pseudotachylyte (*Philpotts*, 1964; *Maddock et al.*, 1987; *Magloughlin*, 1992; *Obata and Karato*, 1995; *Boullier et al.*, 2001) suggests that water fluids must be present at the time of pseudotachylyte generation in some cases. However, it is not clear if water was present as pore water before frictional heating or, instead, if it was released by the breakdown of water-bearing minerals (biotite, chlorite, epidote, amphibole, etc.) during frictional heating (e.g., *Moecher and Sharp*, 2004). The presence of pore water in a slipping zone should impede the achievement of high temperatures, because (1) fluid expansion promoted by frictional heating induces a drastic decrease in the effective normal stress and lubricates the fault (*Sibson*, 1973; *Allen*, 1979; *Lachenbruch*, 1980; *Mase and Smith*, 1987; *Bizzarri and Cocco*, 2006; *Rice*, 2006), (2) water vaporization adsorbs heat, and (3) the expulsion of hot fluids cools the slipping zone. It follows that pseudotachylytes have been commonly assumed to develop in relatively dry rocks (*Sibson*, 1973; *Sibson and Toy*, 2006). On the other hand, water-rich conditions may also promote the production of pseudotachylytes by lowering the rock and mineral melting temperatures (*Allen*, 1979; *Magloughlin*, 1992). Recent field observations support the occurrence of frictional melting under fluid-rich conditions in some cases (*Rowe et al.*, 2005; *Ujiie et al.*, 2007). *Okamoto et al.* (2006) described pseudotachylytes bounded by a carbonate-matrix implosion breccia in the Shimanto accretionary complex (Japan). In this water-rich environment, during seismic slip, frictional heating expanded the pore water (thermal pressurization) in the slipping zone. Thermal pressurization induced (1) fracturing in the wall rocks, (2) expulsion of pressurized water, and (3) carbonate precipitation. Expulsion of pressurized water increased the effective stress (and temperature) leading to frictional melting in the slipping zone (*Okamoto et al.*, 2006). In other words, frictional melting occurred *after* the expulsion of water from the slipping zone during the same seismic rupture.

## 3. A NATURAL LABORATORY OF AN EXHUMED SEISMOGENIC SOURCE

A few studied outcrops have suitable characteristics (large exposures, polished surfaces, presence of pseudotachylytes) that allow a wealth of information about earthquake mechanics to be retrieved from exhumed faults. Examples of such outcrops are from the Outer Hebrides Thrust (Scotland; *Sibson*, 1975), the Homestake Shear Zone (Colorado; *Allen*, 2005), the Fort Foster Brittle Zone (Maine; *Swanson,* 1988, 1989, 2006), and the Woodroffe Thrust (Australia; *Camacho et al.*, 1995). In this chapter we will describe an exceptional outcrop of the Gole Larghe Fault Zone (Southern Alps, Italy) (*Di Toro and Pennacchioni*, 2004, 2005; *Di Toro et al.*, 2005a, 2005b, 2006b; *Pennacchioni et al.*, 2006).

The Gole Larghe Fault Zone is a dextral strike-slip exhumed structure crosscutting the Adamello tonalites (Italian Alps) and forming a southern branch of the Tonale line (a segment of the Periadriatic Lineament, the major fault system of the Alps; *Schmid et al.*, 1989) (Figure 3a). The Gole Larghe Fault Zone is exposed in large glacier-polished unweathered outcrops, which allow a 3D investigation of the structures (Figure 3b) and where single faults can be mapped in detail (Figures 3c and 3d). The fault zone hosts a large number of pseudotachylytes, which have largely escaped alteration and structural reworking during the exhumation to the Earth's surface and therefore preserve an intact record of the coseismic processes that occurred at depth. In addition, the numerous pseudotachylytes present along the different faults are "simple" (a single continuous layer of melt): each pseudotachylyte layer records a single coseismic slip. At the same time, the fault zone contains hundreds of

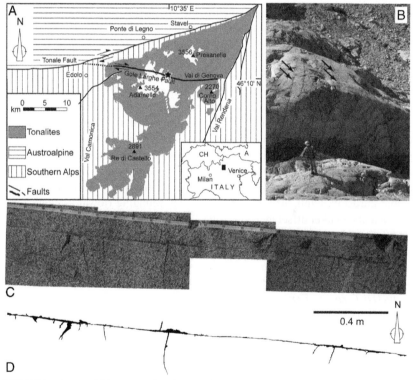

**FIGURE 3**  A natural laboratory of a seismogenic fault zone: the Gole Larghe Fault Zone in the Adamello batholith (Italy). (a) Tectonic sketch map of the Adamello region showing the location of the Gole Larghe Fault and of the glaciated outcrops (star) analyzed in detail in this contribution. (b) Field view of the exposures of the Gole Larghe Fault Zone. Presence of deep creeks allows a 3D view of the fault zone. The fault zone is made of about 200 subparallel strike-slip faults (some indicated by arrows). (c) Photomosaic showing a pseudotachylyte-bearing fault zone. The excellent exposure allows the detailed mapping of the pseudotachylyte vein network and the precise estimate of the melt volume (area) per unit fault length. (d) Drawing of the pseudotachylytes from the photomosaic of Figure 3c. The orientation of the fractures filled by pseudotachylyte was used to reconstruct the seismic rupture directivity (see Section 4.2).

faults, which possibly record different seismic slip increments thus forming a statistically representative population of earthquakes occurring under identical ambient conditions and geological context. The outcrops contain numerous markers (aplite dikes, basic enclaves) crosscut by the faults, which allow the offset to be estimated for each single structure, whereas the host rock (tonalite) is homogeneous over kilometers along the fault. Numerous researchers have investigated the geology of the area since the 1950s, and there are tight constraints on the age and ambient conditions of deformation. All of the features make this fault an exceptional and rather unique natural laboratory for the direct study of an exhumed segment of a seismic zone (the earthquake engine).

The Gole Larghe Fault Zone is exposed for a length of about 12 km in the northern part of the tonalitic Adamello batholith (Figure 3a). In this area, the tonalites consist of plagioclase (48% in volume), quartz (29%), biotite (17%), and K-feldspar (6%) (*Di Toro and Pennacchioni*, 2005), and are dated at 34 Ma (*Del Moro et al.*, 1983). In the upper Genova Valley (star in Figure 3a), the fault zone is 550 m thick and accommodates about 1 km of slip (*Di Toro and Pennacchioni*, 2005; *Pennacchioni et al.*, 2006). The fault zone includes hundreds of subparallel faults striking east-west, and mostly steeply dipping (50° to 80°) to the south, showing a spacing on the range of decimeters to a few meters (some shown in Figure 3b). The presence of subhorizontal roof pendants sunk in the Adamello batholith suggests that the batholith and the fault zone were not tilted during exhumation (*Callegari and Brack*, 2002). The faults exploit a main set of east-west trending joints pervasively developed in the whole intrusion. Fault surfaces have shallowly plunging (toward W) slicken- lines and the marker offset indicates right-lateral strike-slip kinematics (*Di Toro and Pennacchioni*, 2005). Major faults accommodate up to 20 m of slip and are spaced every ~10 m, whereas the minor faults in between accommodate offsets of a few decimeters to a few meters. These subparallel faults are associated with a network of small fault-fractures produced during slip on the major and minor faults (*Di Toro and Pennacchioni*, 2005).

Fault rocks are cataclasites and pseudotachylytes. The Gole Larghe cata- clasites are cohesive fault rocks cemented by the pervasive precipitation of epidote, K-feldspar, and minor chlorite due to fluid-rock interaction along the faults (*Di Toro and Pennacchioni*, 2005). These epidote- and K-feldspar- bearing fault rocks are referred to as cataclasites in the following text. Typ- ically, pseudotachylyte is associated with the last event of slip recorded by each fault segment; pseudotachylyte overprints cataclasite (*Di Toro and Pennacchioni*, 2005). This sequence of events possibly results from the fact that (1) grain size reduction by cataclasis promotes successive melting (e.g., *Spray*, 1995) and (2) quenching of the friction-induced melts welds the fault and favors migration of successive slip events to other subparallel cataclastic faults. This interpretation is consistent with experimental observations: sam- ples containing an artificially generated pseudotachylyte (produced by former high-speed frictional experiments with precut samples) break along the "intact" sample rather than along the welded surfaces (*Hirose and Di Toro*, unpublished data). This indicates that the fault recovers mechanical properties similar to that of the intact rock after sealing off by pseudotachylyte. In the Gole Larghe Fault Zone, the welding process of individual faults by pseudotachylyte formation determines the progressive thickening of the fault zone and results in a low displacement/thickness ratio of the fault zone. This is in stark contrast with what occurs in many mature faults where repeated seismic ruptures localize along the same weak horizons and result in high displacement/ fault thickness ratio (e.g., *Chester et al.*, 1993).

Ar-Ar stepwise dating of the Adamello pseudotachylytes yields ages of ~30 Ma (*Pennacchioni et al.*, 2006), indicating that seismic slip along the Gole Larghe was contemporary to the activity of the Tonale Fault (*Stipp et al.*, 2002). Pseudotachylytes were produced only 3 to 4 Ma after the emplacement of the pluton. Zircon and apatite fission track data (*Viola et al.*, 2001; *Stipp et al.*, 2004) and other geological constraints indicate that faulting in the Adamello occurred after cooling of the pluton at the ambient conditions of 0.25 to 0.35 GPa (corresponding approximately to 9 to 11 km depth) and 250° to 300°C before the uplift of the batholith (*Di Toro and Pennacchioni*, 2004; *Di Toro et al.*, 2005b; *Pennacchioni et al.*, 2006). Therefore, the Gole Larghe pseudotachylytes record events of seismic slip that occurred at the base of the elastico-frictional (brittle) crust.

A main drawback at using the Gole Larghe faults to determine mechanical parameters of a single earthquake is that, in many faults, it is not possible to partition the fault slip between the cataclastic and the coseismic, pseudotachylyte-producing slip (*Di Toro et al.*, 2005b). However, a few fault segments of the Gole Larghe Fault contain only pseudotachylytes and record a single seismic rupture of intact tonalite (*Di Toro et al.*, 2006a; see Section 5.1). The absence of an epidote- and K-feldspar-bearing cataclasite precursor in these fault segments was determined by both field and microstructural observations (*Pittarello et al.*, 2008). Note that rock fragmentation *during* seismic slip is a precursor to frictional melting (*Spray*, 1995) and is part of the *same* seismic rupture that produces the pseudotachylyte (see Figure 2b and Equation 1), but the epidote-bearing cataclasites were produced during a *different* deformation event with respect to the pseudotachylytes (i.e., not the same seismic rupture; see *Di Toro and Pennacchioni*, 2005, for field and microstructural evidence).

A further difficulty to study earthquake mechanics by the use of exhumed faults is that information is needed about the effective stress tensor at the time of seismic slip. In the Adamello Faults, the orientation of principal axis of stress is relatively well constrained by regional and field structural data (*Mittempergher et al.*, 2007; *Pennacchioni et al.*, 2006). The maximum compressive stress is approximately horizontal and nearly at 45° to the faults and the stress field can be reasonably assumed as Andersonian (*Di Toro et al.*, 2005b). The widespread production of pseudotachylytes constrains the pore fluid pressure to hydrostatic or less (*Sibson*, 1973). The friction coefficient at rupture was assumed of 0.75 given the high segmentation of the precursor joints (which involves, in some places, the propagation of the seismic rupture in an intact tonalite) and the pervasive cementation of fault segments by indurated, precursor, cataclasites (*Di Toro and Pennacchioni*, 2005). Given these assumptions, we estimated, for subvertical faults at 10 km depth, a normal stress to the fault ranging between 112 and 184 MPa, and a resolved shear stress ranging between 84 and 138 MPa (*Di Toro et al.*, 2005b, 2006b).

Though the structure, fault rock mineral assemblage, and geochemical composition are identical over 12 km along strike of the Gole Larghe Fault Zone (i.e., *Pennacchioni et al.*, 2006), we focused our attention on the outcrops located at the base of the Lobbia glacier because of their excellent exposure (star in Figure 3a). The results described in the following sections were obtained from this area.

## 4. RUPTURE DYNAMICS

As discussed in Section 2, the presence of pseudotachylyte on exhumed faults is a clear marker of seismic activity. In addition, some of the damage structures are not only involved in the seismic process but are ostensibly of coseismic genesis, in particular, freshly fractured fault branches and lateral fractures permeated with pseudotachylyte that are clearly not associated with any preexisting regional or local deformation trend (e.g., Figure 2b). As a consequence, those coseismically generated structures may be identified as markers of the aggressive dynamic stress transient associated with fast rupture propagation (and, eventually, of the associated fluid pressure and temperature surges). The pattern, distribution, and the intensity of the stress transient associated with fracture propagation are specific of dynamic conditions, namely fracture length, velocity, propagation direction, stress drop, energy dissipation during fracturing, and extension of the weakening process zone at the crack tip (where slip weakening occurs). Therefore, any measurable feature (position, orientation, opening) of the coseismic structures on an exhumed fault is a potential gauge allowing the properties of the paleo-earthquake source to be constrained and characterized, with implications for present-day earthquakes under a similar context. In the following section, some salient features of the fracture transient stress will be illustrated, henceforth related to the marker structures observed in the field, and, finally, used to reconstruct features of the associated earthquake source.

### 4.1. Transient Stress Pattern

Both static and dynamic cracks embedded in an elastic medium, subjected to a remotely applied load, induce a perturbation of the stress field in the surrounding medium. Whereas the pattern associated with a static crack only depends on the slip distribution inside the crack itself, the pattern associated with a dynamically propagating fracture will also critically depend on its tip propagation velocity relative to the wave velocity in the surrounding medium. The differences between static and dynamic fracturing will become significant when the fracture is moving at a substantial fraction of the medium's sound velocity. A determining feature is the stress concentration located at or in the vicinity of the propagating tip, where the yielding strength of the rock is reached and a finite energy flow allows the dissipative process of fracture propagation to occur (*Irwin*, 1957). Some of the simplest mathematical models, assuming cracks with infinitely sharp tips, involve the idea that stress diverges to infinity at the fracture termination (*Kostrov*, 1964). Such a feature is obviously an unphysical consequence of a simplistic mathematical model; however, even more realistic models where the crack tip is allowed to spread over a finite process region or "end zone" (*Ida*, 1972) predict that the stress concentration at the fracture termination can be pronounced. This feature is compatible (1) with the observation of profuse damage observed in the vicinity of crack extremities (*Rice*, 1966) and (2) with the necessity of finite energy flow within a relatively small region around the tip to allow for fracture propagation (*Irwin*, 1957). Theory predicts that, for a fixed amount of energy dissipated in the propagation of fracture, the length of the end zone at the fracture tip collapses as its propagation velocity increases, tending to zero as the velocity approaches its limiting velocity (either the shear or Rayleigh wave velocity of the medium, depending on the configuration). Such a process is analogous to the Lorentz contraction observed in relativistic physics (*Burridge et al.*, 1979). Therefore, spreading of the tip is reduced and the stress concentration tends to reproduce a quasi-singular stress distribution, in agreement with that of the sharp tip mathematical model. Under such circumstances, the perturbation relative to the background stress is much more localized and intense for a crack propagating close to its limiting velocity. Recent crack models take into consideration the dissipative process not only on the crack surface itself but also in a finite volume around the slip zone and in the vicinity of the propagating tip (where the large transient stress is attained) (*Andrews*, 2005). Through such a mechanism, the end zone is not subjected to the shrinking effect discussed earlier; the dissipation increases with crack propagation and is distributed over a wider volume, whereas the stress concentration remains bounded.

## 4.2. Examples of Transient Stress Markers Observed

Markers of transient stress have been described in numerical simulations, laboratory tests, and even in one instance on natural faults, the case of the paleoseismic faults of the Gole Larghe (*Di Toro et al.*, 2005a).

Because rocks and most materials are weaker in tension than in compression, secondary fractures are expected to be produced in the block under tension (e.g., *Andrews*, 2005). This theoretical prediction is confirmed by experiments conducted in photoelastic rock analog materials (homalite). *Samudrala et al.* (2002) fired bullets on a sample assemblage made of a precut homalite sheet glued along a preexisting interface (Figure 4a). The bullet triggered eastward propagation of the crack, dextral slip along the precut surface, and formation of secondary fractures on the sample side experiencing tensile transient stresses (Figure 4b). Though these experiments were performed under no confinement, it can be shown that a state of absolute tension can be reached during dynamic fracturing even under the lithostatic loads expected at  seismogenic depth (*Griffith et al.*, 2008).

In the field example of the Gole Larghe faults, a series of coseismic, secondary fractures was observed to branch from the main faults (Figure 4c). Assuming that the geometry of the observed paleoseismic faults is almost that of a purely vertical, strike-slip fault, and considering a horizontal section through the fault plane, the geometry may be reduced for simplicity to that of an in-plane (mode II) fracture. In this case, the limiting velocity in the subshear wave case is $V_{Ray}$ (the Rayleigh wave velocity, in general slightly above 90% of the shear wave velocity). As the fracture propagation approaches $V_{Ray}$, it can be shown that the transient stress perturbations around the propagating fracture are exacerbated, eventually inducing tension sufficient to surpass the lithostatic load at the estimated depth at the time of activity (10 km) and allowing the opening of tension cracks. A rotation of the axis of major tension and compression also occurs; the angle of principal

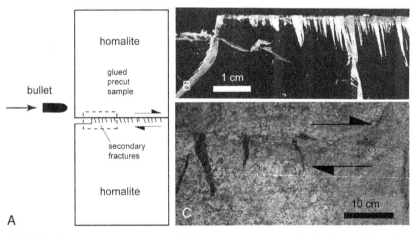

**FIGURE 4**  Markers of transient stress: field and experimental results. (a) A description of the experimental setup, where a dynamic fracture is triggered by shooting a bullet onto a homalite sample (modified from *Samudrala et al.*, 2002). (b) The results of the same experiment, with secondary branching cracks on the side where tensile transient stress is produced. (c) An example of exhumed seismic fault (*Di Toro et al.*, 2005a), where the lateral branching cracks are found in majority on one side and in a dominant direction, yielding constraints on rupture directivity and rupture velocity (see the text for further details).

compressive stress initially at 30° to the fault plane progressively increases until reaching a maximum of about 90° as the fracture velocity increases and approaches $V_{Ray}$. The location of tension is always on the side of the fault where slip is in the direction opposite to fracture propagation, whereas compression is observed on the other side. Accordingly, the presence of open (mode I) pseudotachylyte-filled cracks branching off mainly in the southern wall of the Gole Larghe Faults at roughly 90° (Figure 4c), given that the slip motion was dextral, indicates that rupture was propagating from the east to the west. In addition, to reach absolute tensional conditions at 90°, the fracture velocity ought to be close to $V_{Ray}$.

Branching fractures were filled by pseudotachylyte (e.g., Figures 3c and 4c), confirming that they were produced or opened during the propagation of the seismic rupture. In particular, of 624 fractures filled by pseudotachylyte and measured over a length of 2 to 10 m in 28 different subparallel fault segments, 67.7% intruded the southern bounding block (i.e., the veins injecting into the southern block are more than 70% on 17 fault segments, 60% to 70% on three segments, 50% to 60% on five segments, and less than 50% in only three fault segments) (Figure 5a). The angle of the fracture, measured clockwise starting from the east, had two dominant orientations, at about 30° to 210° (set 1) and 90° to 270° (set 2) with respect to the fault trace (*Di Toro et al.*, 2005a). Set 1 fractures intruded preexisting cataclastic faults, whereas set 2 fractures were produced during rupture propagation along the main fault surface.

The stress perturbations induced by the propagation of the rupture during the 30 Ma old Gole Larghe earthquakes were investigated by means of numerical models (Figures 5b, 5c, and 5d). In these models were considered different coseismic slips (up to 1.5 m, consistent with displacements measured in the field; see Figure 6), rupture propagation modes (crack-like versus self-healing pulses, the latter of 0.1 to 1 km in length—according to the self-healing pulse model, the slipping region is much smaller than the final dimension of the earthquake), stress drops, and slip weakening distances. Because the dynamic stress field, given a fixed prestress direction, depends on the ratio of the rupture propagation velocity $V_{rupt}$ with the shear wave velocity $V_{shear}$, the simulation considered slow rupture velocities ($V_{rupt} = 0.6\ V_{shear}$), typical of large dissipation in the fracture process; high rupture velocities ($V_{rupt} = 0.9\ V_{shear}$), about 98% of the Rayleigh wave velocity, commonly reported for many earthquakes; and supershear rupture velocities ($V_{rupt} = \sqrt{2}V_{shear}$), estimated for some large earthquakes (*Bouchon and Vallée*, 2003). In the case of subsonic rupture velocities ($V_{rupt} < V_{shear}$; Figure 5b), an analytical solution was used, whereas a finite difference solution was used in the case of supersonic rupture speeds (Figure 5d). In the model, the properties of the tonalite were used (Poisson ratio of 0.25, shear modulus of 26 GPa, fracture toughness of 2 MPa $m^{0.5}$, rock density of 2700 kg $m^{-3}$). Whatever the combination of rupture speed and crack mode, the rupture tip was under tension in the southern block for right-lateral faults propagating from the west to the east (consistently with the experimental results shown in Figure 4b). However, by varying

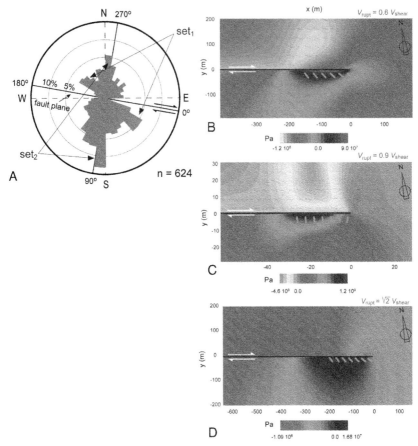

**FIGURE 5** (a) Area-weighted rose diagram showing the orientation of the injection veins filled by pseudotachylyte from 28 fault segments of the Gole Larghe Fault Zone. The fractures are measured clockwise from the east side of the fault (see Figure 3c). Most fractures are toward the south and oriented at about 30° and 85° from the main fault. (b-d) Numerical models of the tensile stress field (positive is tension, negative compression) close to the rupture tip for three different rupture velocities (b: $V_{rupt} = 0.6\ V_{shear}$; c: $V_{rupt} = 0.9\ V_{shear}$; d: $V_{rupt} = \sqrt{2}V_{shear}$). The fracture tip is shown as a black thick line and viewed from above. The fault is dextral, the rupture is propagating eastward and the wall rocks are under tension in the southern side in all models. The planes of maximum tensile stress are indicated by thin black segments: the planes in the southern side and near to the rupture tip are evidenced in orange. For $V_{rupt} = 0.9\ V_{shear}$ (Figure 3c), the planes of maximum tension are oriented at about 85° from the main fault, consistently with the most common orientation of the fractures observed in the Gole Larghe Fault. All figures are from *Di Toro et al.* (2005a). **(See Color Plate 14)**.

the rupture velocity, the magnitude of the stresses varied and the orientation of the planes of maximum tension rotated. In particular, for $V_{rupt}$ approaching the $V_{Ray}$ (i.e., about 0.9 $V_{shear}$, Figure 5c), the solutions yielded the highest values of tensional stress (up to 1.7 GPa, well above the strength under

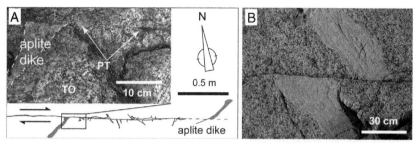

**FIGURE 6** Estimate of dynamic fault strength from field exposures. (a) Fault segment, decorated by (only) pseudotachylyte, separating an aplite dike of about 1.5 m. Most pseudotachylyte is injected in the wall rocks, and pseudotachylyte fault vein thickness is variable along strike. In this case, the pseudotachylyte "thickness" $w_{av}$ appearing in Equation 12 was estimated as the ratio between the pseudotachylyte area in outcrop and the length of the fault segment. (b) Aplite dike is crosscut at an offset of about 30 cm by a fault segment decorated by only pseudotachylyte. The fault displacement was estimated by the measures of the marker separation and of the orientation of the fault, outcrop, marker, and slickenlines (*Di Toro and Pennacchioni*, 2005).

tension for granite, which is about 20 MPa) and an orientation of the planes under maximum tension consistent (80° to 90°) with the most frequent orientation of tensional fractures (injection veins) measured in the field. Some injection veins intrude the northern wall and could be the result of nonplanar geometry of the fault, fracturing due to the volume increase related to the melting of tonalite and cataclasite (about 17%), or some rupture complexities like the interaction of neighbor fault segments. The stress drop associated with the modeled rupture is 42 MPa, which is consistent with the expected stress drop related to the lubrication effects of the frictional melt (see Sections 5 and 6). The estimated fracture energy (the energy dissipated during the propagation of the crack) is between 8 and 67 MJ m$^{-2}$. As a consequence, fieldwork and numerical modeling suggest eastward propagation of the seismic ruptures at Rayleigh wave velocities. Because the Gole Larghe Fault is a dextral branch of the Tonale Fault, which was active at 30 Ma, and the structure, mineralogy, and geochemistry of the fault zone are identical along the whole length of the fault, the interpretation is that the Gole Larghe Fault Zone records hundreds of ruptures propagating from the west (the Tonale Fault) to the east.

Large earthquakes (some fault segments in the Gole Larghe record coseismic slip of 1.5 m (see Figure 6a) which is compatible with M6-7 earthquakes; *Sibson*, 1989) occur over repeated times of the order of 100 to 1000 years (*Scholz*, 2002). A major advantage in using exhumed ancient faults compared to monitoring active faults is the possibility of investigating rupture directivity over the geological timescale and to produce a statistically robust database. The conclusion that some fault zone may record a dominant rupture directivity has implications in earthquake hazard evaluation. Indeed, the radiated wave field from a unilateral propagating rupture is amplified in the direction of propagation (directivity effect) but reduced in the opposite direction, thus inducing increased strong motion in one direction and anisotropy in the potential distribution of damage.

## 5. DYNAMIC FAULT STRENGTH

The determination of the magnitude of the shear stress and traction acting on the fault surface during seismic slip is relevant in earthquake mechanics. For instance, the decrease in shear stress during seismic slip may determine (1) whether dynamic stress drop is larger than static stress drop (*Bouchon*, 1997), (2) the rupture propagation mode (e.g., self-healing pulse versus crack-like; *Heaton*, 1990; *Beeler and Tullis*, 1996; *Nielsen and Carlson*, 2000), (3) the increase in the ratio of radiated energy versus seismic moment with earthquake size (*Mayeda and Walter*, 1996), and (4) the production of heat during seismic slip (e.g., *Lachenbruch*, 1980). However, fault strength during seismic slip cannot be retrieved through seismological methods except for particular cases (*Guatteri and Spudich*, 1998, 2000). In fact, extrapolation from seismic waves may allow the estimate of the static stress drop (*Hanks*, 1977)—and, by means of strong assumptions, of the dynamic stress drop (*Bouchon*, 1997)—but not the absolute values of the shear stress (*Scholz*, 2002).

Experiments conducted with the "conventional" biaxial and triaxial apparatuses (*Lockner and Beeler*, 2002) at slip rates <1 mm/s and slip <1 cm (orders of magnitude lower than seismic deformation conditions) show that rock friction ($\mu = \tau/\sigma$ where $\tau$ is shear stress and $\sigma$ is normal stress to the fault) is about 0.7 in most cohesive and noncohesive rocks (*Byerlee*, 1978) over a large range of temperatures (almost up to 400°C in the case of granite; *Stesky et al.*, 1974) and pressures (up to 2 GPa; *Byerlee*, 1978). Exceptions to Byerlee's law are some clay minerals and phyllosilicates (montmorillonite, talc, vermiculite) showing lower friction values. The "conventional" experiments also show that the friction coefficient is slightly perturbed (few percentage points at most) by variations in slip rate (velocity weakening) and increasing displacement (slip weakening) (*Tullis*, 1988; *Marone*, 1998; *Scholz*, 2002). The evolution of the $\mu$ with slip rate and displacement is described by the Dieterich-Ruina rate and state (R&S) friction law (*Dieterich*, 1978, 1979; *Ruina*, 1983), which has found a broad application in earthquake mechanics (from rupture dynamics to aftershock physics; see *Scholz*, 1998, for a review). However, the R&S friction law was formulated to describe experimental results obtained under deformation conditions significantly different from the seismic ones and is not capable of accounting for some seismological observations such as those listed in points 1 through 4, presented earlier. To explain these observations, several fault-weakening mechanisms were proposed, including thermal pressurization (*Sibson*, 1973), normal interface vibrations (*Brune et al.*, 1993), acoustic fluidization (*Melosh*, 1996), elastohydrodynamic lubrication (*Brodsky and*

*Kanamori*, 2001), silica gel lubrication (*Goldsby and Tullis*, 2002; *Di Toro et al.*, 2004), thermal decomposition (*Han et al.*, 2007), flash heating at asperity contacts (*Rempel*, 2006; *Rice*, 2006), gouge-related weakening (*Chambon et al.*, 2002; *Mizoguchi et al.*, 2007), and melt lubrication (*Spray*, 1993). Among these, experimental evidence was given for silica gel lubrication (*Goldsby and Tullis*, 2002; *Di Toro et al.*, 2004), thermal decomposition (*Han et al.*, 2007), gouge-related weakening (*Mizoguchi et al.*, 2007), and melt lubrication (*Spray*, 2005; *Di Toro et al.*, 2006a). However, because these weakening mechanisms have been found very recently in the laboratory, their occurrence in nature and their physics remain almost unknown (*Beeler*, 2006). The exception is melt lubrication. Artificial pseudotachylytes were produced in high-velocity friction experiments in the late 1980s (*Spray*, 1987), and their mechanical properties were investigated from the late 1990s (*Tsutsumi and Shimamoto*, 1997a). Noteworthy, pseudotachylytes (solidified melts) exist in natural exhumed faults and are similar to artificial pseudotachylytes (*Spray*, 1995; *Di Toro et al.*, 2006a). Then, the physics of melt lubrication of faults has been studied with some detail (*Fialko and Khazan*, 2005; *Nielsen et al.*, 2008). Lastly, it is possible to estimate the dynamic shear stress during seismic slip from the amount of melt produced (*Sibson*, 1975). As a consequence and different from the other weakening mechanisms proposed so far, in the presence of pseudotachylyte it is possible to extrapolate experimental results to natural conditions and validate if there is any weakening effect on fault strength at seismic slip rates. As shown in the next sections, field, experimental, and theoretical work suggests that faults are lubricated by seismic melts.

## 5.1. Field Estimates

*Sibson* (1975) and *Wenk et al.* (2000) suggested estimating the average dynamic shear stress in pseudotachylyte-bearing faults by assuming that all the frictional work during seismic slip is converted to heat (i.e., the process is fully adiabatic) and that the energy expended to produce new surfaces is negligible. Actually, both assumptions are probably valid at 10 km depth (*Lockner and Okubo*, 1983; *Di Toro et al.*, 2006b; *Pittarello et al.*, 2008). From Section 2.5 we note:

$$W_f \approx Q \tag{5}$$

That is, most frictional work during sliding results in fault rock heating and, eventually, in melting (*Jeffreys*, 1942; *McKenzie and Brune*, 1972). The total energy $E_m$ input for unit mass (J kg$^{-1}$) is partitioned in (1) energy exchanged for rock heating until melting ($c_p (T_m - T_{hr})$), (2) energy exchanged during rock melting ($H$), and (3) energy required for further melt heating ($c_{pm} (T_M - T_m)$) if frictional melts are superheated (*Di Toro and Pennacchioni*, 2004; *Fialko and Khazan*, 2005; *Nielsen et al.*, 2008):

$$E_m = c_p(T_m - T_{hr}) + H + c_{pm}(T_M - T_m) \tag{6}$$

where $T_m$ is the rock melting temperature, $T_{hr}$ is the ambient host rock temperature, $T_M$ is the maximum temperature achieved by the melt, $H$ is the latent heat of fusion, and $c_p$ and $c_{pm}$ are the specific heat (J K$^{-1}$ kg$^{-1}$) at constant pressure of the solid phase and of the melt, respectively. In the case of frictional melts, because of the presence of survivor clasts in the matrix that do not exchange latent heat of fusion, Equation 6 becomes

$$E_m = c_p(T_m - T_{hr}) + (1 - \phi)H + (1 - \phi)c_{pm}(T_M - T_m) \\ + \phi \, c_p(T_M - T_m) \tag{7}$$

where $\phi$ is ratio between the volume of lithic clasts and the total volume of pseudotachylyte (i.e., matrix + lithic clasts). Then, assuming that $c_{pm}(T) \approx c_p(T)$:

$$E_m \approx (1 - \phi)H + c_p(T_M - T_{hr}) \tag{8}$$

and the energy input $E$ (J) is:

$$E \approx [(1 - \phi)H + c_p(T_M - T_{hr})]\rho \, A \, w_{av} \tag{9}$$

where $\rho$ is the density (assuming melt density approximately equal to rock density), $A$ is the area of the fault surface, and $w_{av}$ is the average width of friction-induced melt present as veins along the fault plane. The work for unit area (J m$^{-2}$) done to overcome the frictional sliding resistance for a constant $\tau_f$ is

$$W_f = d \, \tau_f \tag{10}$$

where $d$ is the coseismic fault displacement. $W_f$ is converted to heat $E$. Considering the area of the fault surface where frictional work occurs, from Equations 8, 9, and 10,

$$E \approx E_m \, \rho \, A \, w_{av} \approx d \, \tau_f A \tag{11}$$

and the average $\tau_f$ is

$$\tau_f \approx \rho \, E_m \, w_{av}/d \approx \rho[(1 - \phi)H + c_p(T_M - T_{hr})] \, w_{av}/d \tag{12}$$

In the case of the Gole Larghe Faults zone, all the variables in Equation 9 can be determined for pseudotachylyte-bearing faults formed in intact tonalite (i.e., without a precursor, preseismic, cataclasite), and it is possible to estimate the average dynamic shear stress during seismic faulting. The values for $c_p \approx 1200$ J kg$^{-1}$ K$^{-1}$, $H = 3.24 \times 10^5$ J kg$^{-1}$, and $\rho = 2700$ kg m$^{-3}$ are known from the literature (*Di Toro et al.*, 2005a, and references therein). The ambient host rock temperature of $T_{hr} \sim 250°$C is well constrained by geological data, microstructural observations, and the mineralogy of cataclasite coeval with pseudotachylyte production and seismic faulting (*Di Toro and Pennacchioni*, 2004). The peak temperature of the friction-induced melt was

>1200°C, as indicated by the crystallization of plagioclase microlites with andesine composition and the assimilation of plagioclase clasts, and it was estimated at 1450°C based on numerical modeling of microlitic-spherulitic zoned distribution in pseudotachylyte veins (*Di Toro and Pennacchioni*, 2004). A value $\phi = 0.2$ was determined by means of image analysis of scanning electron microscope (SEM) images of the pseudotachylyte veins (*Di Toro and Pennacchioni*, 2004). These values yield $E_m = 1.7 \times 10^6$ J kg$^{-1}$ from Equation 8. The ratio $w_{av}/d$ was determined in fault segments where field and microstructural (SEM, FE-SEM) investigations revealed that the pseudotachylyte layer and seismic slip were produced during a single slip event (i.e., there was no evidence of alteration or precursor cataclasite) and rupture propagated through an intact rock (Figure 6). In the case of thin fault veins, the thickness was measured directly in the field (Figure 6a). In the case of thicker and more complicated pseudotachylyte veins filling larger displacement fault segments, the average thickness was determined from fault profiles (intersection of the fault with the outcrop surface) obtained from photomosaics of the fault (Figures 3c, 3d, and 6b). The average thickness $w_{av}$ was calculated as the area of the pseudotachylyte (including the melt injected in the host rock) divided by the length of the fault profile. The coseismic slip $d$ was determined from separations of dikes. From Equation 12, the calculated dynamic shear stress was in the range between 14.9 and 48.1 MPa.

These values for shear stresses could be high or low depending on the magnitude of the stress normal to the fault. As described in Section 3, the effective stress normal to the fault ranged between 112 MPa (hydrostatic pore pressure) and 182 MPa (no pore pressure). Therefore, the ratio between shear stress and normal stress is in the range between 0.4 and 0.08, well below the friction coefficient typical for tonalites, and suggests fault lubrication by friction-induced melts.

## 5.2. Experimental Results

Fault lubrication by friction-induced melts is supported by nonconventional rock friction experiments performed with high-velocity rotary shear apparatuses on tonalite (described here) and other silicate-built rocks (gabbros, monzodiorites, peridotites; *Hirose and Shimamoto*, 2005a; *Del Gaudio et al.*, 2006). The effective slip velocities in these experiments (1 cm/s to 1.3 m/s for standard cylindrical samples with an external diameter of 22.3 mm) are comparable to slip rates achieved in earthquakes. Therefore, contrary to biaxial and triaxial apparatuses (total displacements <2 cm and slip rates <1 cm s$^{-1}$), it allows the investigation of the seismic slip. The apparatus has a horizontal arrangement, with a stationary shaft on one end and the rotating shaft on the opposite end. The stationary shaft provides up to 1 ton of axial thrust (corresponding to maximum normal stress of 20 MPa for a 22.3 mm in diameter solid sample) by means of a pneumatic axial load actuator. The rotary shaft provides a rotary motion up to 25 rounds per second by means of an electric engine and an electromagnetic clutch (for a detailed description of the apparatus see *Hirose and Shimamoto*, 2005a; *Shimamoto and Tsutsumi*, 1994). Given the cylindrical shape of the samples, the determination of the slip rate and shear stress is problematic, because slip rate and torque (in the experiments are measured with the rotating speed and the torque) increase with sample radius $r$ ($V = \omega r$; $\omega$ is the rotary speed). So the experimental data are obtained in terms of "equivalent slip rate" $V_e$ (*Shimamoto and Tsutsumi*, 1994; *Hirose and Shimamoto*, 2005a, 2005b):

$$V_e = \frac{1}{\tau A} \int_{r_1}^{r_2} 4\pi^2 \, \tau R r_i^2 \, dr_i \qquad (13)$$

where $\tau$ is the shear stress (assumed as constant over the sliding surface), $A$ is the area of the sliding surface, $R$ is the revolution rate of the motor, and $r_1$, $r_2$ are the inner and outer radius of a hollow cylindrical specimen, respectively. For solid cylindrical specimens, where $r_1 = 0$,

$$V_e = \frac{4 \pi R r}{3} \qquad (14)$$

We refer to the equivalent slip rate simply as slip rate $V$ hereafter.

In the experiments conducted on Adamello tonalite, slip rates were of $\sim 1 \text{ ms}^{-1}$ and, when the normal stress was sufficiently high (i.e., 15 to 20 MPa), the total slip was a few meters at most, so comparable with deformation conditions typical of large earthquakes. Given the dramatic decrease in rock strength with increasing heat due to (1) thermal expansion of minerals and (2) the $\alpha - \beta$ transition in quartz, which by increasing the quartz volume by 5% renders the rock weak (thermal fracturing; *Ohtomo and Shimamoto*, 1994), higher normal stresses were achieved by using solid cylinders of rock confined with aluminum rings (Figures 7b and 7c). Because aluminum melts at 660°C, whereas most rock-forming minerals melt at more than 1000°C (*Spray*, 1992), the aluminum ring sustains the sample till bulk melting on the sliding surface occurs.

The experiments consisted of four main steps (1 to 4). In step 1, the axial thrust was applied to the sample. In step 2, the target rotation speed was achieved while an electromagnetic clutch split the rotating column from the specimen. In step 3, when the motor speed reached the target value, the electromagnetic clutch was switched on and the specimens started to slide. In step 4, the motor was switched off to end the experiment. The number of rotations, the axial force, the axial shortening of the specimen, and the torque were recorded by a digital recorder with a data sampling rate of 500 and 1000 Hz.

In the case of the Adamello tonalite, four experiments were conducted at constant slip rate (1.28 m s$^{-1}$) and increasing normal stress (5, 10, 15, and 20 MPa) (experiments with cataclasites yielded similar results; *Di Toro et al.*, 2006b). Similarly to the frictional melting model for the elastico-frictional crust

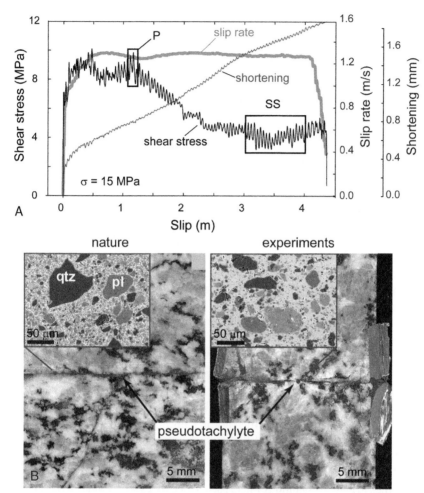

**FIGURE 7**   Results and products of high-velocity rock friction experiments. (a) Shear stress, slip rate, and sample shortening versus slip curves for the Adamello tonalite. Boxes P and SS indicate the displacement intervals used to determine the peak and the steady-state shear stress, respectively, and are plotted in Figure 8a. (b) Comparison between natural (left) and artificial (right) pseudotachylyte. The insets are back-scatter electron scanning electron microscope images of the slipping zones. Both natural and artificial pseudotachylyte have quartz (qtz) and plagioclase (pl) clasts suspended in a cryptocrystalline (natural) and glassy (artificial) matrix. The cryptocrystalline matrix found in nature is explained by the higher ambient temperature in nature (~250°C) with respect to the lab (20°C) and the different cooling (and geological) history. All figures from *Di Toro et al.* (2006a).

(Figure 2b), production of melt in the experiments was preceded by production and extrusion of rock powder from the sample assembly. Melt was also extruded after the powder, similar to what is found in natural faults, where most of the melt is injected into the wall rock (e.g., Figure 4c). Extrusion resulted in a progressive shortening of the sample with increasing slip. At the end of the experiments, a thin layer of melt (fault vein) separated the two opposite sliding surfaces. This artificial pseudotachylyte was remarkably similar to natural pseudotachylytes from the Gole Larghe Fault (Figures 7b and 7c).

In each experiment, the shear stress evolved with increasing slip: after an increase (strengthening stage) to a peak value (P in Figure 7a), shear stress decreased toward a steady-state value (SS in Figure 7a) through a transient stage whose length decreased with increasing normal stress. This evolution of shear stress with slip is typical of all the frictional melting experiments performed so far in different rocks (gabbro, peridotite, etc.; *Tsutsumi and Shimamoto*, 1997a; *Hirose and Shimamoto*, 2005a; *Del Gaudio et al.*, 2006). The evolution of the shear stress with slip in these experiments can be correlated with the microstructural evolution of the melt layer (i.e., amount of clasts) and of the topography of the boundary between the melt and the wall rock (*Del Gaudio et al.*, 2006; *Hirose and Shimamoto*, 2003, 2005a, 2005b). For instance, during the strengthening stage, the slipping zone is discontinuously decorated by melt patches, whereas, during the transient stage and the steady-state stage, the sliding surfaces are separated by a continuous layer of melt (*Hirose and Shimamoto*, 2005a, 2005b; *Del Gaudio et al.*, 2006).

In each experiment with tonalite, the plot of peak and steady-state shear stress versus normal stress shows a slight dependence of shear stress with normal stress which can be roughly described by the equations:

$$\text{Peak shear stress} \quad \tau_p = 4.78 \text{ MPa} + 0.22 \ \sigma \qquad (15)$$

$$\text{Steady-state shear stress} \quad \tau_{ss} = 3.32 \text{ MPa} + 0.05 \ \sigma \qquad (16)$$

The "effective" friction coefficient, or ratio of shear stress versus normal stress, ranges between 0.05 and 0.22, which is well below Byerlee's friction (see Figure 8a). This suggests lubrication of the fault surfaces operated by frictional melts. By extrapolating the effective friction coefficient to seismogenic depth (i.e., to natural conditions where pseudotachylytes were found), experimental data fit well with the natural data (Figure 8b). Of course, the effective friction coefficient corresponds to the best fit line of the experimental data and does not describe the physics of the process. A better extrapolation, which considers the physics of the frictional melting and melt lubrication, is proposed in Section 5.3.

## 5.3. Theoretical Estimates

Theoretical derivations of the shear stress dependence on normal stress and slip rate in the presence of melt have been proposed (*Fialko and Khazan*, 2005; *Nielsen et al.*, 2008). *Nielsen et al.* (2008) argued that a steady-state limit condition exists in case of melt extrusion, based on the balance of heat and mass flow during frictional melt. They derived an expression for shear resistance at the steady-state as a function of applied slip rate and normal

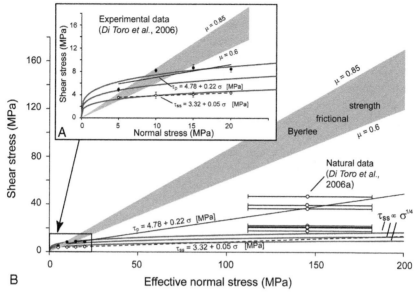

**FIGURE 8**   Shear stress versus normal stress for experimental, natural, and theoretical data compared to solid friction for tonalite (modified from *Di Toro et al.*, 2006a). (a) A zoombox showing the detail of experimental results. (b) The global representation including field estimates. The light gray area indicates the range of typical frictional values for tonalite (*Byerlee*, 1978). Solid circles and open diamonds are experimental values for peak shear stress and steady-state shear stress (Figure 7). Field data have a large range of effective normal stress (as indicated by the length of the horizontal error bar) due to the poorly constrained pore pressure at the time of seismic faulting during the Gole Larghe earthquakes. The solid line is the best linear fit for the peak shear stress data (Equation 15); the dashed line is the best linear fit line for the steady-state shear stress (Equation 16). The three gray curves represent theoretical predictions for viscous strength assuming $\tau_{ss} = \alpha\,\sigma^{0.25}$ (according to Equation 22) and setting $\alpha = 8 \times 10^4$, $11 \times 10^4$, and $14 \times 10^4$, respectively.

stress. In their model, melt thickness, temperature, and viscosity profile are not imposed but result from the coupling of viscous shear heating, thermal diffusion, and melt extrusion. In fact, lubrication does not automatically arise from the simple presence of melt; indeed a thin layer of high-viscosity melt may induce strengthening of the interface (*Koizumi et al.*, 2004; *Scholz*, 2002). Thus, the determination and combination of melt temperature, composition and clast content (which all affect melt viscosity), and melt layer thickness are needed to estimate shear resistance. These parameters themselves depend on the extrusion rate (squeezing of melt under normal stress tending to reduce the thickness of the layer), melting rate (providing new melt tending to increase the layer thickness), and strain rate in the melt layer (affecting heat production, hence altering melting rate, melt temperature, and melt viscosity). In other words, lubrication is the result of a complex feedback between viscosity, normal stress, and slip rate as quantified by *Nielsen et al.* (2008).

The existence of a steady-state is also manifest in the experimental results, which indicate that melt thickness (*Hirose and Shimamoto*, 2005a) and shortening rate tend to a stable value after a variable amount of slip (Figure 7a). From simple mass conservation arguments, constant thickness and shortening rate imply that melting and extrusion rates are constant too. Heat is produced by the viscous shear of the melt layer and removed through (1) extrusion of hot melt, (2) latent heat for melting, and (3) diffusion into the wall rocks. However, because the rock is consumed by the advancement of melting (shortening), the advancement of heat by diffusion into the rock is compensated by the shortening rate. As a consequence, the isotherms are fixed in space and time with respect to the sliding interface; a solid particle belonging to the rock sample moves toward the interface and traverses the isotherms until it melts, and, finally, it is extruded from the sample assembly (Figure 9a). The complex physics of frictional melt lubrication is described by a system of five differential equations (*Nielsen et al.*, 2008), which describe the following:

1. *Melting at the solid interface* (the Stefan problem, *Turcotte and Schubert*, 2002):

$$\rho \, L \, v = \kappa \, \rho \, c_p \frac{\partial T}{\partial z}\bigg|_{w+} - \kappa_m \rho_m c_{pm} \frac{\partial T}{\partial z}\bigg|_{w-} \tag{17}$$

where $L = H \,(1\text{-}\phi)$ is effective latent heat, $v$ is shortening rate, $\kappa$, $\rho$, $c$ are the thermal diffusivity, mass density, and heat capacity of the solid

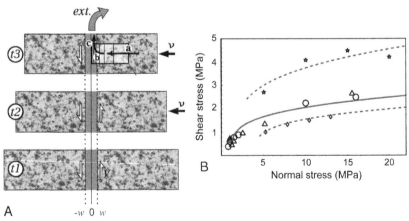

**FIGURE 9**  Estimate of dynamic fault strength: the theoretical model. (a) A cartoon explaining the geometry of the shear/shortening mechanism analyzed, inside a rock volume, at three successive time steps *t1, t2,* and *t3*. The central gray area represents the melt layer of half-thickness *w*. The white arrows represent the shear motion, whereas the lateral arrows represent the convergent motion of the blocks due to the shortening process. (b) Fitting of experimental data according to Equation 22 for peridotite (diamonds), gabbro, monzodiorite (triangles and circles, respectively), and tonalite (stars). Reproduced with permission from *Nielsen et al.* (2008).

(whereas melt parameters are denoted by the subscript $m$), and the two derivatives at $w^-$ and $w^+$ represent temperature gradient on the melt side and on the solid side, respectively;

2. *Heat diffusion at the steady-state*, with a mass flow at velocity v and a heat source density due to shear rate $\dot{\varepsilon}$ (the equation is slightly simplified in either the solid or the melt: heat source is zero inside the solid, whereas it can be shown that heat transfer by flow is negligible inside the melt):

$$\kappa_m \frac{\partial^2 T}{\partial z^2} + \frac{\tau_{ss}\dot{\varepsilon}}{\rho_m c_m} - v\frac{\partial T}{\partial z} = 0 \tag{18}$$

3. *The normal stress $\sigma_n$ supported by a layer of melt* of equivalent viscosity $\eta_e$ being squeezed at a rate (which corresponds to the shortening rate of the specimen) $v$:

$$\sigma_n = C\frac{\eta_e v L_e^2}{w^3} \tag{19}$$

with $C$ a geometrical factor ($C = 3/16$ for a solid cylinder), $w$ the melt layer half thickness and $L_e$ the escaping distance of the frictional melt from the sliding surface (in experiments, $L_e$ is about the radius of the sample: in nature, $L_e$ is the half distance between injection veins, e.g., decimeters to meters).

4. *The strain rate* within the melt layer of thickness $z$ and viscosity $\eta$:

$$\dot{\varepsilon}(z) = \frac{\tau_{ss}}{\eta(z)} \tag{20}$$

5. *The viscosity dependence on temperature $T$* (note that the viscosity law adopted is rather simplified: it is a linearization of the exponential law for a Newtonian fluid):

$$\eta(T) = \eta_c \exp(-(T - T_m)/T_c) \tag{21}$$

where $T_c = T_m^2/B$ and $\eta_c = A\exp(B/T_m)$ are the characteristic temperature and viscosity of the melt of temperature $T_m$. Values of $\eta_c$ and $T_c$ may be estimated based on composition, clast content, and a well-known empirical viscosity law (*Shaw*, 1972; *Spray*, 1993). Note that this only leads to a very rough estimate, given that the result is highly dependent on clast concentration and melt composition; furthermore, it assumes that the melt behaves as a Newtonian fluid (viscosity independent of shear rate), which is probably not the case here. As a consequence, *a priori* estimates of $\eta_c$, $T_c$ are only indicative for the time being and shall require a substantial analysis of experimental and natural samples, which has started.

The coupled of the preceding system of differential equations may be solved analytically, yielding an expression for shear stress in closed form (*Nielsen et al.*, 2008):

$$\tau_{ss} = \sigma^{0.25} \frac{N_f}{\sqrt{L_e}} \sqrt{\frac{\log(2V/W)}{V/W}} \qquad (22)$$

where $N_f$ is a normalizing factor, $L_e$ is function of the geometry of the sample (it is related to the escaping distance of the frictional melt from the sliding surface), $V$ is slip rate, and $W$ is a factor with velocity dimensions, grouping six constitutive parameters:

$$W = \sqrt{\frac{8T_c \kappa_m \rho_m c_{pm}}{\eta_c}} \qquad (23)$$

The corresponding melt half-thickness is also obtained as

$$w = \frac{W\eta_c}{\tau_{ss}} \frac{\text{arctanh}\left(\dfrac{\frac{V}{W}}{\sqrt{\left(\frac{V^2}{W^2}+1\right)}}\right)}{\sqrt{\left(\frac{V^2}{W^2}+1\right)}} \qquad (24)$$

Finally, details of the inhomogeneous viscosity and temperature profiles across the melt layer may be found in *Nielsen et al.* (2008). Note that Equation 19 assumes that all of the normal load is supported by viscous push of the melt layer. A more complex normal stress dependence could arise if hydrodynamic push, bubble formation, or interaction at the asperity contacts or through clast clusters play an important role. In spite of the simplifying assumptions, the shear stress curves obtained are in reasonable agreement with the experimental data. Specific experiments on gabbro tested the relationship between steady-state shear stress and, for instance, normal stress while the other variables (e.g., slip rate $V$) were kept constant (e.g., Figure 8b).

To extrapolate Equation 22 to the ambient conditions of the Gole Larghe paleoearthquakes, we used $\tau_{ss} = \alpha\,\sigma^{0.25}$ (by grouping all parameters other than stress into a single value $\alpha$) and set $\alpha = 8 \times 10^4$. The equation fits reasonably well both the experimental data and the field estimates, corroborating the idea of effective lubrication in the presence of friction melts (Figure 9b); curves obtained for alternative values of $\alpha = 11 \times 10^4$ and $14 \times 10^4$ are also represented.

## 6. DISCUSSIONS AND CONCLUSIONS

The study of the Adamello Faults and associated pseudotachylytes has the potential to unravel aspects of rupture dynamics and fault strength during seismic slip. The "direct" investigation of the exhumed seismic fault network in the Adamello, following the pioneering work of *Sibson* (1975) on pseudotachylytes, has proven an outstanding source of information, complementary to seismological information, if integrated with calibrated experimental and theoretical analysis. This approach deserves future efforts and extension to this and other natural laboratories.

Field description of natural pseudotachylyte-bearing faults as well as theoretical models and physical experiments all need improvements to better describe the process of seismic faulting. Field description needs detailed quantitative methods to map the 3D geometry of the fault zone and the pseudotachylyte distribution between and along the different fault segments. Also, there is the need to further investigate how the results of experimental and theoretical studies, performed under simplified conditions and geometries, can be extrapolated to earthquake source mechanics. For instance, the constitutive equation for frictional melt lubrication (Equation 22), though tested in the laboratory, describes the shear stress during steady-state conditions of slip; however, abrupt variations in slip rate and normal stress (the so-called transient stage) occur in nature (e.g., *Beeler and Tullis*, 1996). In addition, current experiments and theoretical models do not take into account the geometry and complexity of natural faults. The presence of bumps on a fault surface (*Power et al.*, 1988; *Sagy et al.*, 2007) might impede the smooth sliding reproduced in the laboratory, and future studies should consider the effects of fault roughness on frictional sliding. Though this is difficult to investigate in experiments due to the small size of the samples and technical limitations, the problem can be tackled by numerical models, which include the fault roughness and the constitutive law (including the transient stage) for melt lubrication. The input roughness of the fault for the numerical model can be imported directly from quantitative field reconstruction of the fault geometry (e.g., by light detection and ranging [LIDAR]; see also Section 6.1).

The dependence of shear stress with slip rate in the presence of friction melts may have dramatic effects on the rupture propagation process. The outcome of the abrupt decrease in slip rate on the shear stress is shown in an experiment performed at low normal stress (0.72 MPa) on hollow-shaped (24.9 mm and 15.6 mm external and internal diameter, respectively) gabbro samples (Figure 10) (*Di Toro and Hirose*, unpublished data). Once the steady-state shear stress was achieved, the perturbation induced in the system "melt layer plus apparatus" by decreasing the slip rate from 1.6 to 0.8 m s$^{-1}$ resulted in an abrupt increase in strength and the achievement of an unstable regime (see the stick-slip–like behavior after about 115 m of slip). Such abrupt increase in strength may produce self-healing rupture pulses in nature and may result in lower stress drops than those retrieved from the current theoretical models and experimental data.

A main result of the integrated studies shown in this contribution is the recognition of fault lubrication by frictional melts during earthquakes. An expected outcome of melt lubrication is the occurrence of large dynamic stress drops, especially at depth. Field, experimental, and theoretical data summarized in Figure 9 all indicate stress drops in the range of 40 to 120 MPa at 10 km depth. However, these stress drops are larger than those

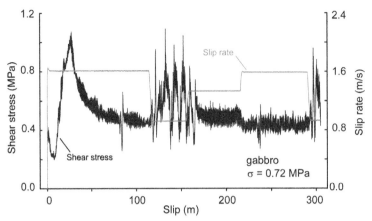

**FIGURE 10** Dependence of shear stress with slip rate in the presence of frictional melts. For a particular combination of slip rate and normal stress (in this case, $V = 0.8$ m/s and $\sigma = 0.72$ MPa), the shear stress is highly unstable (see the seven peaks of high shear stress between 115 and 170 m of slip). Though poorly investigated, the nonlinear dependence of shear stress with slip rate may explain rupture heal and self-healing pulses. The large displacement achieved in this experiment is the result of the application of a small normal stress (compare with Figure 7a).

expected in the upper crust from seismological investigations (between 0.1 and 30 MPa according to *Hanks*, 1977). This discrepancy well illustrates the uncertainties of extrapolating our results to nature and to match our observation with seismological data. The reasons for the "stress drop" discrepancy can be multifold, including the fact that the actual experimental and theoretical models might fail to completely describe the seismic slip process (as underlined earlier). But a reasonable explanation for the discrepancy could be that we estimated *dynamic* stress drops, whereas those considered by *Hanks* (1977) were *static* stress drops (e.g., *Scholz*, 2002, p. 204). *Bouchon* (1997) estimated dynamic stress drop as large as 100 MPa in several discrete fault elements (mostly located at about 10 km depth, so at ambient conditions similar to those in the Gole Larghe Fault Zone) during the Loma Pietra $M_L$ 6.9, 1989 earthquake. Another possible explanation for the lower stress drops retrieved by seismology is that frictional melting is rare in nature and that the stress drops we estimated occur only in some special cases. According to *Sibson and Toy* (2006), pseudotachylytes are not common between fault rocks and, therefore, other coseismic weakening mechanisms (e.g., thermal pressurization, flash heating; *Bizzarri and Cocco*, 2006; *Rice*, 2006) may be more relevant than melt lubrication. We believe, however, that frictional melting is common in silicate-built rocks during faulting in the lower part of the brittle crust as supported, for example, by the widespread occurrence of pseudotachylytes reported in the European Alps. A reason for the apparent scarcity of pseudotachylyte is probably due to the elusive nature of these fault rocks (e.g., veins are thin and pristine "magmatic" structures may easily undergo alteration during postseismic deformation and fault exhumation; for a detailed discussion, see *Di Toro et al.*, 2006b). It follows that pseudotachylytes can be easily missed in the field, even in locales investigated, in detail, for decades (see the case for the pseudotachylyte recently found in the Mount Abbot quadrangle in Sierra Nevada, California; *Kirkpatrick et al.*, 2007; *Griffith et al.*, 2008). In any case, it is worth noting that all friction experiments performed so far on cohesive and noncohesive rocks at seismic slip rates indicate the activation of some weakening mechanisms (thermal decomposition, gouge-related weakening, silica gel lubrication) that, in the absence of frictional melts, produce a dramatic decrease in shear stress similar to that observed in the case of melt lubrication (*Di Toro and Nielsen*, 2007; *Wibberley et al.,* 2008). The discrepancy between experimental results and seismological observations suggests that other factors including the dependence of the shear stress with slip rate and the roughness of the fault surface should be considered in rupture dynamics models in order to extrapolate experimental observations to natural conditions.

## 6.1. A New Approach to the Study of Exhumed Pseudotachylyte-Bearing Faults

The review of previous work shows that it is possible to retrieve information on earthquake mechanics from pseudotachylyte-bearing fault-fracture networks. However, a realistic physical model of the earthquake source should include, for instance, the roughness of natural fault surfaces, because fault geometry controls nucleation, propagation, and arrest of the seismic rupture (e.g., asperity and barriers; *Scholz*, 2002).

The synoptic view of Figure 11 summarizes an ideal multidisciplinary approach for the study of the earthquake mechanics by "direct" investigation of exhumed faults. The 3D architecture of an exhumed seismogenic fault zone is the basic input parameter for modeling coseismic fault slip and rupture propagation and requires a complete 3D mapping of fault geometry and distribution of different fault rocks within and between the different elements of the deformation network. To this aim, classical field structural analysis can be integrated with high-resolution laser scan (light detection and ranging [LIDAR]) mapping, allowing the reconstruction of fault zone architecture with a millimeter scale precision over large outcrops in the case of the Adamello Fault zone (Figure 11a). LIDAR and conventional structural data can be processed by geomodeling platforms to produce "true" 3D models of fault zones (*Jones et al.*, 2004) (Figure 11b). A further step to understanding earthquake processes from the field studies of exhumed fault zones will be the upscaling of analysis from the outcrop ($10^2$ to $10^3$ m) to the large earthquake scale ($10^4$ to $10^5$ m). This can be accomplished using geostatistical techniques developed in the oil industry (*Mallet*, 2002). The mapped 3D geometries of the fault zone will form the reference of calibrated numerical

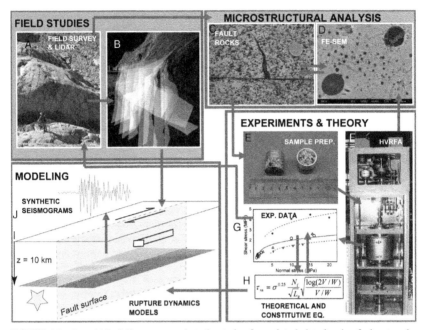

**FIGURE 11**  A multidisciplinary approach to the study of pseudotachylyte-bearing fault networks: synoptic view (see the text for an explanation). The Back Scatter Electron-Field Emission-Scanning Electron Microscope (BSE-FE-SEM) image in Figure 11d is courtesy of INGV. **(See Color Plate 15.)**

simulations for rupture nucleation and propagation (Figure 11i) (see Section 4). On a parallel line of research, samples from the fault zone (Figure 11c) are prepared (Figure 11e) and tested in rock friction experiments (Figure 11f) to investigate their frictional behavior during the seismic cycle (Figure 11g) (Section 5.2). Natural and experimentally produced fault rocks are analyzed in the lab (through microstructural, geochemical, and mineralogical analysis (Figure 11d) and compared to determine deformation processes. Constitutive and theoretically based equations (Figure 11h) that describe the rheology of the fault materials are tested in the nonconventional rock friction apparatus (Figures 11e, 11f, and 11g) (see Section 5.3). The experimental results (e.g., magnitude of the fault shear strength during seismic slip) are also checked in the field (Figures 11a and 11g) (see Section 5.1). The fault rheology constitutive equations (Figure 11h) are then applied in rupture models (Figure 11i) of real faults (Figures 11a and 11b). Earthquake rupture modeling (Figure 11i) will generate synthetic seismograms (Figure 11j) (*Olsen et al.*, 1997) to compare to real seismograms.

The multidisciplinary approach suggested here may exploit the extraordinary wealth of information frozen in large exposures of pseudotachylyte-bearing fault networks and yields a new vision of earthquake mechanics based on the physical processes occurring at seismogenic depth.

## ACKNOWLEDGMENTS

The authors thank Eiichi Fukuyama for editorial work, and we thank Hideo Takagi and an anonymous reviewer for their detailed and accurate comments to the paper. GDT and SN thank Takehiro Hirose and Toshihiko Shimamoto for allowing, helping, and running rock friction experiments at Kyoto University. This research was funded by the Fondazione Cariparo (Progetti di Eccellenza 2006), the University of Padova Scientifica fondi quota ex 60%, the European Research Council Starting Grant Project USEMS (2008–2013), and MIUR (PRIN2005). GDT thanks Elena Narduzzo for field support.

## REFERENCES

Allen, A. R., (1979), Mechanism of frictional fusion in fault zones, *J. Struct. Geol.*, **1**, 231-243.

Allen, J. L., (2005), A multi-kilometer pseudotachylyte system as an exhumed record of earthquake rupture geometry at hypocentral depths (Colorado, USA), *Tectonophys.*, **402**, 37-54.

Andrews, J. D., (2005), Rupture dynamics with energy loss outside the slip zone, *J. Geophys. Res.*, **110**, B01307, doi:10.1029/2004JB003191.

Archard, J. F., (1958), The temperature of rubbing surfaces, *Wear*, **2**, 438-455.

Austrheim, H. and T. M. Boundy, (1994), Pseudotachylytes generated during seismic faulting and eclogitization of the deep crust, *Science*, **265**, 82-83.

Beeler, N. M. and T. E. Tullis, (1996), Self-healing slip pulses in dynamic rupture models due to velocity dependent strength, *Bull. Seismol. Soc. Am.*, **86**(4), 1130-1148.

Beeler, N. M., (2006), Inferring earthquake source properties from laboratory observations and the scope of lab contributions to source physics, In: *Earthquakes: Radiated Energy and the Physics of Faulting*, edited by Abercrombie, R., A. McGarr, G. Di Toro, and H. Kanamori, *Geophysical Monograph Series*, **170**, 99-119, American Geophysical Union, Washington, D.C.

Blenkinsop, T. G., (2000), *Deformation Microstructures and Mechanisms in Minerals and Rocks*, Kluwer Academic Publishers, pp. 1-132.

Bizzarri, A. and M. Cocco, (2006), A thermal pressurization model for the spontaneous dynamic rupture propagation on a three-dimensional fault: 1. Methodological approach, *J. Geophys. Res.*, **111**, B05303, doi:10.1029/2005JB003862.

Bossière, G., (1991), Petrology of pseudotachylytes from the Alpine Fault of New Zealand, *Tectonophys.*, **196**, 173-193.

Bouchon, M., (1997), The state of stress on same faults of the San Andreas system as inferred from near-field strong motion data, *J. Geophys. Res.*, **102**(B6), 11731-11744.

Bouchon, M. and M. Vallée, (2003), Observation of long supershear rupture during the magnitude 8.1 Kunlunshan earthquake, *Science*, **301**, 824-826.

Boullier, A.-M., T. Ohtani, K. Fujimoto, H. Ito, and M. Dubois, (2001), Fluid inclusions in pseudotachylytes from the Nojima fault, Japan, *J. Geophys. Res.*, **106**(B10), 21965-21977.

Brodsky, E. E. and H. Kanamori, (2001), Elastohydrodynamic lubrication of faults, *J. Geophys. Res.*, **106**(B8), 16357-16374.

Brune, J. N., S. Brown, and P. A. Johnson, (1993), Rupture mechanism and interface separation in foam rubber models of earthquakes: A possible solution to the heat flow paradox and the paradox of large overthrusts, *Tectonophys.*, **218**(1-3), 59-67.

Burridge, R., G. Conn, and L. B. Freund, (1979), The stability of a rapid mode II shear crack with finite cohesive traction, *J. Geophys. Res.*, **85**, 2210-2222.

Byerlee, J. D., (1978), Friction of rocks, *Pure Appl. Geophys.*, **116**, 615-626.

Camacho, A., R. H. Vernon, and J. D. Fitz Gerald, (1995), Large volumes of anhydrous pseudo-tachylyte in the Woodroffe Thrust, eastern Musgrave Ranges, Australia, *J. Struct. Geol.*, **17**, 371-383.

Callegari, E. and P. Brack, (2002), Geological map of the Tertiary Adamello batholith (Northern Italy)—Explanatory notes and legend, *Mem. Sci. Geol.*, **54**, 19-49.

Chambon, G., J. Schmittbuhl, and A. Corfdir, (2002), Laboratory gouge friction: Seismic-like slip weakening and secondary rate- and state-effects, *Geophys. Res. Lett.*, **29**(10), 1366, doi:10.1029/2001GL014467.

Chester, F. M., J. P. Evans, and R. L. Biegel, (1993), Internal structure and weakening mechanisms of the San Andreas fault, *J. Geophys. Res.*, **98**(B1), 771-786.

Clarke, G. L. and A. R. Norman, (1993), Generation of pseudotachylite under granulite facies conditions, and its preservation during cooling, *J. Metam. Geol.*, **11**, 319-335.

Cowan, D. S., (1999), Do faults preserve a record of seismic slip? A field geologist's opinion, *J. Struct. Geol.*, **21**(8-9), 995-1001.

Del Gaudio, P., R. Han, G. Di Toro, T. Hirose, T. Shimamoto, and M. Cocco, (2006), Dynamic strength of peridotite at seismic slip rates: Experimental results. *EOS, Trans., AGU*, **87**(52), Fall Meeting Suppl. Abst., S33A-0219.

Del Moro, A., G. Pardini, C. Quercioli, I. Villa, and E. Callegari, (1983), Rb/Sr and K/Ar chronology of Adamello granitoids, Southern Alp, *Mem. Soc. Geol. Ital.*, **26**, 285-299.

Dieterich, J. H., (1978), Time-dependent friction and the mechanics of stick-slip, *Pure Appl. Geophys.*, **116**(4-5), 790-806.

Dieterich, J. H., (1979), Modeling of rock friction 1. Experimental results and constitutive equations, *J. Geophys. Res.*, **84**, 2161-2168.

Di Toro, G. and G. Pennacchioni, (2004), Superheated friction-induced melts in zoned pseudotachylytes within the Adamello tonalites (Italian Southern Alps), *J. Struct. Geol.*, **26**, 1783-1801.

Di Toro, G. and G. Pennacchioni, (2005), Fault plane processes and mesoscopic structure of a strong-type seismogenic fault in tonalites (Adamello batholith, Southern Alps), *Tectonophys.*, **402**, 54-79.

Di Toro, G., D. L. Goldsby, and T. E. Tullis, (2004), Friction falls towards zero in quartz rock as slip velocity approaches seismic rates, *Nature*, **427**, 436-439.

Di Toro, G., S. Nielsen, and G. Pennacchioni, (2005a), Earthquake rupture dynamics frozen in exhumed ancient faults, *Nature*, **436**, 1009-1012.

Di Toro, G., G. Pennacchioni, and G. Teza, (2005b), Can pseudotachylytes be used to infer earthquake source parameters? An example of limitations in the study of exhumed faults, *Tectonophys.*, **402**, 3-20.

Di Toro, G., T. Hirose, S. Nielsen, G. Pennacchioni, and T. Shimamoto, (2006a), Natural and experimental evidence of melt lubrication of faults during earthquakes, *Science*, **311**, 647-649.

Di Toro, G., T. Hirose, S. Nielsen, and T. Shimamoto, (2006b), Relating high-velocity rock friction experiments to coseismic slip in the presence of melts, In: *Earthquakes: Radiated Energy and the Physics of Faulting*, edited by Abercrombie, R., A. McGarr, G. Di Toro, and H. Kanamori, *Geophysical Monograph Series*, **170**, 121-134, American Geophysical Union, Washington, D.C.

Di Toro, G., and S. Nielsen, (2007), Low dynamic fault strength at seismic slip rates: Experimental evidence, *XXIV General Assembly of the International Union of Geodesy and Geophysics*, abstract SW 003.

Ermanovics, I. F., H. Helmstaedt, and A. G. Plant, (1972), An occurrence of Archean pseudo-tachylite from Southern Manitoba, *Can. J. Earth Sci*, **9**, 257-265.

Fabbri, O., A. Lin, and H. Tokushige, (2000), Coeval formation of cataclasite and pseudotachylyte in a Miocene forearc granodiorite, southern Kyushu, Japan, *J. Struct. Geol.*, **22**, 1015-1025.

Fialko, Y., (2004), Temperature fields generated by the elastodynamic propagation of shear cracks in the Earth, *J. Geophys. Res.*, **109**, B01303, doi:10.1029/2003JB002497.

Fialko, Y. and Y. Khazan, (2005), Fusion by earthquake fault friction: Stick or slip?, *J. Geophys. Res.*, **110**, B12407, doi:org/10.1029/2005JB003869.

Francis, P. W., (1972), The pseudotachylyte problem, comments on earth sciences, *Geophys.*, **3**, 35-53.

Goldsby, D. L. and T. E. Tullis, (2002), Low frictional strength of quartz rocks at subseismic slip rates, *Geophys. Res. Lett.*, **29**(17), 1844, doi:10.1029/2002GL01240.

Griffith, W. A., G. Di Toro, G. Pennacchioni, and D. D. Pollard, (2008), Thin pseudotachylytes in faults of the Mt. Abbot Quadrangle, Sierra Nevada: Physical constraints for small seismic slip events. *J. Struct. Geol.*, **30**, 1086-1094.

Griffith, W. A., A. J. Rosakis, D. D. Pollard and C. Ko, (2008), Tensile cracks: A new link between geological observations of faults and seismological models of earthquake dynamics, *Eos Trans.*, AGU, **89**(53), Fall Meeting Abst., T22A-05.

Grocott, J., (1981), Fracture geometry of pseudotachylyte generation zones: A study of shear fractures formed during seismic events, *J. Struct. Geol.*, **3**, 169-178.

Grunewald, U., R. S. J. Sparks, S. Kearns, and J. C. Komorowski, (2000), Friction marks on blocks from pyroclastic flows at the Soufriere Hills volcano, Montserrat: Implication for flow mechanisms, *Geology*, **28**, 827-830.

Guatteri, M. and P. Spudich, (1998), Coseismic temporal changes of slip direction; the effect of absolute stress on dynamic rupture, *Bull. Seismol. Soc. Am.*, **88**, 777-789.

Guatteri, M. and P. Spudich, (2000), What can strong-motion data tell us about slip-weakening fault-friction laws?, *Bull. Seismol. Soc. Am.*, **90**, 98-116.

Han, R., T. Shimamoto, T. Hirose, J.-H. Ree, and J. Ando, (2007), Ultralow friction of carbonate faults caused by thermal decomposition, *Science*, **316**, 878-881.

Hanks, T. C., (1977), Earthquake stress drops, ambient tectonic stresses and stresses that drive plate motions, *Pure Appl. Geophys.*, **143**, 441-458.

Heaton, T. H., (1990), Evidence for and implications of self healing pulses of slip in earthquake rupture, *Phys. Earth Planet. Interi.*, **64**, 1-20.

Hickman, S., M. Zoback, and W. Ellsworth, (2004), Introduction to special section: Preparing for the San Andreas Fault Observatory at Depth, *Geophys. Res. Lett.*, **31**, L12S01, doi:10.1029/2004GL020688.

Hirose, T. and T. Shimamoto, (2003), Fractal dimension of molten surfaces as a possible parameter to infer the slip-weakening distance of faults from natural pseudotachylytes, *J. Struct. Geol.*, **25**, 1569-1574.

Hirose, T. and T. Shimamoto, (2005a), Growth of molten zone as a mechanism of slip weakening of simulated faults in gabbro during frictional melting, *J. Geophys. Res.*, **110**, B05202, doi:10.1029/2004JB003207.

Hirose, T. and T. Shimamoto, (2005b), Slip-weakening distance of faults during frictional melting as inferred from experimental and natural pseudotachylytes, *Bull. Seismol. Soc. Am.*, **95**, 1666-1673.

Holland, T. H., (1900), The charnokite series, a group of Archean hypersthenic rocks in peninsular India, *India Geological Survey Memories*, **28**, 119-249.

Ida, Y., (1972), Cohesive force across the tip of a longitudinal shear crack and Griffith's specific surface energy, *J Geophys. Res.*, **77**, 3796-3805.

Irwin, G. R., (1957), Analysis of stresses and strains near the end of a crack traversing a plate, *J. Appl. Mech.*, **24**, 361-364.

Jeffreys, H., (1942), On the mechanics of faulting, *Geol. Mag.*, **79**, 291-295.

Jones, R. R., K. J. W. McCaffrey, R. W. Wilson, and R. E. Holdsworth, (2004), Digital field data acquisition: Towards increased quantification of uncertainty during geological mapping, In: *Geological Prior Information*, edited by Curtis, A., and R. Wood, *Geological Society Special Publication*, **239**, 43-56.

Kanamori, H., D. L. Anderson, and T. H. Heaton, (1998), Frictional melting during the rupture of the 1994 Bolivian earthquake, *Science*, **279**, 839-842.

Kanamori, H. and E. Brodsky, (2004), The physics of earthquakes, *Rep. Prog. Phys.*, **67**, 1429-1496.

Kirkpatrick, J. D., Z. K. Shipton, J. P. Evans, S. Micklethwaite, S. J. Lim, and P. McKillop (2008), Strike-slip fault terminations at seismogenic depths: The structure and kinematics of the Glacier Lakes fault, Sierra Nevada United States, *J. Geophys. Res.*, **113**, B04304, doi:10.1029/2007JB005311.

Koizumi, Y., K. Otsuki, A. Takeuchi, and H. Nagahama, (2004), Frictional melting can terminate seismic slips: Experimental results of stick-slip, *Geophys. Res. Lett.*, **31**, L21605, doi:10.1029/2004GL020642.

Kostrov, B., (1964), Self-similar problems of propagation of shear cracks, *J. Appl. Mech.*, **28**, 1077-1087.

Kostrov, B. and S. Das, (1988), *Principles of Earthquake Source Mechanics*, Cambridge Univ. Press, London.

Lachenbruch, A. H., (1980), Frictional heating, fluid pressure, and the resistance to fault motion, *J. Geophys. Res.*, **85**(B11), 6249-6272.

Lee, W. H., H. Kanamori, P. C. Jennings, and C. Kisslinger, (2002), *International Handbook of Earthquake & Engineering Seismology*, Part A & B, Academic Press, Amsterdam.

Lin, A., (1994), Glassy pseudotachylytes from the Fuyun Fault Zone, Northwest China, *J. Struct. Geol.*, **16**, 71-83.

Lin, A. and T. Shimamoto, (1998), Selective melting processes as inferred from experimentally generated pseudotachylytes, *J. Asian Earth Sci.*, **16**, 533-545.

Lin, A., A. Chen, C. Liau, C. Lee, C. Lin, P. Lin, S. Wen, and T. Ouchi, (2001), Frictional fusion due to coseismic slip landsliding during the 1999 Chi-Chi (Taiwan) ML 7.3 earthquake, *Geophys. Res. Lett.*, **28**, 4011-4014.

Lin, A., (2007), *Fossil Earthquakes: The Formation and Preservation of Pseudotachylytes*, Springer, Berlin, 348 pp.

Lockner, D. A. and N. M. Beeler, (2002), Rock failure and earthquakes, In: *Earthquake & Engineering Seismology*, edited by Lee, W. H., H. Kanamori, P. C. Jennings, and C. Kisslinger, 505-537, Academic Press, Amsterdam.

Lockner, D. A. and P. G. Okubo, (1983), Measurements of frictional heating in granite, *J. Geophys. Res.*, **88**, 4313-4320.

Ma, K.-F., S.-R. Song, H. Tanaka, C.-Y. Wang, J.-H. Hung, Y.-B. Tsai, J. Mori, Y.-F. Song, E.-C. Yeh, H. Sone, L.-W. Kuo, and H.-Y. Wu, (2006), Slip zone and energetics of a large earthquake from the Taiwan Chelungpu-fault Drilling Project (TCDP), *Nature*, **444**, 473-476.

Maddock, R. H., (1983), Melt origin of fault-generated pseudotachylytes demonstrated by textures, *Geology*, **11**, 105-108.

Maddock, R. H., (1986), Partial melting of lithic porphyroclasts in fault-generated pseudotachylytes, *Neues Jahrbuch für Mineralogie—Abhandlungenr*, **155**, 1-14.

Maddock, R. H., J. Grocott, and M. Van Nes, (1987), Vesicles, amygdales and similar structures in fault-generated pseudotachylytes, *Lithos*, **20**, 419-432.

Maddock, R. H., (1992), Effects of lithology, cataclasis and melting on the composition of fault-generated pseudotachylytes in Lewisian gneiss, Scotland, *Tectonophys.*, **204**, 261-278.

Magloughlin, J. F., (1989), The nature and significance of pseudotachylite from the Nason terrane, North Cascade Mountains, Washington, *J. Struct. Geol.*, **11**, 907-917.

Magloughlin, J. F., (1992), Microstructural and chemical changes associated with cataclasis and frictional melting at shallow crustal levels: The cataclasite-pseudotachylyte connection, *Tectonophys.*, **204**, 243-260.

Magloughlin, J. F. and J. G. Spray, (1992), Frictional melting processes and products in geological materials: Introduction and discussion, *Tectonophys.*, **204**, 197-206.

Magloughlin, J. F., (2005), Immiscible sulfide droplets in pseudotachylyte: Evidence for high temperature (> 1200°C) melts, *Tectonophys.*, **402**, 81-91.

Mallet, J. L., (2002), *Geomodelling*, Oxford Univ. Press, Oxford.

Marone, C., (1998), Laboratory-derived friction laws and their application to seismic faulting, *Ann. Rev. Earth Planet. Sci.*, **26**, 643-696.

Masch, L., R. H. Wenk, and E. Preuss, (1985), Electron microscopy study of hyalomylonites—Evidence for frictional melting in landslides, *Tectonophys.*, **115**, 131-160.

Mase, C. W. and L. Smith, (1987), Effects of frictional heating on the thermal, hydrologic, and mechanical response of a fault, *J. Geophys. Res.*, **92**(B7), 6249-6272.

Mayeda, K. and W. R. Walter, (1996), Moment, energy, stress drop, and source spectra of western United States earthquakes from regional coda envelopes, *J. Geophys. Res.*, **101**, 11, 195-11208.

McKenzie, D. and J. N. Brune, (1972), Melting on fault planes during large earthquakes, *Geophys. J. Roy. Astr. Soc.*, **29**, 65-78.

McPhie, J., M. Doyle, and R. Allen, (1993), *Volcanic Textures: A Guide to the Interpretation of Textures in Volcanic Rocks*, Hobart, Tasmania.

Melosh, H. J., (1996), Dynamical weakening of faults by acoustic fluidization, *Nature*, **397**, 601-606.

Mittempergher, S., G. Di Toro, and G. Pennacchioni, (2007), Effects of fault orientation on the fault rock assemblage of exhumed seismogenic sources (Adamello, Italy), *Geosciences Union Annual Meeting*, Vienna (A) EGU2007-A-05503.

Mizoguchi, K., T. Hirose, T. Shimamoto, and E. Fukuyama, (2007), Reconstruction of seismic faulting by high-velocity friction experiments: An example of the 1995 Kobe earthquake, *Geophys. Res. Lett.*, **34**, L01308, doi:10.1029/2006GL027931.

Moecher, D. P. and A. J. Brearley, (2004), Mineralogy and petrology of a mullite-bearing pseudotachylyte: Constraints on the temperature of coseismic frictional fusion, *Am. Mineral.*, **89**, 1485-1496.

Moecher, D. P. and Z. D. Sharp, (2004), Stable isotope and chemical systematics of pseudotachylyte and wall rock, Homestake shear zone, Colorado, USA: Meteoric fluid or rock-buffered conditions during coseismic fusion?, *J. Geophys. Res.*, **109**, B12206, doi:10.1029/2004JB003045.

Nielsen, S. and J. M. Carlson, (2000), Rupture pulse characterization: Self-healing, self-similar, expanding solutions in a continuum model of fault dynamics, *Bull. Seismol. Soc. Am.*, **90**, 1480-1497.

Nielsen, S., G. Di Toro, T. Hirose, and T. Shimamoto, (2008), Frictional melt and seismic slip, *J. Geophys. Res.*, **113**, B01308, doi:10.1029/2007JB0051222008.

Obata, M. and S. Karato, (1995), Ultramafic pseudotachylite from the Balmuccia peridotite, Ivrea-Verbano zone, northern Italy, *Tectonophys.*, **242**, 313-328.

O'Hara, K., (2001), A pseudotachylyte geothermometer, *J. Struct. Geol.*, **23**, 1345-1357.

O'Hara, K. and Z. D. Sharp, (2001), Chemical and oxygen isotope composition of natural and artificial pseudotachylyte: Role of water during frictional melting, *Earth Planet. Sci. Lett.*, **184**, 393-406.

Ohtani, T., K. Fujimoto, H. Ito, H. Tanaka, N. Tomida, and T. Higuchi, (2000), Fault rocks and past to recent fluid characteristics from the borehole survey of the Nojima fault ruptured in the 1995 Kobe earthquake, southwest Japan, *J. Geophys. Res.*, **105**, 16161-16171.

Ohtomo, Y. and T. Shimamoto, (1994), Significance of thermal fracturing in the generation of fault gouge during rapid fault motion: An experimental verification, *Struct. Geol.*, **39**, 135-144 (in Japanese with English abstract).

Okamoto, S., G. Kimura, S. Takizawa, and H. Yamaguchi, (2006), Earthquake fault rock indicating a coupled lubrication mechanism, *eEarth*, **1**, 23-28.

Olsen, K. B., R. Madariaga, and R. J. Archuleta, (1997), Three-dimensional dynamic simulation of the 1992 Landers earthquake, *Science*, **278**, 834-838.

Otsuki, K., N. Monzawa, and T. Nagase, (2003), Fluidization and melting of fault gouge during seismic slip: Identification in the Nojima fault zone and implications for focal earthquake mechanism, *J. Geophys. Res.*, **108**(B4), doi:10.1029/2001JB001711.

Passchier, C. W., (1982), Pseudotachylyte and the development of ultramylonite bands in the Saint-Barthelémy Massif, French Pyrenees, *J. Struct. Geol.*, **4**, 69-79.

Passchier, C. W. and R. A. J. Trouw, (1996), *Microtectonics*, Springer, Berlin, p. 289.

Pennacchioni, G. and B. Cesare, (1997), Ductile-brittle transition in pre-Alpine amphibolite facies mylonites during evolution from water-present to water-deficient conditions (Mont Mary nappe, Italian Western Alps), *J. Metam. Geol.*, **15**, 777-791.

Pennacchioni, G., G. Di Toro, P. Brack, L. Menegon, and I. M. Villa, (2006), Brittle-ductile-brittle deformation during cooling of tonalite (Adamello, Southern Italian Alps), *Tectonophys.*, **427**, 171-197.

Philpotts, A. R., (1964), Origin of pseudotachylites, *Am. J. Science*, **262**, 1008-1035.

Philpotts, A. R., (1990), *Principles of Igneous and Metamorphic Petrology*, Prentice Hall, Englewood Cliffs.

Pittarello, L., G. Di Toro, A. Bizzarri, G. Pennacchioni, J. Hadizadeh, and M. Cocco, (2008), Energy partitioning during seismic slip in pseudotachylyte-bearing faults (Gole Larghe Fault, Adamello, Italy), *Earth Planet. Sci. Lett.*, **269**, 131-139.

Power, W. L., T. E. Tullis, and J. D. Weeks, (1988), Roughness and wear during brittle faulting, *J. Geophys. Res.*, **93**, 15268-15278.

Price, N. J. and J. W. Cosgrove, (1990), *Analyses of Geological Structures*, Cambridge University Press.

Ray, S. K., (1999), Transformation of cataclastically deformed rocks to pseudotachylyte by pervasion of frictional melt: Inferences from clast size analysis, *Tectonophys.*, **301**, 283-304.

Reid, H. F., (1910), The mechanism of the earthquake, In: *The California Earthquake of April 18, 1906, Report of the State Earthquake Investigation Commission*, **2**, 1-192, Carnegie Institutions, Washington, D.C.

Reimold, W. U., (1998), Exogenic and endogenic breccias: A discussion of major problematics, *Earth Sciences Reviews*, **43**, 25-47.

Rempel, A. W., (2006), The effects of flash-weakening and damage on the evolution of fault strength and temperature, In: *Earthquakes: Radiated Energy and the Physics of Faulting*, edited by Abercrombie, R., A. McGarr, G. Di Toro, and H. Kanamori, *Geophysical Monograph Series*, **170**, 263-270, American Geophysical Union, Washington, D.C.

Rice, J. R., (1966), Contained plastic deformation near cracks and notches under longitudinal shear, *Int. J. Frac. Mech.*, **2**, 426-447.

Rice, J. R., (2006), Heating and weakening of faults during earthquake slip, *J. Geophys. Res.*, **111**, B05311, doi:10.1029/2005JB004006.

Rowe C. D., J. C. Moore, F. Meneghini, and A. W. McKeirnan, (2005), Large-scale pseudotachylytes and fluidized cataclasites from an ancient subduction thrust fault, *Geology*, **33**(12), 937-940.

Ruina, A., (1983), Slip instability and state variable friction laws, *J. Geophys. Res.*, **88**(B12), 10359-10370.

Sagy, A., E. Brodsky, and G. J. Axen, (2007), Evolution of fault-surface roughness with slip, *Geology*, **35**, 283-286.

Samudrala, O., Y. Huang, and A. J. Rosakis, (2002), Subsonic and intersonic shear rupture of weak planes with a velocity weakening cohesive zone, *J. Geophys. Res.*, **107**, 2170, doi:10.1029/2001JB000460.

Schmid, S. M., H. R. Aebli, F. Heller, and A. Zingg, (1989), The role of the Periadratic Line in the tectonic evolution of the Alps, In: *Alpine Tectonics*, edited by Coward M. P., D. Dietrich, and R. G Park, *Geological Society Special Publication*, **45**, 153-171.

Scholz, C. H., (1998), Earthquakes and friction laws, *Nature*, **391**, 37-42.

Scholz, C. H., (2002), *The Mechanics of Earthquakes and Faulting, 2$^{nd}$ ed.*, Cambridge University Press, Cambridge, 471 pp.

Scott, J. S. and H. I. Drever, (1953), Frictional fusion along an Himalayan thrust, *Proceedings of the Royal Society of Edinburgh, sect. B*, **65**, 121-140.

Shand, S. J., (1916), The pseudotachylyte of Parijs (Orange Free State) and its relation to "trap-shotten gneiss" and "flinty crush rock," *Quart. J. Geol. Soc. London*, **72**, 198-221.

Shaw, H., (1972), Viscosities of magmatic silicate liquids; an empirical method of prediction, *Am. J. Sci.*, **272**, 870-893.

Shimada, K., Y. Kobari, T. Okamoto, H. Takagi, and Y. Saka, (2001), Pseudotachylyte veins associated with granitic cataclasite along the Median Tectonic Line, eastern Kii Peninsula, Southwest Japan, *J. Geol. Soc. Japan*, **107**, 117-128.

Shimamoto, T. and H. Nagahama, (1992), An argument against the crush origin of pseudotachylytes based on the analysis of clast size distribution, *Struct. Geol.*, **14**, 999-1006.

Shimamoto, T. and A. Tsutsumi, (1994), A new rotary-shear high-speed frictional testing machine: Its basic design and scope of research, *J. Tectonic Res. Group of Japan*, **39**, 65-78 (in Japanese with English abstract).

Sibson, R. H., (1973), Interactions between temperature and pore fluid pressure during earthquake faulting—A mechanism for partial or total stress relief, *Nature*, **243**, 66-68.

Sibson, R. H., (1975), Generation of pseudotachylyte by ancient seismic faulting, *Geophys. J. Roy. astr. Soc.*, **43**(3), 775-794.

Sibson, R. H., (1977), Fault rocks and fault mechanisms, *J. Geol. Soc. London*, **133**, 191-213.

Sibson, R. H., (1980), Transient discontinuities in ductile shear zones, *J. Struct. Geol.*, **2**, 165-171.

Sibson, R. H., (1989), Earthquake faulting as a structural process, *J. Struct. Geol.*, **11**, 1-14.

Sibson, R. H., (2003), Thickness of the seismic slip zone, *Bull. Seismol. Soc. Am.*, **93**(3), 1169-1178.

Sibson, R. H. and V. Toy, (2006), The habitat of fault-generated pseudotachylyte: Presence vs. absence of friction melt, In: *Earthquakes: Radiated Energy and the Physics of Faulting*, edited by Abercrombie, R., A. McGarr, G. Di Toro, and H. Kanamori, *Geophysical Monograph Series*, **170**, 153-166, American Geophysical Union, Washington, D.C.

Snoke, A. W., J. Tullis, and V. Todd, (1998), *Fault-related Rocks: A Photographic Atlas*, Princeton University Press, New Jersey, 617 pp.

Spray, J. G., (1987), Artificial generation of pseudotachylyte using friction welding apparatus: Simulation of melting on a fault plane, *J. Struct. Geol.*, **9**, 49-60.

Spray, J. G., (1988), Generation and crystallization of an amphibolite shear melt: An investigation using radial friction welding apparatus, *Contrib. Mineral. Petrol.*, **99**, 464-475.

Spray, J. G., (1992), A physical basis for the frictional melting of some rock-forming minerals, *Tectonophys.*, **204**(3-4), 205-221.

Spray, J. G., (1993), Viscosity determinations of some frictionally generated silicate melts: Implications for fault zone rheology at high strain rates, *J. Geophys. Res.*, **98**, 8053-8068.

Spray, J. G., (1995), Pseudotachylyte controversy: Fact or friction?, *Geology*, **23**, 1119-1122.

Spray, J. G., (1997), Superfaults, *Geology*, **25**, 305-308.

Spray, J. G., (2005), Evidence for melt lubrication during large earthquakes, *Geophys. Res. Lett.*, **32**, L07301, doi:10.1029/2004GL022293.

Stesky, R. M., W. F. Brace, D. K. Riley, and P.-Y.F. Robin, (1974), Friction in faulted related rock at high temperature and pressure, *Tectonophys.*, **23**, 177-203.

Stipp, M., H. Stünitz, R. Heilbronner, and S. Schmid, (2002), The eastern Tonale fault zone: A 'natural laboratory' for crystal plastic deformation of quartz over a temperature range from 250 to 700°C, *J. Struct. Geol.*, **24**, 1861-1884.

Stipp, M., B. Fügenschuh, L. P. Gromet, H. Stünitz, and S. M. Schmid, (2004), Contemporaneous plutonism and strike-slip faulting: A case study from the Tonale fault zone north of the Adamello pluton (Italian Alps), *Tectonics*, **23**, TC3004, doi:10.1029/2003TC001515.

Swanson, M. T., (1988), Pseudotachylyte-bearing strike-slip duplex structures in the Fort Foster Brittle Zone, S. Maine, *J. Struct. Geol.*, **10**, 813-828.

Swanson, M. T., (1989), Side wall ripouts in strike-slip faults, *J. Struct. Geol.*, **11**, 933-948.

Swanson, M. T., (1992), Fault structure, wear mechanisms and rupture processes in pseudotachylyte generation, *Tectonophys.*, **204**, 223-242.

Swanson, M. T., (2006), Pseudotachylyte-bearing strike-slip faults in mylonitic host rocks, Fort Foster Brittle Zone, Kittery, Maine, In: *Earthquakes: Radiated Energy and the Physics of Faulting*, edited by Abercrombie, R., A. McGarr, G. Di Toro, and H. Kanamori, *Geophysical Monograph Series*, **170**, 167-179, American Geophysical Union, Washington, D.C.

Takagi, I., K. Goto, and N. Shigematsu, (2000), Ultramylonite bands derived from cataclasite and pseudotachylyte in granites, northeast Japan, *J. Struct. Geol.*, **22**, 1325-1339.

Toyoshima, T., (1990), Pseudotachylite from the Main Zone of the Hidaka metamorphic belt, Hokkaido, northern Japan, *J. Metam. Geol.*, **8**, 507-523.

Tsutsumi, A. and T. Shimamoto, (1997a), High-velocity frictional properties of gabbro, *Geophys. Res. Lett.*, **24**, 699-702.

Tsutsumi, A. and T. Shimamoto, (1997b), Temperature measurements along simulated faults during seismic fault motion, *Proc. 30$^{th}$ Int'l Geol. Congr.*, **5**, 223-232.

Tsutsumi, A., (1999), Size distribution of clasts in experimentally produced pseudotachylyte, *J. Struct. Geol.*, **21**, 305-312.

Tullis, T. E., (1988), Rock friction constitutive behavior from laboratory experiments and its implications for an earthquake prediction field monitoring program, *Pure Appl. Geophys.*, **126**, 556-588.

Turcotte, D. L., and G. Schubert, (2002), *Geodynamics*, $2^{nd}$ *ed.*, Cambridge University Press, Cambridge, USA, 472 pp.

Ueda, T., M. Obata, G. Di Toro, K. Kanagawa, and K. Ozawa, (2008), Mantle earthquakes frozen in mylonitized ultramafic pseudotachylytes of spinel-lherzolite facies, *Geology*, **36**(8), 607-610, doi:10.1130/G24739A.1.

Ujiie, K., H. Yamaguchi, A. Sakaguchi, and T. Shoichi, (2007), Pseudotachylytes in an ancient accretionary complex and implications for melt lubrication during subduction zone earthquakes, *J. Struct. Geol.*, **29**, 599-613.

Viola, G., N. S. Mancktelow, and D. Seward, (2001), Late Oligocene–Neogene evolution of Europe–Adria collision: New structural and geochronological evidence from the Giudicarie fault system (Italian Eastern Alps), *Tectonics*, **20**, 999-1020.

Wenk, H. R., (1978), Are pseudotachylites products of fracture or fusion?, *Geology*, **l6**, 507-511.

Wenk, H. R., L. R. Johnson, and L. Ratschbacher, (2000), Pseudotachylites in the Eastern Peninsular Ranges of California, *Tectonophys.*, **321**, 253-277.

White, J. C., (1996), Transient discontinuities revisited: Pseudotachylyte, plastic instability and the influence of low pore fluid pressure on the deformation processes in the mid-crust, *J. Struct. Geol.*, **18**, 1471-1486.

Wibberley. C. A. J, Y., Graham and G. Di Toro, (2008), Recent advances in the understanding of fault zone internal structure: A review, In: *The Internal Structure of Fault Zones: Implications for Mechanical and Fluid Flow Properties*, edited by Wibberley, C. A. J., W. Kurz, J. Imber. R. E. Holdsworth, and C. Collettini, *Geological Society Special Publication*, **299**, 5-33, The Geological Society Publishing House, Bath, UK.

## Chapter 6

# The Critical Slip Distance for Seismic and Aseismic Fault Zones of Finite Width

## Chris Marone
*Istituto Nazionale di Geofisica e Vulcanologia, Rome, Italy*
*Department of Geosciences, Penn State University, University Park, Pennsylvania, U.S.A*

## Massimo Cocco
*Istituto Nazionale di Geofisica e Vulcanologia, Rome, Italy*

## Eliza Richardson
*Istituto Nazionale di Geofisica e Vulcanologia, Rome, Italy*
*Department of Geosciences, Penn State University, University Park, Pennsylvania, U.S.A*

## Elisa Tinti
*Istituto Nazionale di Geofisica e Vulcanologia, Rome, Italy*

We present a conceptual model for the effective critical friction distance for fault zones of finite width. A numerical model with 1D elasticity is used to investigate implications of the model for shear traction evolution during dynamic and quasi-static slip. The model includes elastofrictional interaction of multiple, parallel slip surfaces, which obey rate and state friction laws with either Ruina (slip) or Dieterich (time) state evolution. A range of slip acceleration histories is investigated by imposing perturbations in slip velocity at the fault zone boundary and using radiation damping to solve the equations of motion. The model extends concepts developed for friction of bare surfaces, including the critical friction distance $L$, to fault zones of finite width containing wear and gouge materials. We distinguish between parameters that apply to a single frictional surface, including $L$ and the dynamic slip weakening distance $d_o$, and those that represent slip for the entire fault zone, which include the effective critical friction distance, $D_{cb}$, and the effective dynamic slip weakening distance $D_o$. A scaling law for $D_{cb}$ is proposed in terms of $L$ and the fault zone width. Earthquake source parameters depend on net slip across a fault zone and thus scale with $D_{cb}$, $D_o$, and the slip at yield strength $D_a$. We find that $D_a$ decreases with increasing velocity jump size for friction evolution via the Ruina law, whereas it is independent of slip acceleration rate for the Dieterich law. For both laws, $D_a$ scales with fault zone width and shear traction exhibits prolonged hardening before reaching a yield strength. The parameters $D_{cb}$ and $D_o$ increase roughly linearly with fault zone thickness. This chapter and Chapter 7 in

Fault-Zone Properties and Earthquake Rupture Dynamics

the volume discuss the problem of reconciling laboratory measurements of the critical friction distance with theoretical and field-based estimates of the effective dynamic slip weakening distance.

---

## 1. INTRODUCTION

The nature of the transition from interseismic fault creep to transiently accelerating slip and dynamic rupture propagation is a central problem in earthquake science. Laboratory experiments, studies of frictional instability, and theoretical models confirm that strength breakdown and transient frictional behavior are fundamental to phenomena ranging from aseismic slip and earthquake nucleation to dynamic earthquake triggering and postseismic fault slip (e.g., *Dieterich*, 1972, 1979, 1992; *Ida*, 1972; *Scholz et al.*, 1972; *Ohnaka*, 1973; *Palmer and Rice*, 1973; *Andrews*, 1976a, 1976b; *Rice and Ruina*, 1983; *Ruina*, 1983; *Tullis and Weeks*, 1986; *Tullis*, 1988, 1996; *Ohnaka and Kuwahara*, 1990; *Marone et al.*, 1991; *Beeler et al.*, 1996; *Boatwright and Cocco*, 1996; *Sleep*, 1997; *Marone*, 1998; *Scholz*, 1998; *Richardson and Marone*, 1999; *Bizzarri et al.*, 2001; *Cocco and Bizzarri*, 2002; *Bizzarri and Cocco*, 2003; *Lapusta and Rice*, 2003; *Boettcher and Marone*, 2004; *Tinti et al.*, 2005; *Bhat et al.*, 2007; *Liu and Rice*, 2007; *Savage and Marone*, 2007; *Ampuero and Rubin*, 2008). In the context of rate and state friction laws, a key parameter determining transient slip behavior is the critical friction distance $L$ for breakdown of frictional strength. Rapid fault slip acceleration and earthquakes are favored by large ratios of dynamic stress drop to $L$, whereas transient slip, slow earthquakes, and aseismic phenomena are favored by near zero and negative values of this ratio. The physical mechanisms governing the critical slip distance are poorly understood, and the few available observations come from laboratory experiments on rock fracture and friction. Strength breakdown during preseismic and aseismic slip likely involves multiple processes, each with its own characteristic length and timescale. This implies that rupture nucleation and dynamic propagation are inherently scale-dependent processes. In the present study, we focus on the length scale parameters and, in particular, on the slip associated with the breakdown phase.

In this chapter, we distinguish between parameters that apply to a single frictional surface, including $L$ and the dynamic slip weakening distance $d_o$, and those that represent slip for the entire fault zone, which include the effective fault zone critical friction distance $D_{cb}$ and the effective dynamic slip weakening distance $D_o$. Earthquake source parameters depend on net slip across a fault zone and thus scale with $D_{cb}$, $D_o$, and the slip at yield strength $D_a$. We propose a scaling relation between $D_{cb}$ and the intrinsic critical friction distance for frictional contact junctions $L$. We posit that $D_{cb}$ is a key parameter determining rupture acceleration, dynamic rupture propagation, and the mechanical energy absorbed within the fault zone. In Chapter 7,

*Cocco et al.* (2009) investigate the dynamic slip weakening parameter $D_c$, which we refer to as $D_o$ in this chapter.

The pioneering laboratory friction experiments of *Rabinowicz* (1951) showed that the transition between static and dynamic friction occurs over a characteristic slip ($s_k$; Figure 1a). This concept has been incorporated in frictional instability models (*Palmer and Rice*, 1973) and theoretical slip weakening laws (*Ida*, 1972; *Andrews*, 1976a, 1976b). However, only the rate- and

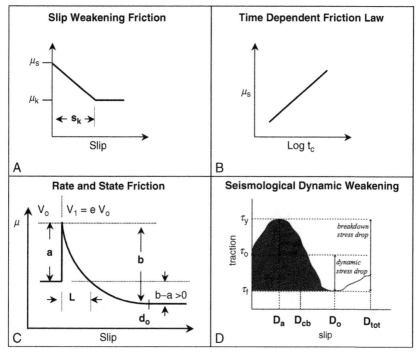

**FIGURE 1**    (a) Slip weakening friction law in which friction decreases from a static value $\mu_s$ to a kinetic value $\mu_k$ linearly over slip distance $s_k$. Kinetic friction is independent of slip velocity. (b) Idealized representation of the increase in static friction with the time of contact $t_c$, so-called *frictional healing*, showing approximately linear healing with the logarithm of contact age. (c) Rate and state friction law in which static and kinetic friction and frictional healing are special cases of a more general behavior. Friction varies with slip velocity, time of stationary contact, and recent memory of sliding conditions including velocity, normal stress, chemical environment, and temperature. The parameters $L$ and $d_o$ are the critical friction distance and the dynamic slip weakening distance, respectively, for a single frictional surface. (d) Idealized shear traction evolution curve for a dynamic rupture on a fault zone of finite width. The parameters $\tau_o$ (initial stress), $\tau_y$ (yield stress), and $\tau_f$ (dynamic frictional strength) define stress drop and strength excess. The shaded region represents the seismological fracture energy, which is referred to as breakdown work $W_b$. The parameters $D_a$, $D_{cb}$, $D_o$, and $D_{tot}$ are the slip at peak strength, the fault zone critical friction distance, the effective dynamic slip weakening distance, and the total slip, respectively, for a fault zone of finite thickness.

state-dependent friction laws (*Dieterich*, 1979; *Rice and Ruina*, 1983; *Ruina*, 1983) distinguish between the characteristic length scale of the constitutive formulation ($L$, see Figure 1c) and the slip weakening distance of shear traction evolution ($d_o$, see Figure 1c). By modeling the dynamic propagation of a shear crack governed by rate- and state-dependent friction, *Cocco and Bizzarri* (2002) and *Bizzarri and Cocco* (2003) proposed a scaling law for $L$ and $d_o$.

The overall behavior of shear stress breakdown and friction evolution is commonly named dynamic fault weakening (e.g., *Rice and Cocco*, 2007; *Cocco and Tinti*, 2008). The physical interpretation of the length scale parameters characterizing dynamic fault weakening depends on the adopted fault mechanical model and the processes controlling friction evolution. In the framework of a fault zone model consisting of two surfaces in contact, the parameter $L$ is commonly interpreted as the slip necessary to renew the population of contacts, and $d_o$ is a parameter contributing to the energy necessary to maintain dynamic crack propagation (i.e., fracture energy). However, in a more complex fault zone model of finite thickness and heterogeneous strain localization, as indicated by geological observations of faults (e.g., *Chester et al.*, 1993; *Arboleya and Engelder*, 1995; *Cowan*, 1999; *Sibson*, 2003; *Billi and Storti*, 2004; *Cashman et al.*, 2007), the physical interpretation of the constitutive parameter $L$ and of the critical slip weakening distance is not straightforward (e.g., *Ando and Yamashita*, 2007; *Bhat et al.*, 2007).

At seismic rupture speeds, breakdown weakening processes may include failure of adhesive contact junctions, yielding of grain cements, flash heating, particle fracture, and shear-induced melting (e.g., *Storti et al.*, 2003; *Di Toro et al.*, 2005, 2006; *Rice*, 2006; *Nielsen et al.*, 2008). At slower rates, such as appropriate for earthquake nucleation and aseismic creep, breakdown weakening may include mineral growth, shear localization, pressure solution creep, and reorganization of fault zone microstructures (e.g., *Rutter et al.*, 1986; *van der Pluijm et al.*, 2001; *Solum et al.*, 2003; *Niemeijer and Spiers*, 2006; *Ikari et al.*, 2007). In laboratory experiments on bare rock surfaces sheared at slow rates ($< 0.01$ to $0.1$ m/s) $L$ can be related to asperity contact junctions via concepts of Hertzian contacts (*Rabinowicz*, 1951; *Scholz*, 2002). When wear material is involved, processes involving granular physics become important, and higher strain rates mean that processes involving shear heating must be included (e.g., *Di Toro et al.*, 2006).

We emphasize here that although the parameter $L$ is the intrinsic length scale for rate- and state-dependent friction, seismically determined earthquake source parameters will depend on net slip across the fault zone and thus the effective fault zone critical friction distance $D_{cb}$ and the effective dynamic slip weakening distance $D_o$ (Figure 1d). $L$ is a constitutive parameter of laboratory inferred rate and state constitutive laws and describes the transition from one frictional state (e.g., static contact or steady creep) to another, for example, as measured during velocity step experiments (Figure 1c).

Comparison of field, laboratory, and numerical estimates of slip weakening distances reveals significant discrepancies (*Marone and Kilgore*, 1993; *Ide and Takeo*, 1997; *Guatteri et al.*, 2001; *Mikumo et al.*, 2003; *Ohnaka*, 2003). Although laboratory friction data are reasonably well described by the rate and state friction constitutive laws, many of the underlying processes are poorly understood. This complicates application of laboratory data to *in situ* fault conditions, which span timescales from seismic to interseismic and include complex thermal, chemical, and hydraulic processes. For example, *Bizzarri and Cocco* (2006a, 2006b) used rate- and state-dependent constitutive laws to investigate frictional heating and thermal pressurization during dynamic fault weakening. Their results clearly show that $D_o$ depends strongly on the hydraulic and thermal parameters of the fault zone. Moreover, analysis of the effective critical slip distance determined from seismological data reveals fundamental problems associated with resolution and scaling (*Guatteri et al.*, 2001; *Cocco et al.*, 2009).

The purpose of this chapter is to describe a frictional model for fault zones of finite width and summarize recent progress on understanding the critical slip distance. We focus in particular on the problem of reconciling laboratory measurements of the critical slip distance with theoretical- and field-based estimates. In addition, we compare and discuss laboratory friction measurements for slow slip rates ($< 0.01$ to $0.1$ m/s) with those derived from experiments at high slip rates ($>1$ m/s), which yield significantly larger values of $L$ and $D_o$.

## 2. FRICTION LAWS AND THE TRANSITION FROM STATIC TO KINETIC FRICTION

*Rabinowicz* (1951) introduced the slip weakening parameter, $s_k$, in his seminal paper on the transition from static to kinetic friction. He was motivated by the implausibility of simple static-kinetic friction models, which assume that friction changes instantaneously with the onset of slip, and he showed that $s_k$ could be understood in terms of asperity contact properties. *Rabinowicz* (1951, 1958) also rejected purely velocity-dependent friction models on the basis of their inability to describe the transition from static to kinetic friction (Figure 1a) and available data showing that friction was not a single valued function of sliding velocity (*Bowden and Tabor*, 1950; for a summary, see *Dowson*, 1979). The simplest model representing strength evolution that is consistent with both laboratory data and theoretical work on crack tip stresses, and elastodynamic rupture propagation is one in which details of the transient weakening are simplified to a linear trend over slip distance $s_k$ (Figure 1a).

A consequence of slip weakening behavior is the existence of a breakdown zone, which represents the spatial scale over which slip weakening occurs. *Barenblatt* (1959) and *Ida* (1972) introduced the breakdown zone for tensile and shear cracks, respectively, to avoid infinitely large stress concentrations on the fracture plane. One form of the breakdown zone model involving linear slip weakening, such as in Figure 1a, has been widely adopted in the literature (*Ida*, 1972; *Palmer and Rice*, 1973; *Andrews*, 1976a, 1976b).

Modern friction laws account for differences between static and kinetic friction in addition to (1) variation in kinetic friction with slip velocity and (2) frictional aging (Figure 1b) in which the static friction varies with waiting time (e.g., *Dieterich*, 1972; *Beeler et al.*, 1994; *Karner and Marone*, 2001). The rate and state friction laws (for reviews, see *Tullis*, 1996; *Marone*, 1998) combine aging, velocity-dependent kinetic friction, and observations of instantaneous friction rate effects (Figure 1c). These laws reproduce a wide range of laboratory observations and are capable of describing the full spectrum of fault behaviors, ranging from aseismic slip events to slow earthquakes and dynamic rupture (e.g., *Scholz*, 2002).

In the context of rate and state friction, slip weakening represents a memory effect that fades with time and slip following a perturbation in slip velocity, normal stress, or physical properties of the fault zone that are described by a friction state variable (Figure 1c). Rate and state friction behavior is characterized by three or more empirical parameters, which may be loosely thought of as material properties. These parameters vary with rock mineralogy, physicochemical conditions, and stressing rate of the frictional surface or shear zone, reflecting changes in the atomic and larger-scale processes of strain accommodation.

A major challenge in earthquake science is that of determining the *in situ* conditions and frictional properties appropriate for a given fault zone and seismic or aseismic phenomena. It is important to relate the friction parameters to seismological parameters of stress drop, slip weakening, and breakdown work (*Tinti et al.*, 2005). Figure 1d shows these parameters in the context of dynamic rupture modeling inferred from seismological data. A companion paper (*Cocco et al.*, 2009) focuses on these parameters and the problem of measuring slip weakening using seismic data and interpreting its scaling with other earthquake source parameters.

## 3. CONTACT MODEL FOR THE CRITICAL SLIP DISTANCE OF SOLID SURFACES AND SHEAR ZONES

*Rabinowicz* (1951, 1958) showed that the slip weakening distance $s_k$ could be understood in terms of asperity contact properties. He proposed a model in which $s_k$ is proportional to contact junction size (Figure 2a) and slip weakening results from progressive reduction of the real area of contact or replacement of older, stronger contact area with newly created, weaker contact (Figure 2b). This model is consistent with data showing that the rate-state friction parameter $L$ is proportional to roughness of clean solid surfaces and with measurements showing that $L$ scales with particle diameter of sheared granular layers (*Dieterich*, 1981; *Marone and Kilgore*, 1993). The asperity model

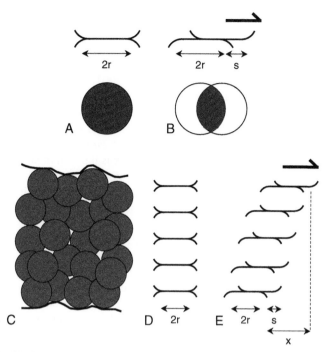

**FIGURE 2**   Idealized representations of friction contact junctions for clean surfaces (a and b) and surfaces separated by wear (gouge) materials (c through e). (a, b) Section and plan views of an asperity contact junction of diameter $2r$ formed during (a) static loading and (b) after sliding a distance $s$. Shaded areas represent the true area of contact, which decreases with slip and is reformed after a critical time. (c) Cross section of a gouge zone between rough surfaces showing 3D contact geometry of idealized wear particles. (d) Section view of idealized, boundary-parallel contact junctions within the model fault zone. Each contact has diameter $2r$. (e) Contact geometry after the upper boundary is displaced by a slip distance $x$; each contact slips a distance $s$. For clean surfaces, the transition from static to kinetic friction is determined by contact junction properties, whereas for a wear zone, the transition is determined by the number of contacts, contact junction properties, and particle interactions.

for $s_k$ and $L$ forms the basis of a common interpretation of the state variable in rate and state friction laws as the average age of a contact given by the ratio of $L$ to slip velocity.

The contact junction model for $L$ (Figures 2a and b) can be extended to a granular shear zone (Figure 2c). We focus here on the case of a mature fault zone of finite width, where the width is many times larger than the average particle diameter. Shearing of the fault boundary is accomplished by slip increments within the zone, along surfaces between contact junctions. Figure 2c shows a simplified geometry in which contacts are subparallel to the shear zone boundaries (Figure 2d). The model is based on field and laboratory studies showing (1) that shear becomes localized along boundary parallel planes that migrate throughout the zone with continued shear (e.g., *Mandl et al.*, 1977; *Logan et al.*, 1979; *Beeler et al.*, 1996; *Marone*, 1998); (2) the macroscopic coefficient of kinetic friction for a sheared granular layer is linearly related to particle contact dimensionality, such that friction of 1D grains is a factor of 3 lower than that for an assemblage of 3D spherical particles (*Frye and Marone*, 2002; *Knuth and Marone*, 2007); (3) the observation that the effective critical friction distance scales with shear zone thickness (*Marone and Kilgore*, 1993); and (4) theoretical studies showing the necessity to identify finite length and timescale parameters for dynamic rupture (*Ando and Yamashita*, 2007; *Bhat et al.*, 2007; *Cocco and Tinti*, 2008). Frictional shear of the layer boundaries by a slip distance $x$ is the sum of slip increments $s$ on the number of contacts within the shear zone (Figures 2c, 2d, and 2e). This model links frictional behavior to dynamic rupture by defining a scaling of $L$ with breakdown work and related seismic parameters (*Cocco et al.*, 2009).

For a fault zone of thickness $T$, the effective critical slip distance is given by the sum of contributions from individual contacts within the zone. We define the critical friction distance for a single contact as $L$. Then, $D_{cb} = n L \chi$, where $\chi$ is a geometric factor to account for contact orientation and $n$ is the number of surfaces in the shear zone (Figure 2). Particle diameter $D$ can be related to $L$ via contact properties as $L = D \zeta$, where $\zeta$ is a constant including elastic and geometric properties and the slip needed for fully developed sliding at the contact (*Boitnott et al.*, 1992). Combining these relations and the constants, we can define a linear relation between $D_{cb}$ and shear zone thickness as

$$D_{cb} = T \, \gamma_c \tag{1}$$

where $\gamma_c$ is the critical strain derived from slip increments on individual surfaces within the shear zone. This parameter is given by the product of $\chi$ and $\zeta$, and we expect that it varies with particle size, angularity, fault gouge mineralogy, and perhaps fault zone roughness. *Marone and Kilgore* (1993) determined $\gamma_c = 0.01$ using laboratory data on granular shear zones of varying thickness. Rewriting Equation 1 in terms of $L$ yields

$$\gamma_c = n \, L\chi/T \tag{2}$$

Thus, $\gamma_c$ is directly proportional to $L$.

Our proposed granular model for $D_{cb}$ retains the connection between slip velocity and contact lifetime suggested by *Rabinowicz* (1951). If the boundary shearing velocity is $v$, the average interparticle slip velocity on surfaces within the shear zone is $v/n$, and the average contact lifetime is given by $L/v$. This result is consistent with studies showing that frictional memory effects—and the friction state variable—depend on contact lifetime and also internal granular structure such as packing density, porosity, shear fabric, and granular force chains (e.g., *Marone et al.*, 1990; *Segall and Rice*, 1995; *Beeler et al.*, 1996; *Mair et al.*, 2002; *Mair and Hazzard*, 2007).

## 4. MODEL FOR A SHEAR ZONE OF FINITE THICKNESS

We investigate implications of a shear zone model in which slip occurs on multiple surfaces in a zone of finite thickness (Figure 2) and focus, in particular, on scaling of the critical slip distance with fault zone properties. Consider an idealized fault of width $T$ within a fractured host rock at Earth's surface (Figure 3). To account for both internal shear behavior and macroscopic properties of the fault zone, we assume the fault is composed of multiple, subparallel slip surfaces, each of which exhibits rate and state frictional characteristics (Figure 3). We focus on mature faults that contain boundary parallel shear localization, and we do not consider oblique shear bands such as Riedel shears. The fault zone is assumed to be symmetric about its midpoint $T/2$. Slip surfaces are separated by a distance $h$ and coupled elastically to adjacent surfaces via a stiffness $K_{int}$ (shown schematically as a leaf spring in Figure 3). The shear zone boundary is driven by remote loading at a constant tectonic displacement rate $v_{pl}$, and this elastic coupling is characterized by stiffness $K_{ext}$ (Figure 3). Our model is conceptually similar to a Burridge-Knopoff model (*Burridge and Knopoff*, 1967) for a series of slider blocks

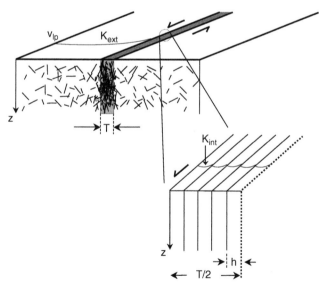

**FIGURE 3**   Fault zone model. Idealized crustal fault zone of thickness $T$. Fracture density increases toward fault zone, which includes multiple subparallel slip surfaces. The fault zone boundary is driven by remote loading $v_{lp}$ and 1D stiffness $K_{ext}$. (Inset) Enlargement of fault zone showing slip surfaces separated by distance $h$; half-thickness of fault zone is $T/2$. The left side is the shear zone boundary, the dotted line is the shear zone center. Each slip surface obeys the friction constitutive law. Fault zone shear is driven by boundary shear and elastofrictional interaction between surfaces. Thin lines denote elastic connection between surfaces (equivalent to leaf-springs) characterized by $K_{int}$.

linked by springs, except that the slider blocks in our model are in frictional contact and connected elastically in parallel rather than in series. Thus, slip on any surface within the fault zone reduces shear stress on all surfaces (Figure 3 inset). We assume stress equilibrium such that shear and normal stresses are equal on each surface. This simplified elastic model allows attention to focus on the role of friction parameters and fault zone width; future studies will incorporate models with internal stress/strength heterogeneity and variations in fault zone properties along strike.

For a given shear zone, slip is distributed along $n_s$ parallel surfaces, and we index these surfaces with the parameter $i$, starting with $i = 0$ at the boundary. Each slip surface $i$ is coupled elastically to its neighbors via stiffness $K_{int}$. We take $K_{int} = G/h$, where $G$ is shear modulus and $h$ is layer spacing (Figure 3) and use $G = 30$ GPa and $h = 6$ mm. We explore a range of fault zone widths but keep surface spacing $h$ constant. In the model results shown here, we assume that remote tectonic loading of the shear zone boundary is compliant relative to $K_{int}$ and fix the ratio $K_{int}/K_{ext}$ at 10. Another possibility, for future studies, would be to take $K_{int}/K_{ext}$ equal to the number of surfaces in the shear zone. Table 1 contains other details of the parameters used.

Each surface $i$ in the shear zone exhibits rate and state frictional behavior, such that friction $\mu_i$ is a function of state $\theta_i$ and slip velocity $v_i$ according to

$$\mu_i(\theta_i, v_i) = \mu_o + a \ln\left(\frac{v_i}{v_o}\right) + b \ln\left(\frac{v_o\theta_i}{L}\right) \qquad (3)$$

where $\mu_o$ is a reference friction value at slip velocity $v_o$, and the parameters $a$, $b$, and $L$ are empirically derived friction constitutive parameters, which we

**TABLE 1** Model Parameters for All Cases, $G = 30$ GPa, $\sigma = 100$ MPa, $K_{int} = G/h$; $K_{int}/K_{ext} = 10$; $v_o = $ 1e-6 m. $n_s/2$ Is the Number of Surfaces in the Fault Zone Half Width $T/2$

| a | b | L (m) | h (m) | $K_{ext}/\sigma_n$ (m$^{-1}$) | $n_s/2$ | T (m) | v (m/s) | Model; Figure(s) |
|---|---|---|---|---|---|---|---|---|
| 0.012 | 0.016 | 1e-5 | 6e-3 | 5e4 | 20 | 0.24 | 0.01 | 11.12_3; 4 |
| 0.012 | 0.016 | 1e-5 | 6e-3 | 5e4 | 10 | 0.12 | 0.01 | 11.12_4; 5 |
| 0.012 | 0.016 | 1e-5 | 6e-3 | 5e4 | 50 | 0.60 | 0.01 | 11.12_4; 6 |
| 0.012 | 0.016 | 1e-5 | 6e-3 | 5e4 | 60 | 0.72 | 0.01 | 13.12_4; 7 |
| 0.012 | 0.016 | 1e-5 | 6e-3 | 5e4 | 10-100 | Var. | Var. | 13.12; 8, 9, 10 |
| 0.012 | 0.016 | 1e-5-1e2 | 6e-3 | 5e4 | 20-40 | Var. | Var. | 3.2; 11 |
| 0.012 | 0.016 | 3e-5 | 6e-3 | 5e4 | 20 | 0.24 | 0.01 | 16.12; 12 |

assume to be equal on all surfaces. Tectonic fault zones are likely to include spatial variations of the friction constitutive parameters within the shear zone, but these are beyond the scope of the present study. We analyze friction state evolution according to either

$$\frac{d\theta_i}{dt} = 1 - \frac{v_i\theta_i}{L} \text{ (Dieterich law)} \tag{4}$$

or

$$\frac{d\theta_i}{dt} = -\frac{v_i\theta_i}{L}\ln\left(\frac{v_i\theta_i}{L}\right) \text{ (Ruina law)} \tag{5}$$

Frictional slip on each surface satisfies the quasi-dynamic equation of motion with radiation damping

$$\mu_i = \frac{\tau_o}{\sigma_n} - \frac{G}{2\beta\sigma_n}(v_i - v_{pl}) + k(v_{pl}\, t - v_i\, t) \tag{6}$$

where $\mu_i$ is the frictional stress, $\tau_o$ is an initial stress, $\beta$ is shear wave speed, $\sigma_n$ is normal stress, $k$ is stiffness divided by normal stress, and $t$ is time. We assume a normal stress of 100 MPa and define stiffness $k$ using this value (Table 1). Differentiating Equations 3 and 6 with respect to time and solving for $dv_i/dt$ yields

$$\frac{dv_i}{dt} = \frac{k(v_{pl} - v_i) - \frac{b\frac{d\theta_i}{dt}}{\theta_i}}{\frac{a}{v_i} + \frac{G}{2\beta\sigma_n}} \tag{7}$$

which applies for each surface within the shear zone. Our approach for including radiation damping (*Rice*, 1993) is similar to that described in previous works (*Perfettini and Avouac*, 2004; *Ziv*, 2007).

We assume steady creep during the interseismic period, and thus each surface of the fault zone undergoes steady-state slip at velocity $v_i = v_o$ with $\mu_o = 0.6$ and $\theta_{ss} = L/v_o$. The effective stiffness $k_i$ between the load point and surface $i$ within the fault zone is given by

$$\frac{1}{k_i} = \frac{1}{K_{\text{ext}}} + \sum_{j=1}^{i}\frac{1}{K_{\text{int}_j}} \tag{8}$$

To determine shear motion within the fault zone, we solve the coupled Equations 4 and 7 or 5 and 7 using stiffness from Equation 8 and a fourth-order Runge-Kutta numerical scheme. A perturbation in slip velocity is imposed at the shear zone boundary, via the remote loading stiffness $K_{\text{ext}}$ and for each time step in the calculation, the surface with the lowest frictional strength is allowed to slip. The initial conditions are that shear and normal stress are the same on each surface, and thus we assume slip occurs where frictional strength is lowest. We ensure that time steps are small compared to the ratio of slip surface separation, $h$, to elastic wave speed. Thus, within a given time step, only one surface slips and it is coupled elastically to the remote loading velocity via the spring stiffness given in Equation 8.

## 5. RESULTS

We adopt the constitutive parameters, elastic properties, and slip velocities used by *Cocco and Bizzarri* (2002). These parameters are consistent with laboratory friction data for bare rock surfaces and fault gouge for which shear is localized (*Marone et al.*, 1990; *Mair and Marone*, 1999). *Cocco and Bizzarri* (2002) solved the elastodynamic equations for a 2D in-plane crack using a finite difference approach. They reported traction evolution and slip histories for several cases. We use their peak dynamic slip velocities as a proxy for dynamic slip in our models. To initiate slip in the model, we apply a step change in loading velocity, as a proxy for arrival of dynamic rupture at a point on a fault undergoing steady slip. We compute the traction evolution as a function of slip at the shear zone boundary by summing the shear displacement on each internal surface. For reference, the traction evolution is also computed for a single surface with the same frictional properties (thin lines in Figure 4). Traction evolution as a function of slip for shear zones that are 24 cm wide and obey either the Ruina or Dieterich state evolution laws are shown in Figure 4; thick lines show traction at the center of the fault as a function of net slip at the boundary. Figure 4a shows the distinction between parameters that apply to a single frictional surface, including $L$ and the dynamic slip weakening distance $d_o$, and those that represent slip for the entire fault zone, which include the effective critical friction distance, $D_{cb}$, and the effective dynamic slip weakening distance $D_o$.

In our model, shear strength exhibits prolonged hardening before reaching a yield strength (thick lines in Figure 4). This arises in part because a perturbation in slip velocity, applied at the boundary, causes each surface in succession to accelerate and strengthen via the friction direct effect, given by the term $(a \ln(v/v_o))$ in Equation 3. At each time step in the calculation, shear traction is equal on all surfaces as required by stress equilibrium, but only the weakest surface slips. Therefore, during the slip acceleration phase, any surface that slips slightly more than surrounding surfaces has higher friction and the next slip increment occurs elsewhere. This ensures that shear is pervasive, and not localized, before peak yield strength. The peak yield strength is slightly lower for fault zones of finite width compared to the reference case due to greater state evolution in the shear zone compared to the reference fault with zero width.

We track the slip at peak friction, defined as $D_a$, and use it to characterize the slip hardening phase (*Cocco et al.*, 2009). For shear zones that obey the Dieterich evolution law, the value of $D_a$ is much larger than that for the reference case (Figure 4a), although the postpeak strength evolution is not a

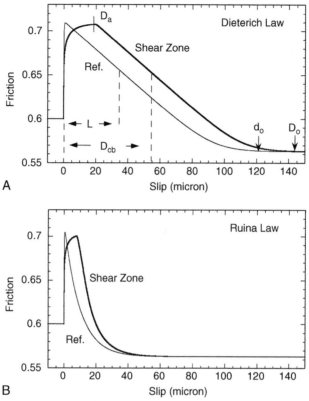

**FIGURE 4** Slip versus frictional traction for a single surface (thin line labeled Ref. in each panel) and for the center of a 24-cm-thick shear zone composed of 40 surfaces. In each case, the loading velocity $v_{lp}$ (Figure 3) is subject to a step increase from 1 micron/s to 1cm/s. Friction parameters are the same for all surfaces (see Table 1), and frictional state evolves following the Dieterich law (a) or the Ruina law (b). Panel a shows that friction parameters $L$ and $d_o$ for a single surface are the corresponding parameters for a shear zone of finite width. $D_{cb}$ is the effective critical friction distance, and $D_o$ is the effective dynamic slip weakening distance, respectively, defined by net slip across a shear zone.

function of shear zone thickness. The parameter $D_o$ corresponds to the slip during the dynamic stress drop (Figure 1d) for an elastodynamic model. We note that our parameter $D_o$ differs from studies of planar surfaces of zero thickness and some laboratory-based values (*Ohnaka*, 2003) for which the hardening phase and the parameter $D_a$ are near zero. In our notation, *Ohnaka*'s (2003) parameter $D_c$ is given by $D_c = D_o - D_a$. In our model, $D_o$ is given by the point at which friction reaches a minimum value after dynamic weakening. The hardening phase that occurs before reaching peak friction is shorter for the Ruina law than for the Dieterich law, and the postpeak traction evolution also requires less slip (Figure 4b).

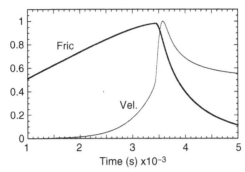

**FIGURE 5**   Time versus friction (thick line) and velocity (thin line) following an increase in loading velocity from 1 micron/s to 1 cm/s for a shear zone 12 cm thick composed of 20 surfaces. $D_c'$ is the slip at peak velocity.

Figure 5 shows temporal evolution of friction and slip velocity for a shear zone that obeys the Ruina friction law and is 12 cm wide (10 surfaces in half width). The parameter $D_c'$ is defined as the slip at peak velocity (*Fukuyama et al.*, 2003; *Mikumo et al.*, 2003) and consistent with previous work (*Tinti et al.*, 2004) is larger than $D_a$. The primary period of slip acceleration occurs during rapid weakening after the yield strength has been reached. Note that the friction and velocity curves of Figure 5 are normalized so that their maximum values are equal to 1.

Friction evolution varies systematically as a function of position within the shear zone (Figure 6). The shear stress is equal on all surfaces of the fault zone at a given time; however, frictional strength evolves according to slip velocity and state. Figure 6 shows details of the friction evolution for surfaces at distances of 6, 12, 18, 24, and 30 cm from the boundary (surface numbers 10, 20, 30, 40, and 50) for a shear zone of thickness 60 cm that contains 100 surfaces each of which follows the Ruina evolution law. The degree of hardening before reaching a yield strength increases with distance from the shear zone boundary and the yield strength decreases as the center of the shear zone is approached. These observations have important implications for seismological breakdown work, which is a more robust measure of fracture energy (*Cocco and Tinti*, 2008), because traction evolution in the preyield stress region represents energy that derives from dynamic stress concentration at the rupture tip and is required to overcome local strength excess (*Cocco et al.*, 2009). Note that slip velocity for the surface at the center of the shear zone, number 50 in Figure 6, exhibits slight overshoot and thus friction approaches the steady-state value from below.

For a point on a fault plane, slip velocity acceleration depends on several factors including rupture velocity, maximum particle velocity for the earthquake, and distance from the nucleation region. We consider a range of acceleration histories by applying velocity steps of different size to a model fault

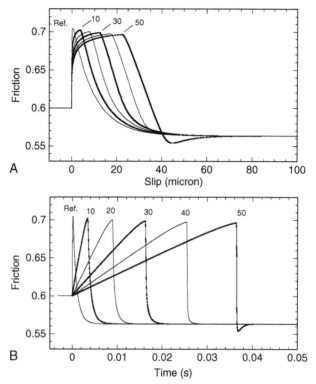

**FIGURE 6**   Slip (a) and time (b) versus frictional traction following an increase in loading velocity from 1 micron/s to 1 cm/s for a single surface (curve labeled Ref.) and for a shear zone 60 cm thick containing 100 surfaces. Friction evolution is shown for various positions within the shear zone. Labels denote surface number: 10 is near the shear zone boundary and 50 is at the center (see Figure 3). Friction parameters are same for all surfaces (see Table 1), and frictional state evolves following the Ruina law.

zone 72 cm thick with 120 surfaces. The far field load point velocity $v_{lp}$ is subject to step increases of ratio 10 to 10,000 relative to the initial value. The yield strength increases with velocity step size, as expected (Figure 7a). For Dieterich frictional state evolution, the parameter $D_a$ is constant, independent of acceleration history, whereas for Ruina state evolution, $D_a$ decreases with increasing magnitude of the slip velocity perturbation (Figure 7b). These observations are consistent with results from solutions of the full elastodynamic equations for a 2D crack (*Bizzarri and Cocco*, 2003).

Details of the relationship between $D_a$, shear zone thickness, and slip velocity perturbation are shown in Figure 8. Symbols represent shear zones of different thickness; the numbers at right represent the number of slip surfaces in the half width (Figure 8a). For the Ruina law, $D_a$ is nearly constant (ranging from 0 to 30 microns) with acceleration history for shear zones up

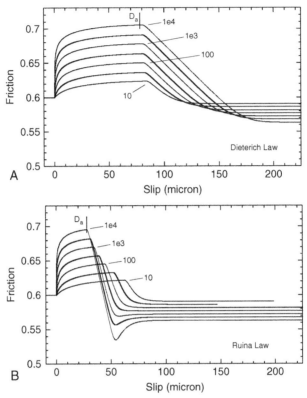

**FIGURE 7** Slip versus frictional traction following velocity steps of different sizes for a 72-cm-thick shear zone containing 120 surfaces. $D_a$, the slip at peak friction, is shown for one case in each panel. $D_a$ is roughly independent of acceleration history for the Dieterich law (a) but decreases with increasing velocity step size for the Ruina evolution law (b). Labels denote ratio of final to initial slip velocity.

to 36 cm wide (30 surfaces in the half width) and decreases with increasing velocity step size at a rate that scales directly with shear zone thickness (Figure 8b). Note that Figure 7 shows details of the traction evolution for one of the cases ($T = 72$ cm, 120 surfaces) shown in Figure 8. For the Dieterich law, $D_a$ is essentially independent of velocity step size for all shear zone thicknesses (Figure 8a).

Because seismological measurements of breakdown work include the contribution of traction evolution in the preyield stress region (*Cocco et al.*, 2009), the differences between Dieterich and Ruina style frictional state evolution have important implications for scaling of dynamic rupture parameters with fault zone thickness. For Dieterich law state evolution on a fault zone that is 1.2 m wide, the ratio $D_a/D_o$ ranges from 0.57 to 0.75 for slip velocity jump ratios from 10 to 10,000, respectively, whereas for a 0.012 m wide fault zone, this ratio ranges from 0.05 to 0.10. For the same range of velocities and

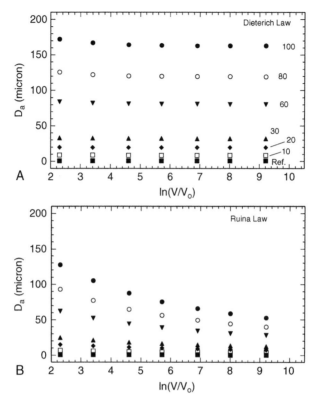

**FIGURE 8**   Slip at peak friction $D_a$ versus velocity jump size $\ln(v/v_o)$ for a variety of shear zone thicknesses. Labels in panel a denote number of surfaces in half width; Ref. indicates a single frictional surface. The same symbols are used to indicate shear zone width in both panels. For the Dieterich law, $D_a$ is nearly independent of velocity step size (a). For the Ruina law, $D_a$ decreases with increasing velocity step size for thick (>30 surfaces) shear zones (b).

Ruina state evolution, a fault zone that is 1.2 m wide has $D_a/D_o$ ranging from 0.32 to 0.69, whereas a 0.012 m wide fault zone has ratio from 0.04 to 0.10. This shows that preyield stress hardening increases with fault zone width and that the hardening phase can contribute up to 75% of the seismological breakdown work. Thus, seismic energy radiation, which is proportional to dynamic stress drop, decreases with increasing breakdown work, because increasing breakdown work reduces the energy left to be radiated.

## 6. IMPLICATIONS FOR SCALING OF THE DYNAMIC SLIP WEAKENING DISTANCE

*Cocco and Bizzarri* (2002) proposed a scaling relation between the critical friction distance $L$ and the dynamic slip weakening distance $d_o$. Written in terms of our variables, their scaling relation is $d_o = L \ln(v/v_o)$. Although the absolute values of $d_o$ obtained in our model are approximate (because we do not solve the full elastodynamic equations of motion), the relative values are meaningful because our model includes radiation damping and uses model parameters obtained by *Cocco and Bizzarri* (2002), who solved the elastodynamic equations for a 2D in-plane crack.

For a fault zone of finite width, the critical friction distance and the dynamic slip weakening distance are given by $D_{cb}$ and $D_o$, respectively. These parameters are proportional to the net slip across the fault zone (Figure 4). Thus, we extend the scaling relation of *Cocco and Bizzarri* (2002) to obtain a relation between $D_o$ and $D_{cb}$:

$$D_o = D_{cb} \ln(v/v_o) \qquad (9)$$

This relation can be modified to include the effects of fault zone thickness by noting that $D_{cb}$ scales with $T/h$: $D_o \approx L \, T/h$. Written in terms of the fault zone thickness and the intrinsic critical friction distance, the scaling relation of *Cocco and Bizzarri* (2002) predicts a linear scaling relation between $D_o$ and fault zone thickness for a given intrinsic critical friction distance.

We may test the validity of this scaling relation in the context of our fault zone model by evaluating the scaling of $D_o$ with velocity step size and fault zone thickness (Figure 9). The plot symbols distinguish the number of shear zone surfaces (proportional to thickness) with the numbers at right in Figure 9a denoting surface number relative to the shear zone boundary (see Table 1 for model details). For the Dieterich law, $D_o$ increases with shear zone thickness and velocity step size (Figure 9a). This is consistent with the observation that $D_a$ is independent of velocity step size (Figure 7) because the rate of postyield stress weakening is independent of acceleration history in this case. For Ruina friction state evolution, $D_o$ exhibits more complex behavior. For thin shear zones (0 to 30 surfaces in half width $T/2$), $D_o$ is nearly independent of velocity step size, whereas for thicker zones (60+ surfaces in half width $T/2$), $D_o$ first decreases and then increases with increasing velocity step size. For the parameters of our model, the minimum $D_o$ in the Ruina case occurs for a $300\times$ velocity jump (Figure 9c). Thus, when frictional state evolves according to the Dieterich law, the linear scaling between fault zone slip weakening distance and $D_{cb}$ predicted by Equation 9 is confirmed. In contrast, when state evolves with the Ruina law, $D_o$ depends on $T$ as predicted but the relationship between $D_o$ and velocity perturbation is more complex than indicated in Equation 9. We find that both $D_o$ and $D_a$ vary with shear zone thickness and slip acceleration history (Figures 6 and 7). It is of interest to distinguish the proportion of the change in dynamic slip weakening distance associated with hardening, as measured by $D_a$ (Figure 9). For the Dieterich law, the difference $D_o - D_a$ increases with velocity jump size (Figure 9b) whereas for the Ruina law, this difference is relatively insensitive to velocity jump size and shows more complex behavior (Figure 9d). These comparisons are useful for understanding the connection between our modeling results and

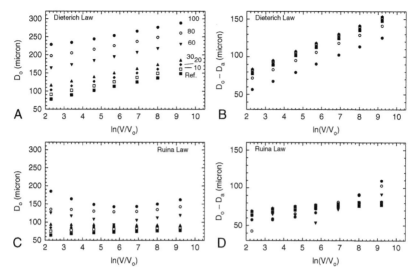

**FIGURE 9**   Effective dynamic slip weakening $D_o$ versus velocity jump size $\ln(v/v_o)$ for a variety of shear zone thicknesses. Labels in (a) denote number of surfaces in half width; Ref. indicates a single frictional surface. The same symbols are used to indicate shear zone width in both panels. (a) For the Dieterich law, $D_o$ increases with shear zone thickness and velocity step size. (b) Note that the difference $D_o - D_a$ increases with velocity jump size. (c) For the Ruina law, $D_o$ is nearly independent of velocity step size for thin (<30 surfaces) shear zones but exhibits complex behavior for thick shear zones. (d) The difference $D_o - D_a$ is relatively insensitive to velocity jump size and shows complex behavior.

laboratory measurements of the critical friction distance and slip weakening distance (e.g., *Marone and Kilgore*, 1993; *Ohnaka*, 2003).

We can further evaluate the scaling between fault zone thickness and the parameters $D_o$ and $D_a$. Figure 10 shows this relationship for velocity steps of $10\times$, $300\times$, and $10,000\times$ and fault zone thicknesses from 12 cm to 1.2 m (see Table 1 for other model parameters). For the Dieterich law, both $D_o$ and $D_a$ increase with thickness and $D_o$ increases with velocity jump size (Figures 10a and 10c). This is consistent with Equation 9 and expected from the traction evolution curves shown in Figure 7a. Friction evolution via the Ruina law shows that the dynamic slip parameters increase with fault zone thickness (Figures 10b and 10d) but smaller velocity jumps lead to greater $D_a$ (Figure 10b) and the dependence of $D_o$ on velocity step size is complicated and nonlinear (Figure 10d). Equation 9 also predicts scaling among $L$, $T$, and velocity jump ratio. Figure 11 shows this relationship for a range of $L$ values for two fault zone widths and two velocity jump ratios (see Table 1 for other model parameters). For the Ruina law, both $D_o$ and $D_a$ scale with $L$ and the magnitude of the velocity perturbation (Figures 11a and 11c). For the Dieterich law, $D_o$ and $D_a$ scale with $L$, but velocity perturbation magnitude has no

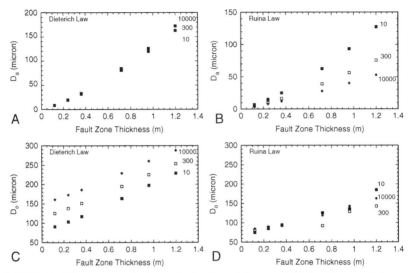

**FIGURE 10** Effective dynamic slip weakening $D_o$ and slip at peak friction $D_a$ versus fault zone thickness for three different velocity jump sizes. For the Dieterich law, $D_a$ is nearly independent of jump size and increases with fault zone thickness (a), whereas $D_o$ increases with both fault zone thickness and increasing velocity jump size (c). For the Ruina law, $D_a$ decreases with increasing jump size and increases with increasing fault zone thickness (b), whereas $D_O$ exhibits complicated behavior with respect to both velocity jump size and fault zone thickness (d).

effect (Figures 11b and 11d). For both state evolution laws, the effects of fault zone thickness and velocity perturbation are small compared to the effect of $L$.

## 7. DISCUSSION

The ontogeny of large tectonic faults, coupled with wear and gouge formation during subsequent offset, means that earthquakes occur within fault zones of finite width. Although there is significant uncertainty about the width of active slip during dynamic rupture, many lines of evidence suggest that average fault zone width is 10's or 100's of cm or more. Our modeling results show that the critical slip distance, as measured at the fault zone boundary, scales with fault zone thickness. This result is robust for fault zones that obey rate and state friction because the only requirement is positive instantaneous friction rate dependence (positive $a$ in Equation 3). Indeed, positive values of this parameter are one of the most consistent observations from laboratory friction studies (e.g., *Marone*, 1998; *Beeler et al.*, 2008) and are confirmed by frictional stability analyses.

Due to the difficulty of measuring the shear zone critical friction distance, $D_{cb}$ directly, we focus on the closely related parameters $D_a$ and $D_o$ (Figure 1). Each of these slip distances is expected to scale with fault width $T$, independent of the friction state evolution law. We studied the role of slip

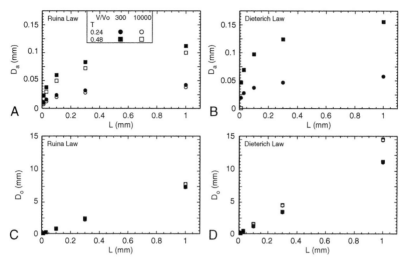

**FIGURE 11**   Effective dynamic slip weakening $D_o$ and slip at peak friction $D_a$ versus intrinsic critical friction distance $L$ for two fault zone thicknesses and two different velocity jump sizes. For the Dieterich law, both $D_a$ and $D_o$ are independent of velocity jump size, whereas $D_a$ and $D_o$ scale with fault zone width. For the Ruina law, both $D_a$ and $D_o$ scale weakly with velocity jump size and more strongly with fault zone width.

acceleration by imposing step changes in slip velocity of varying magnitude at the shear zone boundary and found marked differences between the Dieterich and Ruina friction evolution laws (see Figures 7 and 8). Our results are consistent with those of an elastodynamic model (*Cocco and Bizzarri*, 2002; *Bizzarri and Cocco*, 2003) even though we use a simplified 1D elastic model. We show that $D_a$ decreases with increasing velocity jump size for friction evolution via the Ruina law (Figures 7b and 8b), whereas it is independent of slip acceleration rate for the Dieterich law (Figures 7a and 8a). This is consistent with expectations from Equations 4 and 5 because frictional weakening requires a constant slip, independent of acceleration time, for the Dieterich law, whereas the rate of weakening scales strongly with slip velocity for the Ruina law (e.g., *Ampuero and Rubin*, 2008). For our purposes, the scaling of $D_a$ with velocity perturbation size means that breakdown work and traction evolution during dynamic rupture will differ for the two friction laws. This further complicates the problem of relating seismic estimates of the slip weakening distance to physical models and laboratory measurements of $L$ (*Cocco et al.*, 2009).

Another goal in evaluating the mechanics and scaling of slip weakening is that of relating laboratory data and seismic measurements of $D_{cb}$ to field observations of faulting. An understanding of shear zone width, particle size distribution, slip distribution, strain rate, and shear-induced melting would be a major advance in understanding earthquakes. Our model provides a connection between friction constitutive parameters and fault strain profiles (Figure 12). In this example, the slip distribution across the shear zone is determined by summing slip increments on each of the surfaces within a zone containing 40 slip surfaces. This snapshot in time represents the slip distribution when boundary slip had just reached $D_a$ and thus shear traction was at the yield stress. In our model, subsequent slip after this point is concentrated on a single surface at the center of the fault zone.

The strain rate profile represents the average value of strain rate over the time interval from zero to the point at which stress reached the yield stress

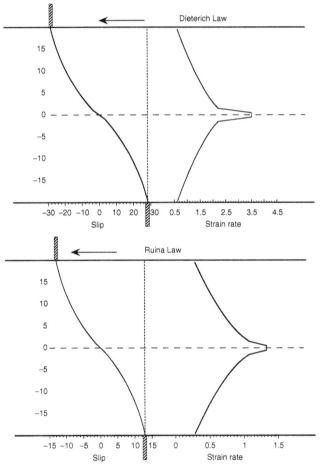

FIGURE 12   Slip profile and strain rate profile across the model fault zone for the Dieterich law (upper plots) and Ruina law (lower plots). Horizontal dashed lines represent center of the fault zone. Vertical dashed line and markers at shear zone boundary provide slip reference. Note the nonlinear slip distribution as a function of position, within the shear zone. Slip and strain rate are shown for the point at which the boundaries slid by an amount equal to $D_a$.

(Figure 12). Because of shear localization, the strain rate peak seen at the center of the shear zone (Figure 12) becomes larger with additional slip after the yield stress. Our model results for fault zone strain rate are generally consistent with field observations and indicate the importance of the transition from pervasive to localized shear within a fault zone. There is clearly a need to extend this approach, to investigate the strain distribution after the full seismic slip. That would require some form of predetermined localization dimension, for example, by coupling slip on a subset of surfaces, or the inclusion of post-yield stress slip hardening, so as to inhibit localized slip under some conditions. Shear heating or hydromechanical effects are obvious directions for future work.

Our simulations show a scaling of $D_o$ versus fault zone thickness $T$ (Figure 10). For $T$ ranging between 0.1 and 2 m, $D_o$ ranges between 0.01 and 0.2 mm. The latter range is smaller than estimates based on seismological investigations (*Ide and Takeo*, 1997; *Guatteri et al.*, 2001; *Fukuyama et al.*, 2003; *Mikumo et al.*, 2003; *Spudich and Guatteri*, 2004; *Fukuyama and Mikumo*, 2007; *Cocco et al.*, 2009) and theoretical models (*Tinti et al.*, 2004) as well as those inferred from high-velocity friction experiments (*Di Toro et al.*, 2006; *Hirose and Shimamoto*, 2005), where $D_o$ is on the order of meters. There are two explanations for this discrepancy. The first relies on the effects of other processes that produce larger $D_o$ values from slip weakening curves. Thermal pressurization, for instance (see *Bizzarri and Cocco*, 2006a, 2006b), produces larger values of $D_o$ than those resulting from a simple rate and state friction model. Indeed, it is interesting to note that in order to have $D_o$ values or the order of meters, following our scaling law of Equation 9, the fault zone critical friction distance is required to be in the range of cm for expected values of the velocity jump ratio. This is larger than values commonly inferred from low-speed velocity stepping friction experiments on gouge (*Mair and Marone*, 1999). Other processes such as melting or silica gel creation might also contribute to larger values of $L$ and $D_{cb}$. It is also possible that $D_{cb}$ is significantly larger in real dynamic fault weakening episodes, because slip velocity is far from a simplistic step function. The second explanation relates to the definition of fault zone thickness. In our study, we define the fault zone thickness as the zone where contacts among gouge grains are distributed. Thus, this would correspond to the fault core thickness. In this case, the proposed range of variability for $T$ agrees with geological and field observations. However, if we define the fault zone thickness as the region where strain rate varies from a nearly constant value to a localized bell-shaped profile, the damage zone surrounding the fault core should be included. This means that would effectively increase the range of $T$ by at least a factor of 100.

The scaling between $D_o$ and fault zone thickness has interesting implications for interpretation of high-velocity friction experiments. Indeed, the experimental setup of these tests is very different from velocity stepping experiments carried out with bare or relatively thin surfaces. Results from laboratory high-velocity friction experiments show that fault zone thickness increases with slip velocity, but they are unable to constrain this relation because gouge escapes from the testing machine in most cases. Finally, in high-velocity friction experiments, damage is created off-fault (see *Hirose and Bystricky*, 2007), and therefore the definition itself of *T* is not straightforward. We emphasize that the likely range of the critical slip weakening distance inferred from laboratory experiments remains poorly constrained. Bare surface, low-velocity, and high-velocity friction experiments represent different proxies for a realistic fault zone. The paucity of detailed laboratory data on fault gouge and the incomplete understanding of laboratory observations from high- and lower-speed friction complicates attempts of bridging the gap between laboratory experiments and natural fault zones.

## ACKNOWLEDGMENTS

We thank two anonymous reviewers and E. Fukuyama for comments that helped to improve the manuscript.

## REFERENCES

Ampuero, J.-P. and A. Rubin, (2008), Earthquake nucleation on rate and state faults: Aging and slip laws, *J. Geophys. Res.*, **113**, B01302, doi:10.1029/2007JB005082.

Ando, R. and T. Yamashita, (2007), Effects of mesoscopic-scale fault structure on dynamic earthquake ruptures: Dynamic formation of geometrical complexity of earthquake faults, *J. Geophys. Res.*, **112**, B09303, doi:10.1029/2006JB004612.

Andrews, D. J., (1976a), Rupture propagation with finite stress in antiplane strain, *J. Geophys. Res.*, **81**(20), 3575-3582.

Andrews, D. J., (1976b), Rupture velocity of plane strain shear cracks, *J. Geophys. Res.*, **81**(32), 5679-5687.

Arboleya, M. L. and T. Engelder, (1995), Concentrated slip zones with subsidiary shears: Their development on three scales in the Cerro Brass fault zone, Appalachian valley and ridge, *J. Struct. Geol.*, **17**(4), 519-532.

Barenblatt, G. I., (1959), The formation of brittle cracks during brittle fracture. General ideas and hypotheses. Axially-symmetric cracks, *Appl. Math. Mech.*, **23**, 1273-1282.

Beeler, N. M., T. E. Tullis, and J. D. Weeks, (1994), The roles of time and displacement in the evolution effect in rock friction, *Geophys. Res. Lett.*, **21**, 1987-1990.

Beeler, N. M., T. E. Tullis, M. L. Blanpied, and J. D. Weeks, (1996), Frictional behavior of large displacement experimental faults, *J. Geophys. Res.*, **101**(B4), 8697-8715.

Beeler, N. M., T. E. Tullis, and D. L. Goldsby, (2008), Constitutive relationships and physical basis of fault strength due to flash heating, *J. Geophys. Res.*, **113**, B01401, doi:10.1029/2007JB004988.

Bhat, H. S., M. Olives, R. Dmowska, and J. R. Rice, (2007), Role of fault branches in earthquake rupture dynamics, *J. Geophys. Res.*, **112**, B11309, doi:10.1029/2007JB005027.

Billi, A. and F. Storti, (2004), Fractal distribution of particle size in carbonate cataclastic rocks from the core of a regional strike-slip fault zone, *Tectonophys.*, **384**, 115-128.

Bizzarri, A., M. Cocco, D. J. Andrews, and E. Boschi, (2001), Solving the dynamic rupture problem with different numerical approaches and constitutive laws, *Geophys. J. Int.*, **144**(3), 656-678.

Bizzarri, A. and M. Cocco, (2003), Slip-weakening behavior during the propagation of dynamic ruptures obeying to rate- and state-dependent friction laws, *J. Geophys. Res.*, **108**, 2373, doi:10.1029/2002JB002198.

Bizzarri, A. and M. Cocco, (2006a), A thermal pressurization model for the spontaneous dynamic rupture propagation on a three-dimensional fault: 1. Methodological approach, *J. Geophys. Res.*, **111**, B05303, doi:10.1029/2005JB003862.

Bizzarri, A. and M. Cocco, (2006b), A thermal pressurization model for the spontaneous dynamic rupture propagation on a three-dimensional fault: 2. Traction evolution and dynamic parameters, *J. Geophys. Res.*, **111**, B05304, doi:10.1029/2005JB003864.

Boatwright, J. and M. Cocco, (1996), Frictional constraints on crustal faulting, *J. Geophys. Res.*, **101**(B6), 13895-13909.

Boettcher, M. S. and C. Marone, (2004), The effect of normal force vibrations on the strength and stability of steadily creeping faults, *J. Geophys. Res.*, **109**, B03406, doi:10.1029/2003JB002824.

Boitnott, G. N., R. L. Biegel, C. H. Scholz, N. Yosioka, and W. Wang, (1992), Micromechanics of rock friction 2: Quantitative modeling of initial friction with contact theory, *J. Geophys. Res.*, **97**, 8965-8978.

Bowden, F. P. and D. Tabor, (1950), *The Friction and Lubrication of Solids*, Part I, Oxford, Clarenden Press.

Burridge, R. and L. Knopoff, (1967), Model and theoretical seismicity, *Bull. Seismol. Soc. Am.*, **57**, 341-371.

Cashman, S. M., J. N. Baldwin, K. V. Casman, K. Swanson, and R. Crawford, (2007), Microstructures developed by coseismic and aseismic faulting in near-surface sediments, San Andreas fault, *California Geology*, **35**, 611-614, doi:10.1130/G23545A.1.

Chester, F. M., J. P. Evans, and R. L. Biegel, (1993), Internal structure and weakening mechanisms of the San Andreas fault, *J. Geophys. Res.*, **98**(B1), 771-786.

Cocco, M. and A. Bizzarri, (2002), On the slip-weakening behavior of rate- and state-dependent constitutive laws. *Geophys. Res. Lett*, **29**(11), 1-4.

Cocco, M. and E. Tinti, (2008), Scale dependence in the dynamics of earthquake propagation: Evidence from seismological and geological observations, *Earth Planet. Sci. Lett.*, **273**(1-2), 123-131.

Cocco, M., E. Tinti, C. Marone, and A. Piatanesi, (2009), Scaling of slip weakening distance with final slip during dynamic earthquake rupture, In: *Fault-Zone Structure and Earthquake Rupture Dynamics*, edited by E. Fukuyama, *International Geophysics Series*, **94**, 163-186, Elsevier.

Cowan, D. S., (1999), Do faults preserve a record of seismic slip? A field geologist's opinion, *J. Struct. Geol.*, **21**(8-9), 995-1001.

Di Toro, G., S. Nielsen, and G. Pennacchioni, (2005), Earthquake rupture dynamics frozen in exhumed ancient faults, *Nature*, **436**, 1009-1012.

Di Toro, G., T. Hirose, S. Nielsen, G. Pennacchioni, and T. Shimamoto, (2006), Natural and experimental evidence of melt lubrication of faults during earthquakes, *Science*, **311**, 647-649, doi:10.1126/science.1121012.

Dieterich, J. H., (1972), Time-dependent friction in rocks, *J. Geophys. Res.*, **77**, 3690-3697.

Dieterich, J. H., (1979), Modeling of rock friction 1. Experimental results and constitutive equations, *J. Geophys. Res.*, **84**, 2161-2168.

Dieterich, J. H., (1981), Constitutive properties of faults with simulated gouge, In: *Mechanical Behavior of Crustal Rocks*, edited by Carter, N. L., M. Friedman, J. M. Logan, and D. W. Stearns, *Geophysical Monograph Series*, **24**, 103-120, American Geophysical Union, Washington, D.C.

Dieterich, J. H., (1992), Earthquake nucleation on faults with rate- and state-dependent strength, *Tectonophys.*, **211**, 115-134.

Dowson, D., (1979), *History of Tribology*, Longman, New York.

Frye, K. M. and C. Marone, (2002), The effect of humidity on granular friction at room temperature, *J. Geophys. Res.*, **107**(B11), 2309, doi:10.1029/2001JB000654.

Fukuyama, E. and T. Mikumo, (2007), Slip-weakening distance estimated at near-fault stations, *Geophys. Res. Lett.*, **34**, L09302, doi:10.1029/2006GL029203.

Fukuyama, E., T. Mikumo, and K. B. Olsen, (2003), Estimation of the critical slip-weakening distance: Theoretical background, *Bull. Seismol. Soc. Am.*, **93**, 1835-1840, doi:10.1785/0120020184.

Guatteri, M., P. Spudich, and G. C. Beroza, (2001), Inferring rate and state friction parameters from a rupture model of the 1995 Hyogo-ken Nanbu (Kobe) Japan earthquake, *J. Geophys. Res.*, **106**, 26511-26521.

Hirose, T. and M. Bystricky, (2007), Extreme dynamic weakening of faults during dehydration by coseismic shear heating, *Geophys. Res. Lett.*, **34**, L14311, doi:10.1029/2007GL030049.

Hirose, T. and T. Shimamoto, (2005), Slip-weakening distance of faults during frictional melting as inferred from experimental and natural pseudotachylytes, *Bull. Seismol. Soc. Am.*, **95**, 1666-1673, doi:10.1785/0120040131.

Ida, Y., (1972), Cohesive force across the tip of a longitudinal-shear crack and Griffith's specific surface energy, *J. Geophys. Res.*, **77**, 3796-3805.

Ide, S. and M. Takeo, (1997), Determination of constitutive relations of fault slip based on seismic wave analysis, *J. Geophys. Res.*, **102**(B12), 27379-27391.

Ikari, M., D. M. Saffer, and C. Marone, (2007), Effect of hydration state on the frictional properties of montmorillonite-based fault gouge, *J. Geophys. Res.*, **112**, B06423, doi:10.1029/2006JB004748.

Karner, S. L. and C. Marone, (2001), Frictional restrengthening in simulated fault gouge: Effect of shear load perturbations, *J. Geophys. Res.*, **106**, 19319-19337.

Knuth, M. and C. Marone, (2007), Friction of sheared granular layers: The role of particle dimensionality, surface roughness, and material properties, *Geochem. Geophys. Geosyst.*, **8**, Q03012, doi:10.1029/2006GC001327.

Lapusta, N. and J. R. Rice, (2003), Nucleation and early seismic propagation of small and large events in a crustal earthquake model, *J. Geophys. Res.*, **108**, 2205, doi:10.1029/2001JB000793.

Liu, Y and J. R. Rice, (2007), Spontaneous and triggered aseismic deformation transients in a subduction fault model, *J. Geophys. Res.*, **112**, B09404, doi:10.1029/2007JB004930.

Logan, J. M., M. Friedman, N. Higgs, C. Dengo, and T. Shimamodo, (1979), Experimental studies of simulated fault gouges and their application to studies of natural fault zones, In: *Analysis of Actual Fault Zones in Bedrock, US Geol. Surv. Open File Rep.*, **1239**, 305-343.

Mair, K., K. M. Frye, and C. Marone, (2002), Influence of grain characteristics on the friction of granular shear zones, *J. Geophys. Res.*, **107**(10), 2219, doi:10.1029/2001JB000516.

Mair, K. and J. F. Hazzard, (2007), Nature of stress accommodation in sheared granular material: Insights from 3D numerical modelling, *Earth Planet. Sci. Lett.*, **259**, 469-485, doi:10.1016/j.epsl.2007.05.006.

Mair, K. and C. Marone, (1999), Friction of simulated fault gouge for a wide range of velocities and normal stress, *J. Geophys. Res.*, **104**(B12), 28899-28914.

Mandl, G., L. N. J. de Jong, and A. Maltha, (1977), Shear zones in granular material, an experimental study of their structure and mechanical genesis, *Rock Mech.*, **9**, 95-144.

Marone, C., (1998), Laboratory-derived friction laws and their application to seismic faulting, *Ann. Rev. Earth Planet. Sci.*, **26**, 643-696.

Marone, C. and B. Kilgore, (1993), Scaling of the critical slip distance for seismic faulting with shear strain in fault zones, *Nature*, **362**, 618-622.

Marone, C., C. B. Raleigh, and C. H. Scholz, (1990), Frictional behavior and constitutive modeling of simulated fault gouge, *J. Geophys. Res.*, **95**(B5), 7007-7025.

Marone, C., C. H. Scholz, and R. Bilham, (1991), On the mechanics of earthquake afterslip, *J. Geophys. Res.*, **96**, 8441-8452.

Mikumo, T., K. B. Olsen, E. Fukuyama, and Y. Yagi, (2003), Stress-breakdown time and slip-weakening distance inferred from slip-velocity functions on earthquake faults, *Bull. Seismol. Soc. Am.*, **93**(1), 264-282.

Nielsen, S., G. Di Toro, T. Hirose, and T. Shimamoto, (2008), Frictional melt and seismic slip, *J. Geophys. Res.*, **113**, B01308, doi:10.1029/2007JB005122.

Niemeijer, A. R. and C. J. Spiers, (2006), Velocity dependence of strength and healing behaviour in simulated phyllosilicate-bearing fault gouge, *Tectonophys.*, **427**(1-4), 231-253, doi:10.1016/j.tecto.2006.03.048.

Ohnaka, M., (1973), Experimental studies of stick-slip and their application to the earthquake source mechanism, *J. Phys. Earth*, **21**, 285-303.

Ohnaka, M., (2003), A constitutive scaling law and a unified comprehension for frictional slip failure, shear fracture of intact rock, and earthquake rupture, *J. Geophys. Res.*, **108**(B2), 2080, doi:10.1029/2000JB000123.

Ohnaka, M. and Y. Kuwahara, (1990), Characteristic features of local breakdown near a crack-tip in the transition zone from nucleation to dynamic rupture during stick-slip shear failure, *Tectonophys.*, **175**, 197-220.

Palmer, A. C. and J. R. Rice, (1973), The growth of slip surfaces in the progressive failure of over-consolidated clay, *Proc. Roy. Soc. London* Ser. A, **332**, 527-548.

Perfettini, H. and J.-P. Avouac (2004), Stress transfer and strain rate variations during the seismic cycle, *J. Geophys. Res.*, **109**, doi:10.1029/2003JB002917.

Rabinowicz, E., (1951), The nature of static and kinetic coefficients of friction, *J. Appl. Phys.*, **22**, 1373-1379.

Rabinowicz, E., (1958), The intrinsic variables affecting the stick-slip process, *Proc. Phys. Soc. (London)*, **71**, 668-675.

Rice, J. R., (1993), Spatio-temporal complexity of slip on a fault, *J. Geophys. Res.*, **98**, 9885-9907.

Rice, J. R., (2006), Heating and weakening of faults during earthquake slip, *J. Geophys. Res.*, **111**, B05311, doi:10.1029/2005JB004006.

Rice, J. R. and M. Cocco, (2007) Seismic fault rheology and earthquake dynamics, In: *Tectonic Faults: Agents of Change on a Dynamic Earth*, edited by M. R. Handy, G. Hirth and N. Hovius (Dahlem Workshop 95, Berlin, January 2005, on *The Dynamics of Fault Zones*), 99-137, The MIT Press, Cambridge, MA, USA.

Rice, J. R. and A. L. Ruina, (1983), Stability of steady frictional slipping, *J. Appl. Mech.*, **105**, 343-349.

Richardson, E. and C. Marone, (1999), Effects of normal stress vibrations on frictional healing, *J. Geophys. Res.*, **104**(B12), 28859-28878.

Ruina, A., (1983), Slip instability and state variable friction laws, *J. Geophys. Res.*, **88**(B12), 10359-10370.

Rutter, E. H., R. H. Maddock, S. H. Hall, and S. H. White, (1986), Comparative microstructure of natural and experimentally produced clay bearing fault gouges, *Pure Appl. Geophys.*, **24**, 3-30.

Savage, H., and C. Marone, (2007), The effects of shear loading rate vibrations on stick-slip behavior in laboratory experiments, *J. Geophys. Res.*, **112**, B02301, doi:10.1029/2005JB004238.

Scholz, C. H., (1998), Earthquakes and friction laws, *Nature*, **391**, 37-42.

Scholz, C. H., (2002), *The Mechanics of Earthquakes and Faulting, 2nd ed.*, Cambridge University Press, Cambridge, 471 pp.

Scholz, C. H., P. Molnar, and T. Johnson, (1972), Detailed studies of frictional sliding of granite and implications for the earthquake mechanism, *J. Geophys. Res.*, **77**, 6392-6406.

Segall, P. and J. R. Rice, (1995), Dilatancy, compaction, and slip instability of a fluid-infiltrated fault, *J. Geophys. Res.*, **100**(B11), 22155-22173.

Sibson, R. H., (2003), Thickness of the seismic slip zone, *Bull. Seismol. Soc. Am.*, **93**(3), 1169-1178.

Sleep, N. H., (1997), Application of a unified rate and state friction theory to the mechanics of fault zones with strain localization, *J. Geophys. Res.*, **102**, 2875-2895.

Solum, J. G., B. A. van der Pluijm, D. R. Peacor, and L. N. Warr, (2003), Influence of phyllo-silicate mineral assemblages, fabrics, and fluids on the behavior of the Punchbowl fault, southern California, *J. Geophys. Res.*, **108**(B5), 2233, doi:10.1029/2002JB001858.

Spudich, P. and M. Guatteri, (2004), The effect of bandwidth limitations on the inference of earth-quake slip-weakening distance from seismograms, *Bull. Seismol. Soc. Am.*, **94**, 2028-2036, doi:10.1785/0120030104.

Storti, F., A. Billi, and F. Salvini, (2003), Particle size distributions in natural carbonate fault rocks: Insights for non-self-similar cataclasis, *Earth Planet. Sci. Lett*, **206**, 173-186.

Tinti, E., A. Bizzarri, A. Piatanesi, and M. Cocco, (2004), Estimates of slip weakening distance for different dynamic rupture models, *Geophys. Res. Lett.*, **31**, L02611, doi:10.1029/2003GL018811.

Tinti, E., P. Spudich, and M. Cocco, (2005), Earthquake fracture energy inferred from kinematic rupture models on extended faults, *J. Geophys. Res.*, **110**, B12303, doi:10.1029/2005JB003644.

Tullis, T. E., (1988), Rock friction constitutive behavior from laboratory experiments and its implications for an earthquake prediction field monitoring program, *Pure Appl. Geophys.*, **126**, 555-588.

Tullis, T. E., (1996), Rock friction and its implications for earthquake prediction examined via models of Parkfield earthquakes, In: *Earthquake Prediction: The Scientific Challenge*, edited by Knopoff, L., *Proc. Natl. Acad. Sci. USA*, **93**, 3803-3810.

Tullis, T. E. and J. D. Weeks, (1986), Constitutive behavior and stability of frictional sliding of granite, *Pure Appl. Geophys.*, **124**, 383-414.

van der Pluijm, B. A., C. M. Hall, P. J. Vrolijk, D. R. Pevear, and M. C. Covey, (2001), The dat-ing of shallow faults in the Earth's crust, *Nature*, **412**, 172-175, doi:10.1038/35084053.

Ziv, A., (2007), On the nucleation of creep and the interaction between creep and seismic slip on rate- and state-dependent faults, *Geophys. Res. Lett.*, **34**, L15303, doi:10.1029/2007GL030337.

## Chapter 7

# Scaling of Slip Weakening Distance with Final Slip during Dynamic Earthquake Rupture

**Massimo Cocco**
*Istituto Nazionale di Geofisica e Vulcanologia, Rome, Italy*

**Elisa Tinti**
*Istituto Nazionale di Geofisica e Vulcanologia, Rome, Italy*

**Chris Marone**
*Istituto Nazionale di Geofisica e Vulcanologia, Rome, Italy*
*Department of Geosciences, Penn State University, University Park, Pennsylvania, U.S.A*

**Alessio Piatanesi**
*Istituto Nazionale di Geofisica e Vulcanologia, Rome, Italy*

We discuss physical models for the characteristic slip weakening distance $D_c$ of earthquake rupture with particular focus on scaling relations between $D_c$ and other earthquake source parameters. We use inversions of seismic data to investigate the breakdown process, dynamic weakening, and measurement of $D_c$. We discuss limitations of such measurements. For studies of breakdown processes and slip weakening, it is important to analyze time intervals shorter than the slip duration and those for which slip velocity is well resolved. We analyze the relationship between $D_c$ and the parameters $D_c'$ and $D_a$, which are defined as the slip at the peak slip velocity and the peak traction, respectively. We discuss approximations and limitations associated with inferring the critical slip weakening distance from $D_c'$. Current methods and available seismic data introduce potential biases in estimates of $D_c$ and its scaling with seismic slip due to the limited frequency bandwidth considered during typical kinematic inversions. Many published studies infer erroneous scaling between $D_c$ and final slip due to inherent limitations, implicit assumptions, and poor resolution of the seismic inversions. We suggest that physical interpretations of $D_c$ based on its measurement for dynamic earthquake rupture should be done with caution and the aid of accurate numerical simulations. Seismic data alone cannot, in general, be used to infer physical processes associated with $D_c$, although the estimation of breakdown work is reliable. We emphasize that the parameters $T_{acc}$ and peak slip velocity contain the same

**Fault-Zone Properties and Earthquake Rupture Dynamics**
**163**

dynamic information as $D_c$ and breakdown stress drop. This further demonstrates that inadequate resolution and limited frequency bandwidth impede to constrain dynamic rupture parameters.

## 1. INTRODUCTION

Understanding shear traction evolution during nucleation and dynamic propagation of earthquakes is one of the major tasks for seismologists and Earth scientists. Earthquakes are the most important expression of faulting, and knowledge of the processes controlling dynamic fault weakening during propagation of a seismic rupture is a crucial goal. This should be achieved by collecting geological and geophysical observations of natural faults, from laboratory experiments of friction and fracture, and by modeling seismic data and earthquake rupture using theoretical models and numerical simulations. Dynamic fault weakening can be fully described by the total shear traction evolution at a target point on the fault plane as a function of time or slip (*Rice and Cocco*, 2006). Figure 1 shows an example of dynamic traction, slip, and slip velocity evolution as a function of time (panel a) and dynamic shear traction as a function of slip (panel b) for a target point on the fault plane; the latter is commonly called the slip weakening curve. The dynamic traction evolution of Figure 1 was obtained using a numerical procedure, discussed later, in which the rupture history derived from a kinematic inversion of seismic recordings is used as a boundary condition on the assumed fault plane (*Bouchon*, 1997; *Ide and Takeo*, 1997; *Tinti et al.*, 2005b). We prefer to focus

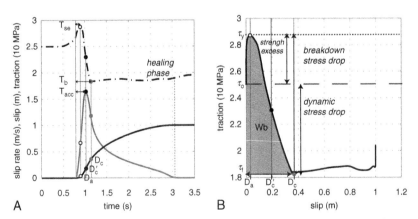

**FIGURE 1**   (a) Comparison of slip velocity, slip, and traction time histories at a target point on a fault plane using a smoothed Yoffe function as a source time function. Black solid circles indicate the time of peak slip velocity ($T_{\mathrm{acc}}$) and the gray solid circles indicate the end of weakening ($T_b$). (b) Corresponding traction versus slip behavior; the same circles of panel a are used to indicate the parameters $D_a$, $D_c'$, and $D_c$.

on dynamic traction evolution and dynamic rupture parameters that we derive from seismological observations, which allow us to constrain the rupture history including final slip, rupture time, slip duration, and details of the source time function.

Dynamic fault weakening is characterized by the stress degradation near the propagating rupture front. We discuss models in which shear stress drops from an upper yield value ($\tau_y$) to a residual level ($\tau_f$) in an extremely short time, called the breakdown time ($T_b$), and over a characteristic slip, called the slip weakening distance ($D_c$). The spatial extent of the breakdown zone ($X_b$), defined as the region of shear stress degradation near the tip of a propagating rupture, depends on the slip weakening distance. Slip velocity reaches its peak in a time $T_{acc}$ (see *Tinti et al.*, 2005b), corresponding to the duration of positive slip acceleration. Generally, $T_{acc}$ is shorter or equal to the breakdown time ($T_{acc} \leq T_b$), as shown in Figure 1. Constraining slip velocity and traction evolution via analysis of seismically radiated waveforms is an extremely important but difficult task for modeling high-frequency radiation of the earthquake source.

The seismic slip duration (i.e., the rise time) is another important parameter characterizing the rupture history, and it is physically associated with healing mechanisms. In the literature, numerous studies represent dynamic rupture propagation through either a cracklike rupture mode or a self-healing pulse propagation (e.g., *Cochard and Madariaga*, 1994, 1996; *Zheng and Rice*, 1998; *Cocco et al.*, 2004). The healing of slip, which may cause short slip durations or slip-velocity pulse-mode rupture propagation, has been attributed to heterogeneity of initial stress or strength on the fault plane (e.g., *Beroza and Mikumo*, 1996; *Bizzarri et al.*, 2001) or to properties of the constitutive law adopted to represent fault friction (e.g., *Perrin et al.*, 1995; *Beeler and Tullis*, 1996). Self-healing ruptures have been documented during rupture propagation between dissimilar materials and in other cases (*Weertman*, 1980; *Andrews and Ben-Zion*, 1997; *Cochard and Rice*, 2000). Seismological models often assume a source time function of finite duration (see, for instance, the slip velocity plotted in Figure 1a), and therefore they may be considered more consistent with self-healing slip rather than with cracklike models. Indeed, the traction evolution shown in Figure 1 exhibits restrengthening associated with healing of slip.

The purpose of this study is to elucidate the physical interpretation and seismic measurement of the characteristic slip weakening distance ($D_c$) with particular focus on the breakdown process and dynamic weakening. A key feature of our approach is the focus on the timescale of the breakdown process. We analyze time intervals shorter than the slip duration and ensure that periods of large slip velocity are well resolved (e.g., Figure 1).

The class of shear traction evolution models for dynamic fault weakening represented by Figure 1 are required to radiate seismic waves and to release the applied tectonic stress. Several stress parameters can be defined from the traction evolution shown in Figure 1: the strength excess ($\tau_y - \tau_o$), the dynamic stress drop ($\tau_o - \tau_f$), and the breakdown stress drop ($\tau_y - \tau_f$), where $\tau_o$ is the initial value of stress for a particular position on the fault plane. The area below the slip weakening curve and above the residual stress level ($\tau_f$) is traditionally identified with the fracture energy ($G$) (see *Palmer and Rice*, 1973; *Andrews*, 1976; *Rice et al.*, 2005), although several authors have proposed that a similar quantity called the breakdown work ($W_b$) is more appropriate (*Tinti et al.*, 2005a; *Cocco et al.*, 2006), at least for interpreting seismological observations.

The breakdown work is a more general definition of seismological fracture energy (*Cocco et al.*, 2006), and it is different from fracture energy as defined in classic fracture mechanics (see *Abercrombie and Rice*, 2005; *Cocco and Tinti*, 2008). The seismological breakdown work accounts for (1) the portion of the mechanical work dissipated within the fault zone, including surface area production, heat, and other factors; (2) traction evolution in the preyield stress region, which represents the energy lost during the initial slip-hardening phase (Figure 1); and (3) the effects of spatial and temporal variations in slip direction. Because it is a more realistic representation of the earthquake process, we use the breakdown work as a proxy for seismological fracture energy in this study. The use of breakdown work also provides a means of studying spatial variations in seismological fracture energy because it can be defined at each point on the fault plane. Breakdown work represents the only measurable portion, through seismological observations, of the energy absorbed on the rupture plane for fracture and frictional dissipation (see *Cocco et al.*, 2006). Therefore, measuring breakdown work is quite important for understanding the earthquake energy balance and for constraining the energy to be radiated as seismic waves.

In this study, we focus on the physical interpretation and measurement of the critical slip weakening distance $D_c$ via modeling of seismological data. Our approach requires knowledge of the rupture history in order to image slip or slip velocity evolution and to constrain dynamic traction evolution on the fault plane. The parameter $D_c$ is commonly measured from the same slip weakening curves (see Figure 1b) used to measure breakdown stress drop and breakdown work (or fracture energy). *Mikumo et al.* (2003) and *Fukuyama et al.* (2003) have proposed an alternative approach that allows estimation of the slip weakening distance directly from seismic observations. They proposed to measure the slip at the time of peak slip velocity, called $D_c{'}$, and to use this as a proxy for $D_c$. As shown in Figure 1a, $D_c{'}$ differs from $D_c$, and, as we will discuss later, their ratio depends on fault constitutive properties (*Tinti et al.*, 2004). However, $D_c{'}$ has been considered a reliable approximation of $D_c$ in some cases. *Fukuyama et al.* (2003) stated that this approximation works well for smoothly propagating ruptures. In this chapter, we further discuss the validity of using $D_c{'}$ as an alternative seismological estimate of $D_c$.

## 2. RUPTURE HISTORY FROM KINEMATIC SOURCE MODELS

Earthquake rupture history is often imaged through inverse approaches using a kinematic source parameterization. Geophysical data (seismograms and geodetic data) are inverted using nonlinear algorithms to infer the spatiotemporal distribution of fault slip, slip direction (rake angle), and slip duration (rise time). The range of numerical approaches in use assume either an analytical source time function (single-window approach) or represent the source time function as the superposition of several triangular functions (multiwindow approach) (see the detailed discussion in *Cohee and Beroza*, 1994). The latter method has the advantage of avoiding the selection of an analytical source time function, but limitations include poor resolution and a sparse sampling of the slip velocity time history. On the contrary, the single window approach has the limitation of the *a priori* choice of source time function but allows higher-resolution sampling of the source time function and thus can provide better constraint on the breakdown process and the rupture history on the fault plane.

Figure 2 displays several examples of slip velocity source time functions currently adopted in the literature. They have different parameterization, and for each model slip velocity reaches its peak in a different time interval. *Piatanesi et al.* (2004) discussed the effect of using different source time functions for imaging the distribution of dynamic parameters on the fault plane (dynamic and breakdown stress drop, strength excess, and critical slip weakening distance). They pointed out that the choice of the slip velocity function affects the inferred dynamic parameters; in particular, as we will discuss, different source time functions yield different values of critical slip weakening distance and a different scaling with final slip.

Figure 3 shows the final slip distribution for the 1979 Imperial Valley earthquake obtained by *Hartzell and Heaton* (1993) by inverting strong motion data. The authors used an asymmetric triangular function having a rise time of 0.7 s and the time to peak slip velocity $T_{acc}$ equal to 0.2 s. The five panels on the bottom display the slip and slip velocity time histories at selected positions on the fault plane indicated by letters in the upper panel. This kinematic model is an example in which the rise time is assumed

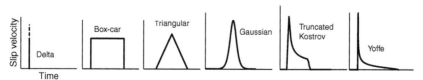

**FIGURE 2**   Several analytical source time functions proposed in the literature to model the slip velocity evolution on the fault plane: delta, box-car, triangular, Gaussian, Kostrov, and Yoffe functions (modified from *Tinti et al.*, 2005b).

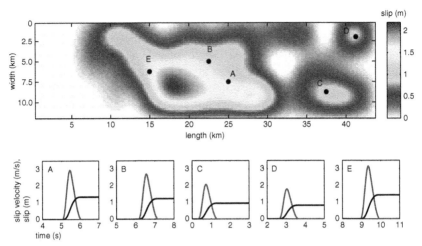

**FIGURE 3**　Upper panel: Slip distribution of kinematic model by *Hartzell and Heaton* (1983) for the 1979 Imperial Valley earthquake. Bottom panel: Slip velocity (gray lines) and slip time histories (black lines) for five subfaults as indicated by the capital letters above. (**See Color Plate 16.**)

constant on the fault (and therefore it is not inverted) and the slip velocity time histories have a constant shape in all subfaults and vary only in amplitude.

As noted by *Piatanesi et al.* (2004), the use in kinematic modeling of source time functions not compatible with the dynamic rupture propagation could bias the estimate of $D_c$ and hence the inferred ratio of $D_c$ to the total slip $D_{tot}$. Based on these results, *Tinti et al.* (2005b) proposed the use of a new source time function to infer kinematic source models consistent with earthquake dynamics, which they named the regularized Yoffe function (shown in Figure 1a, see *Tinti et al.*, 2005b, for the details of its analytical parameterization). Although other candidate source time functions are available in the literature (see, for instance, *Nakamura and Miyatake*, 2000; *Dreger et al.*, 2007), we emphasize that the regularized Yoffe function is consistent with dynamic solutions of the elastodynamic equation (*Nielsen and Madariaga*, 2003) and allows a flexible parameterization for our purposes.

To image the rupture history on the fault plane, robust kinematic inversions have been proposed to improve resolution. Various smoothing constraints have been adopted to ensure stable solutions of the inverse problem (see *Hartzell et al.*, 2007, and references therein). It is generally accepted to use positivity and smoothing constraints (*Hartzell and Heaton*, 1983; *Henry et al.*, 2000, among many others) for reducing the instability and the complexity of the inverted models to levels consistent with resolution of the filtered data. However, depending on the choice of assumed spatial or temporal constraints, the results of the inversions (in terms of kinematic parameters) may change dramatically (see *Beresnev*, 2003). Moreover, other factors can also strongly affect results such as signal preprocessing (*Boore*, 2005; *Boore and Bommer*, 2005), model parameterization (*Piatanesi et al.*, 2007), and inversion schemes (see *Hartzell et al.*, 2007, and references therein). Because of the difficulties in computing accurate Green's functions at high frequencies ($f > 2\text{Hz}$), approaches based on waveform inversion model seismograms in a limited frequency bandwidth. Applying filters to recorded seismograms helps in imaging the slip distribution, but reduces the available resolution of the source process at small wavelengths. *Spudich and Guatteri* (2004) highlighted the effects of the limited frequency bandwidth of modeled data on the inferred dynamic and frictional parameters. Despite these limitations, rupture history can be retrieved only through kinematic source models, and therefore they represent a unique resource of information to better explain earthquake dynamics.

## 3. INFERRING TRACTION EVOLUTION

*Tinti et al.* (2005a) have implemented a 3D finite difference code based on the *Andrews* (1999) approach to calculate the stress time histories on the earthquake fault plane from kinematic rupture models. The fault is represented by a surface containing double nodes, and the stress is computed through the fundamental elastodynamic equation (*Ide and Takeo*, 1997; *Day et al.*, 1998). Each node belonging to the fault plane is forced to move with a prescribed slip velocity time series, which corresponds to imposing the slip velocity as a boundary condition on the fault and determining the stress-change time series everywhere on the fault. This numerical approach does not require specification of any constitutive law relating total dynamic traction to friction. The dynamic traction evolution is a result of the calculations. The numerical model is consistent with the analytical model proposed by *Fukuyama and Madariaga* (1998), where stress change [$\sigma(\boldsymbol{x}, t)$] at a position $\boldsymbol{x}$ of the fault plane is related to slip velocity time history [$v(\boldsymbol{x}, t)$] at time $t$ by means of the following relation:

$$\sigma(\boldsymbol{x},t) = -\frac{\mu}{2\beta}v(\boldsymbol{x},t) + \int_{\Sigma}\int_0^t K(\boldsymbol{x}-\xi;t-t')v(\xi,t)dt'dS \qquad (1)$$

where $\beta$ is the shear wave velocity, $\mu$ the shear rigidity, and $K$ the dynamic load associated to those points that are already slipping (that is, those within the cone of causality around the rupture front). *Piatanesi et al.* (2004) used the same approach to infer dynamic parameters from kinematic models.

The slip velocity time histories at each point on the fault plane are obtained from the kinematic rupture models inferred by inverting geophysical data. To convert the slip model to a continuously differentiable slip-rate function, the original kinematic models have to be interpolated and smoothed both in time and space (see *Day et al.*, 1998; *Tinti et al.*, 2005a, for details). The free surface is included in these computations, and the Earth models are simplified assuming homogeneous half-spaces. As discussed earlier, the inadequate resolution and the limited frequency bandwidth, which characterize inverted kinematic models, reduce the ability to infer the real dynamic traction evolution everywhere on the fault plane. Many recent papers have investigated the limitations of using poorly resolved kinematic source models (*Guatteri and Spudich*, 2000; *Piatanesi et al.*, 2004; *Spudich and Guatteri*, 2004). We will discuss these issues in further detail.

Despite the limitations noted earlier, the methods proposed by *Ide and Takeo* (1997), *Day et al.* (1998), and *Tinti et al.* (2005a) provide the dynamic shear stress time histories on the fault plane. This is an important step in using seismological observations to understand the breakdown process during earthquake ruptures. Moreover, source time functions compatible with earthquake dynamics and suitable for waveform inversions are becoming commonly available (see *Piatanesi et al.*, 2004; *Tinti et al.*, 2005b; *Cirella et al.*, 2006; *Dreger et al.*, 2007). Finally, *Tinti et al.* (2005a, 2008a) have discussed in detail the fidelity of calculations of breakdown work and $D_c$ and concluded that, in agreement with *Guatteri and Spudich* (2000), the estimates of $W_b$ are quite stable despite the limited available resolution in kinematic source models, whereas the $D_c$ parameter is more difficult to constrain.

For these reasons, we use the approach of *Tinti et al.* (2005a) in the present study. Figure 4 shows the traction evolution as a function of slip for the 1979 Imperial Valley earthquake. Numbers along the axes represent the relative position along strike and dip on the fault plane of each target point. The intervals shown for the lower left panel indicate traction (in MPa) and slip (in m) and apply to each panel of Figure 4. From the inferred slip weakening curves, we can measure the $D_c$ breakdown stress drop, and the breakdown work at each grid point on the fault plane. Figure 4 clearly shows high values of stress drop in correspondence to the large slip patch. The slip weakening curves also

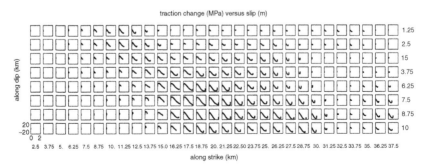

**FIGURE 4** Traction versus slip curves of the 1979 Imperial Valley earthquake for several subfaults. The relative position on the fault plane (dip and strike) is indicated for each subfault. The intervals shown for the lower left panel ([0 2], [−20 20]) indicate traction (in MPa) and slip (in m) and are the scales for each panel.

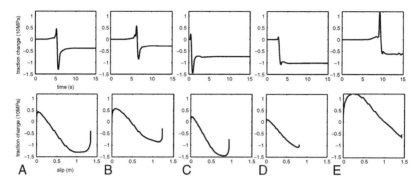

**FIGURE 5**   Traction change time histories (upper panels) and traction change versus slip curves (bottom panels) for the same five subfaults of Figure 3.

exhibit a slow weakening rate due to the smoothed source time function of the kinematic model. Figure 5 displays shear traction time histories (top panels) and the slip weakening curves (bottom panels) for the five points noted in Figure 3. As expected, the dynamic parameters of strength excess and break-down stress drop vary over the fault plane (Figure 5). Moreover, $D_c$ differs for different positions on the fault, varying from $\sim$0.7 to 1.5 m for the selected positions. We also emphasize that the duration of the breakdown phase and the subsequent restrengthening phase vary over the fault plane, even though the rise time (0.7 s) and $T_{acc}$ (0.2 s) are assumed to be constant.

The dynamic traction evolution curves inferred from kinematic rupture histories display an initial increase before reaching the upper yield stress value $\tau_y$, as clearly evident in the examples shown in Figures 1 and 5. This initial slip hardening phase precedes the dynamic weakening phase, and it is associated with relatively small slip amplitudes. We define $D_a$ as the slip at the upper yield stress: Figure 1a shows that, at least in this case, it is much smaller than both $D_c'$ and $D_c$. In the framework of classic dynamic models, this initial hardening phase is associated with the strength excess, and it is an important feature for modeling rupture propagation through spontaneous dynamic models. However, it is important to note that classic slip weakening models implicitly assume $D_a = 0$. This implies that the duration of initial slip hardening ($T_{se}$) is negligible and much shorter than the dynamic weakening phase ($T_{se} << T_{acc} \leq T_b$). It is important to note that the initial slip hardening phase has been observed in laboratory experiments (*Ohnaka and Yamashita,* 1989; *Ohnaka,* 2003) and modeled using rate and state friction (*Bizzarri et al.,* 2001; *Marone et al.,* 2009). We emphasize that the breakdown work estimates done by *Tinti et al.* (2005a) and *Cocco et al.* (2006) include both the initial slip hardening and the subsequent slip weakening phases, and consequently their $D_c$ estimates include a contribution from $D_a$.

## 4. MEASURING $D_c'$ FROM PEAK SLIP VELOCITY

Following the approach proposed by *Mikumo et al.* (2003), we have measured $D_c'$ from the kinematic source model of the 1979 Imperial Valley earthquake, whose slip distribution is shown in Figure 3. Figure 6 displays the distribution of $D_c'$ inferred from the slip history imaged by *Hartzell and Heaton* (1983) by inverting strong motion accelerograms. $D_c'$ ranges between 0.5 and 1 m in the central, high slip patch; these values can be considered as a lower bound for $D_c$. As expected, the spatial distribution of $D_c'$ is strongly correlated with the slip distribution. This is even more evident in Figure 7a, which shows the perfect linear scaling of $D_c'$ with final slip $D_{\text{tot}}$. This scaling is simply the result of the initial hypothesis adopted in kinematic inversion of a fixed source time function (an asymmetric triangular function) with a constant rise time and $T_{\text{acc}}$ (0.7 and 0.2 s, respectively). Therefore, in this kinematic model, the heterogeneity of inferred slip completely dictates the heterogeneity of peak slip velocity and hence $D_c'$.

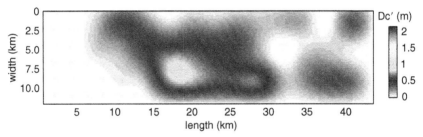

**FIGURE 6** $D_c'$ distribution for the 1979 Imperial Valley earthquake inferred from the kinematic model of *Hartzell and Heaton* (1983). (**See Color Plate 17.**)

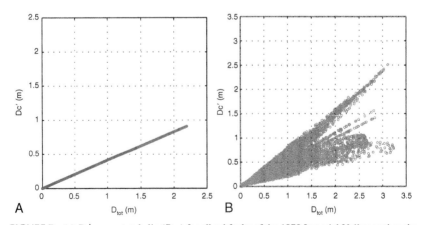

**FIGURE 7** (a) $D_c'$ versus total slip ($D_{\text{tot}}$) for all subfaults of the 1979 Imperial Valley earthquake using the *Hartzell and Heaton* (1983) kinematic model. (b) $D_c'$ versus total slip ($D_{\text{tot}}$) for all subfaults of 1994 Northridge earthquake using the *Wald et al.* (1996) kinematic model.

Nowadays, thanks to computational tools, finite-fault inversions are commonly performed using nonlinear formulations that allow all the kinematic parameters to be inverted (e.g., *Delouis et al.*, 2002; *Ji et al.*, 2002; *Piatanesi et al.*, 2007). To account for the actual rupture complexity, slip or peak slip velocity, rupture time, rise time, and slip direction are simultaneously inverted. However, the parameter $T_{\text{acc}}$ is usually fixed *a priori* because of data frequency bandwidth limitations and, as a consequence, strongly affects estimates of $D_c{}'$. Independent inversion for all of the kinematic parameters should produce a natural scatter of the correlation between $D_c{}'$ and $D_{\text{tot}}$. However, there are still other problems and limitations in measuring $D_c{}'$ from seismological observations. *Spudich and Guatteri* (2004) pointed out that low-pass filtering of strong motion seismograms can affect the estimates of $D_c{}'$ because it biases the inverted rupture models causing an artificial correlation between $D_c{}'$ and $D_{\text{tot}}$. These authors claimed that slip models derived from band-limited ground motion data might not resolve periods shorter than the breakdown time, and therefore the models do not contain periods shorter than $T_{\text{acc}}$ ($\leq T_b$). They defined these inverted models as "temporally unresolved." This means that the process of low-pass filtering ground motion data can remove information about $D_c$ and $D_c{}'$. *Spudich and Guatteri* (2004) concluded that filtering ground motion data or the inferred slip models tends to bias upward the $D_c$ values inferred from the slip weakening curves and to generate artificial correlations with final slip. The effect of filtering is to shift the peak slip velocity later in time, which means that $T_{\text{acc}}$ is overestimated or that postpeak energy is effectively repositioned to a time before the peak slip velocity (*Spudich and Guatteri*, 2004). These issues raise the question of whether estimates of $D_c{}'$ from modeling ground motion waveforms are tenable and corroborated by data.

*Yasuda et al.* (2005) performed a particularly interesting test for the subject discussed here. They simulated a spontaneous dynamic earthquake rupture, by assuming constant $D_c$, and computed the synthetic waveforms that would be observed at actual receivers. Therefore, they inverted these synthetic seismograms to image the kinematic rupture history and to constrain the slip rate function on each subfault. As expected, the spontaneous forward dynamic model has a spatial and temporal resolution much higher than the inverted model (nearly 10 times larger). This numerical test allows the comparison of inferred values of both $D_c$ and $D_c{}'$ with those of the target model. The results of *Yasuda et al.* (2005) clearly show that the $D_c{}'$ values measured from the dynamically generated slip rate functions range between $0.25 \cdot D_c{}'$ and $D_c$ (as expected because $T_{\text{acc}} \leq T_b$). Also, for a constant $D_c$ model (see Figure 3 in *Yasuda et al.*, 2005), the inferred values of $D_c{}'$ scale with final slip. These results are consistent with the findings of *Tinti et al.* (2004), who also demonstrated that the $D_c{}'$ values measured from the inverted slip rate functions scale with final slip. The values of $D_c{}'$ and $D_{\text{tot}}$ so derived define a roughly linear scaling with slope equal to 1/2, although they exhibit larger scatter than values measured from the dynamically generated slip rate functions. *Spudich and Guatteri* (2004) explained the scaling $D_c' = 0.5 \cdot D_{tot}$ as a consequence of the central limit theorem: after the filtering operation, a source time function tends to a Gaussian in which half of the slip occurs before peak slip velocity. *Yasuda et al.* (2005) also concluded that both of their estimates of $D_c'$ exhibit an apparent correlation with final slip.

The correlation between $D_c'$ and final slip $D_{tot}$ has also been obtained in other inversions of kinematic models. Figure 7b shows the scaling of $D_c'$ with $D_{tot}$ for the 1994 Northridge earthquake. These values derive from the calculations performed by *Tinti et al.* (2005a) using the kinematic model results of *Wald et al.* (1996). The latter authors used a multiwindows approach, and the source time function consists of three overlapping isosceles triangles each with duration of 0.6 s and initiations separated by 0.4 s. This allows a rise time lasting up to 1.4 s. We show here values estimated for the dip-slip component of the slip vector (Figure 7b). $D_c'$ exhibits a larger scatter around a linear scaling, which arises in part because in this case the slip rate function can contain multiple peaks at different times (see Figure 2 in *Tinti et al.*, 2005b). Nevertheless, the correlation of $D_c'$ with final slip is still evident (Figure 7b).

A conclusion from all of these models is that $D_c'$ is an accurate estimate of $D_c$ only if the inversion method retains information on the breakdown process. Unfortunately, this condition is not met, due to inherent limitations on spatial and temporal resolution of kinematic source models. Moreover, although one might expect to be able to use $D_c'$ as a proxy for $D_c$ in numerical models of dynamic rupture, it is quite difficult to constrain $D_c'$ from the rupture history imaged from kinematic inversions because in these approaches, the slip velocity function is chosen *a priori* (single window) or it is imaged with a poor resolution insufficient to resolve $T_{acc}$. *Fukuyama et al.* (2003) pointed out that also for a smoothly propagating rupture the standard deviation of the measured $D_c'$ values can be larger than 30%. We therefore conclude that the correlation of $D_c'$ with final slip is often an unavoidable consequence of *a priori* assumptions and limited resolution. In the next sections, we discuss (1) a physical explanation of the deviation of $D_c'$ from $D_c$ and (2) the problem of poor resolution of the breakdown process.

## 5. MEASURING $D_C$ FROM INFERRED TRACTION EVOLUTION CURVES

We have discussed approximations and limitations associated with inferring the critical slip weakening distance from the slip at peak slip rate. Here, we discuss the estimation of $D_c$ directly from the slip weakening curves obtained from the inverted kinematic rupture histories. As noted previously, we follow the approach proposed by *Ide and Takeo* (1997), *Day et al.* (1998), and *Tinti et al.* (2005a). Figure 8 shows the $D_c$ values inferred from the traction evolution curves displayed in Figure 4 and obtained from the slip history imaged by

*Hartzell and Heaton* (1983) for the 1979 Imperial Valley earthquake. A visual comparison between Figures 3 and 8 suggests that $D_c$ is correlated with final slip. This correlation is illustrated in Figure 9, which shows the scaling of $D_c$ with final slip for all the subfaults of this model (see *Tinti et al.*, 2005a, for details). The proportionality between $D_c$ and final slip has been obtained by several other authors (*Dalguer et al.*, 2003; *Pulido and Kubo*, 2004; *Burjanek*

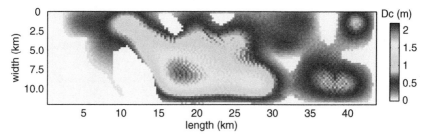

**FIGURE 8**   $D_c$ distribution for the 1979 Imperial Valley earthquake using the *Hartzell and Heaton* (1983) kinematic model as a boundary condition to compute traction history. (**See Color Plate 18.**)

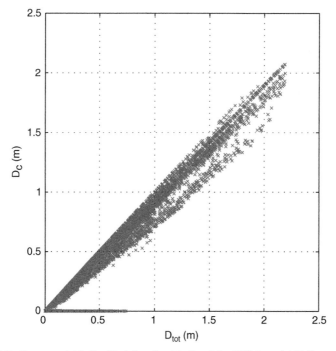

**FIGURE 9**   $D_c$ versus total slip ($D_{tot}$) for all subfaults of the 1979 Imperial Valley earthquake using the *Hartzell and Heaton* (1983) kinematic model as a boundary condition on the fault plane to compute traction change.

*and Zaharadnik*, 2007) and imaged for other earthquakes. Figure 11 of *Tinti et al.* (2005a) displays a similar scaling also for the 1995 Kobe and the 1997 Colfiorito earthquakes. Therefore, we conclude that, similarly to $D_c'$, the estimates of $D_c$ scale with final slip according to a nearly linear relationship.

Another interesting feature emerges from Figure 9. It shows that most of the subfaults have a $D_c$ value quite close to the final slip ($D_c \approx D_{tot}$). This is even clearer in Figure 10, which displays the spatial distribution of the ratio

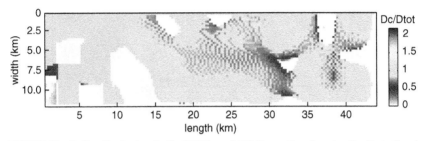

**FIGURE 10** $D_c/D_{tot}$ distribution for the 1979 Imperial Valley earthquake using the *Hartzell and Heaton* (1983) kinematic model. **(See Color Plate 19.)**

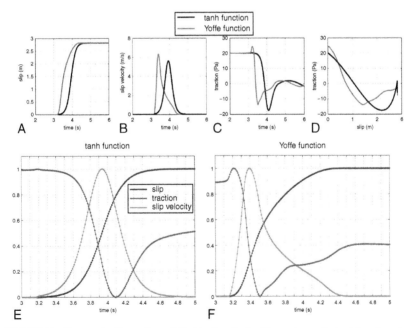

**FIGURE 11** Top panels: Comparison of slip (a), slip velocity (b), traction evolution (c), and traction versus slip curves (d) at a target point for the two source time functions $f_1$ (in black) and $f_2$ (in gray). Bottom panels: Normalized time histories of slip, slip velocity, and dynamic traction calculated with $f_1$ (tanh function) (e) and $f_2$ (Yoffe function) (f) for the same target point. **(See Color Plate 20.)**

$D_c/D_{tot}$ on the fault plane. The ratio is close to the unity over much of the fault plane (Figure 10). This is due to the source time function selected for modeling observed ground motions (*Piatanesi et al.*, 2004) and to the limited spatial and temporal resolution employed in numerical calculations (*Spudich and Guatteri*, 2004; *Yasuda et al.*, 2005).

To better understand the effects of adopting different source time functions in retrieving dynamic traction evolution, we show in Figure 11 the dynamic traction evolution inferred for two slip rate functions:

$$f_1 = f^{\dot{Y}}(t) = H(t)\frac{2}{T_R}\left[1 - \tanh^2\left(\frac{2}{T_R}(2t - \frac{3}{2}T_R)\right)\right]$$

$$f_2 = f^{\dot{Y}}(t) = H(T_R - t)\frac{2}{\pi \cdot T_R}\left(\frac{-t + T_R}{t}\right)^{1/2} \tag{2}$$

where $H(t)$ is the Heaviside function and $T_R$ is the rise time. The first function $f_1$ is a smoothed ramp function in slip resulting in a Gaussian slip rate function; the second function $f_2$ is the Yoffe function, which is singular at the rupture time (see *Piatanesi et al.*, 2004). The numerical representation of the latter is obtained by smoothing the function in time with a moving triangular window (0.37 s) (see *Tinti et al.*, 2005b, for details of the analytical regularization of the Yoffe function). These slip and slip rate functions are shown in panels a and b of Figure 11. Traction changes of panels c and d of Figure 11 are computed for the heterogeneous slip model shown in Figure 12a in a point with 2.7 m of final slip. This is the rupture history imaged by *Iwata and Sekiguchi* (2002) for the 2000 western Tottori, Japan, earthquake, and it is characterized by a nonuniform slip distribution and an extremely variable rupture velocity. Figure 11 allows us to point out the main differences of slip rates and inferred traction evolution curves retrieved by using these two source time functions. To this goal, panels e and f of this figure display the slip, slip rate, and shear traction time histories for the selected target point. It is evident that although $T_{acc}$ for the Yoffe function is much shorter than the rise time, the same parameter is half of the rise time for the smoothed ramp. Moreover, function $f_1$ is characterized by a smooth onset of slip velocity, whereas in function $f_2$ it is quite sharp. All these features, which reflect peculiarities of the dynamics of earthquake rupture, yield quite different evolution curves. In particular, we emphasize that the Yoffe function yields (1) a much shorter duration of breakdown process than the smoothed ramp, (2) smaller values of $D_c$, and $D_c/D_{tot}$ than $f_1$, (3) smaller values of $D_c'$ than those of function $f_1$ (nearly half of the latter), and (4) higher peak slip velocity values. We note that the two rupture histories used to compute the traction evolution curves illustrated in Figure 11 only differ for the selected slip rate function, whereas all the other parameters are the same (the rupture model is that shown in Figure 12a).

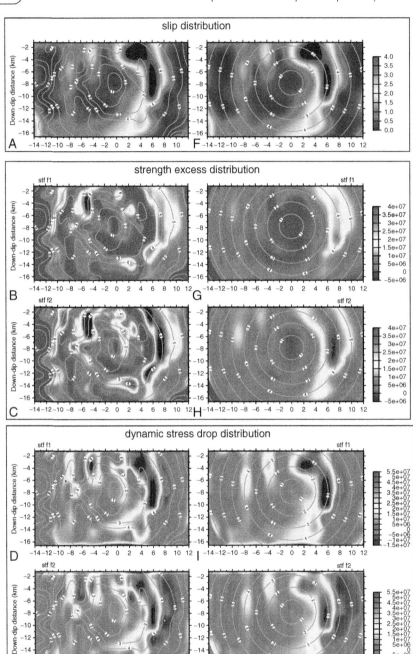

**FIGURE 12**    Distribution of slip (a, f), strength excess (b, c, g, h), and dynamic stress drop (d, e, i, j) on the fault plane retrieved for the two source time function $f_1$ and $f_2$ and for heterogeneous (left panels) and constant (right panels) rupture velocity models. (panels a-e are from *Piatanesi et al.*, 2004) **(See Color Plate 21.)**

Figure 12 shows the strength excess and the dynamic stress drop distributions computed from the slip weakening curves inferred using the two source time functions defined earlier ($f_1$ and $f_2$) and two slip models differing only for the rupture velocity. Left panels depict the slip, strength excess, and dynamic stress drop distributions for the variable rupture velocity model of *Iwata and Sekiguchi* (2002), whereas right panels display the same parameters for a uniform rupture velocity model (equal to 2.1 km/s). This figure clearly shows that, as expected, the variability of rupture velocity controls the heterogeneity of strength excess and dynamic stress drop on the fault plane, but it also shows that the adopted source time function affects the values of these dynamic parameters. This is in agreement with *Guatteri and Spudich* (2000), who concluded that there is a trade-off between strength excess and $D_c$ in controlling the rupture velocity.

## 6. SCALING BETWEEN $D_C$ AND FINAL SLIP

In previous sections, we discussed estimation methods of both $D_c{}'$ and $D_c$ as well as the reliability of their scaling with final slip. The main limitations arise from (1) limitations in our ability to model seismic waveforms given the narrow frequency bandwidths available (i.e., lack of high-frequency components) (see *Spudich and Guatteri*, 2004, and references therein), (2) adoption of source time functions that are not compatible with dynamic rupture propagation (*Piatanesi et al.*, 2004), and (3) the lack of causality constraints on spatial and temporal evolution of slip velocity in seismic inversions (i.e., poor constraints on spatial gradient of slip velocity) (see *Tinti et al.*, 2008a, for details). The duration of the breakdown process and the peak slip velocity depend on the frictional and constitutive properties of the fault. Therefore, we expect that differences between $T_{acc}$ (timing of peak slip velocity) and $T_b$ (breakdown duration)—and consequently between $D_c{}'$ and $D_c$—are controlled by the constitutive properties of the source. *Tinti et al.* (2004) have discussed the difference between $D_c{}'$ and $D_c{}'$ for a 2D rupture governed by either a slip weakening or a rate- and state-dependent friction law. They demonstrated that such a difference is controlled by the strength parameter $S$ (i.e., the ratio between strength excess and dynamic stress drop). These authors have also shown that the rate dependence of the friction law affects slip acceleration and the slip weakening parameter.

We have discussed several interpretations of the biases affecting estimates of the slip weakening distance $D_c$. However, the most important issue concerns the scaling of this parameter with the final slip. Although we cannot exclude the possibility that such a scaling might be physically tenable for real earthquakes, we present evidence showing that the inferred proportionality between $D_c$ and $D_{tot}$ is artificial. The same is true for the parameter $D_c{}'$, whose estimate from kinematic slip models is controlled by the adopted source time function and by other *a priori* assumptions when inverting ground motion data. *Tinti et al.* (2005b) performed several simulations using a Yoffe function to represent slip velocity time history and the traction at split nodes approach to retrieve dynamic shear traction changes caused by coseismic slip. These authors propose that the following relation holds:

$$D_c \propto \sqrt{\frac{T_{acc}}{T_R}} D_{tot} \qquad (3)$$

suggesting that, when the duration of positive slip acceleration $T_{acc}$ (i.e., time of peak slip velocity) and the rise time $T_R$ are both constant or their ratio is constant, there is a direct proportionality between slip weakening distance and final slip. The values of $T_{acc}$ and $T_R$ control peak slip velocity, and therefore the same factors, which bias the estimate of $D_c$, explain the difficulties in constraining both final slip and slip velocity.

These findings are summarized in Figure 13, which shows the dynamic traction changes (panels c, d, e, and f) computed for source models having a uniform slip of 1 m with a constant rupture velocity of 2.0 km/s and a slip velocity time history represented by a Yoffe function (panels a and b). Left panels display the results of calculations obtained for different $T_{acc}$ values and a constant rise time (1.0 s), whereas right panels show those computed for different rise times and a constant $T_{acc}$ equal to 0.225 s (see *Tinti et al.*, 2005b, for further details about these calculations). Panels g and h show the scaling of peak slip velocity with $D_c$ for the two test cases. The dynamic traction histories and the slip weakening curves displayed in Figure 13 clearly illustrate that $D_c$ and the weakening rate vary with the parameter $T_{acc}$. Inverting ground motion data with a limited temporal resolution overestimates the real $T_{acc}$ and consequently produces an overestimate of $D_c$. The weakening rate varies to maintain the same value of the final slip. As a consequence, the peak slip velocity decreases for increasing $T_{acc}$ and $D_c$, as clearly shown in panel g of Figure 13. The effect of using different rise times for the same $T_{acc}$ is to affect the $D_c$ value, but very smoothly affecting the weakening rate. Figure 13 points out that increasing the rise time causes both $D_c$ and peak slip velocity to decrease. Figure 13 summarizes in a simple way all the difficulties in imaging dynamic traction evolution and measuring the slip weakening parameters. It is also useful to illustrate in a schematic way all the limitations in measuring the slip weakening parameters by modeling observed ground motion data.

## 7. DISCUSSION AND CONCLUDING REMARKS

We discuss physical models for the characteristic slip weakening distance $D_c$ and the scaling between $D_c$ and total slip. We show that current methods and available seismic data introduce potential biases in estimates of $D_c$ and its scaling with seismic slip due to the limited frequency bandwidth considered during typical kinematic inversions. For studies of dynamic slip weakening, it is important to analyze time intervals shorter than $T_{acc}$ and $T_b$ to obtain

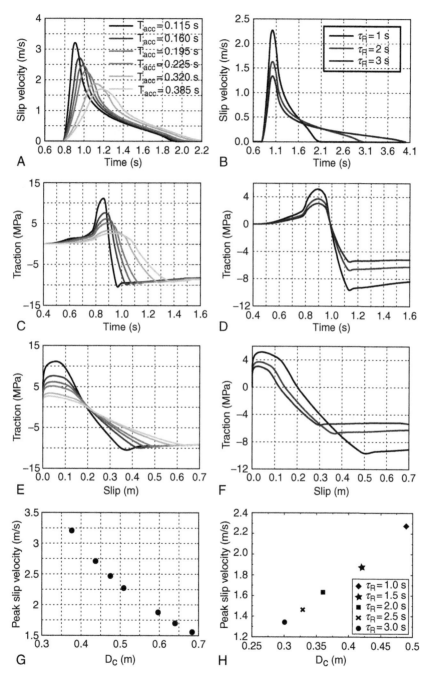

**FIGURE 13**   The dynamic traction changes (panels c, d, e, and f) for source models having uniform slip of 1 m, constant rupture velocity (2.0 km/s), and slip velocity time histories represented by the Yoffe function (panels a and b). Left panels: Calculations obtained for different $T_{acc}$ values and constant rise time (1.0 s). Right panels: Calculations for different rise times and a constant $T_{acc}$ (0.225 s). (modified from *Tinti et al.*, 2005b) **(See Color Plate 22.)**

accurate estimates of $D_c$ and its scaling with total slip. Unfortunately, $T_{acc}$ is usually fixed *a priori* in kinematic inversions due to poor data resolution. The same is true for the rise time, which is poorly constrained in kinematic inversions and often assumed to be spatially uniform on the fault plane. Therefore, in such cases, $T_{acc}$ (and $T_{acc}/T_R$) strongly affect estimates of $D_c{}'$ as well as $D_c$, and, as shown in Equation 3, this generates artificial correlation of $D_c$ with final slip. In this study, we extend previous works (*Guatteri and Spudich*, 2000; *Tinti et al.*, 2008a) and show that $D_c$ inferred through seismological data can be biased unless a proper modeling of high-frequency waves and source parameterization is adopted. *Fukuyama and Mikumo* (2007) estimated the slip weakening distance from seismograms recorded at near field stations. These authors claimed that the proposed approach is not significantly affected by spatiotemporal smoothing and resolution limitations. However, this approach also depends on several *a priori* assumptions concerning the rupture history (rupture velocity, rise time, etc.) and source time function.

We cannot exclude the existence of physical mechanisms controlling the scaling of $D_c$ with final slip $D_{tot}$. *Ohnaka and Yamashita* (1999) and *Ohnaka* (2003) suggested that $D_c$ scales with the roughness of the fault plane in the direction of slip and that $D_c$ has a fractal distribution on the fault plane. The idea that $D_c$ scales with fault maturity was also proposed by *Marone and Kilgore* (1993) and *Abercrombie and Rice* (2005). On the other hand, *Scholz* (1988) argued that rough fault surfaces under lithostatic load will develop a characteristic contact dimension and hence exhibit a constant value of $D_c$ rather than one that scales with roughness (see also *Brown and Scholz*, 1985; *Aviles et al.*, 1987; *Power et al.*, 1987). Nevertheless, the assumption of a fractal distribution of $D_c$ and the scaling with fault roughness might imply a scaling with final slip. Moreover, because slip is heterogeneously distributed on the fault plane, constant $D_c$ models are not physically consistent (that is, they could predict $D_c$ values larger than final slip in locked patches). A full discussion of the physical mechanisms that could yield causal scaling between $D_c$ and final slip is beyond the scope of the present study. We emphasize that the linear scaling between $D_c$ and $D_{tot}$ inferred from kinematic source models is caused by the poor resolution of the breakdown process.

In this study, we have shown that the estimates of breakdown work from kinematic source models are more reliable. The breakdown work is a more appropriate quantity for assessing the earthquake energy budget than the fracture energy as defined in classic fracture mechanics (*Cocco et al.*, 2006; *Cocco and Tinti*, 2008). Figure 14 shows the distribution of $W_b$ (J/m$^2$) obtained by *Tinti et al.* (2005a, 2008b) for a kinematic model of the 1979 Imperial Valley earthquake (*Hartzell and Heaton*, 1983). The spatial distribution of breakdown work is strongly correlated with the slip distribution (see Figure 3). The correlation between the distributions of $W_b$ and slip is due primarily to the correlation of $D_c$ with slip but also secondarily to the correlation of stress drop with total slip.

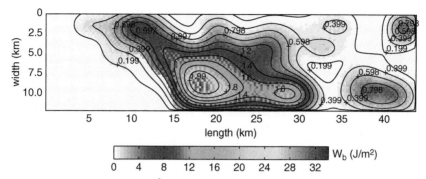

**FIGURE 14**   $W_b$ distribution (J/m$^2$) of the 1979 Imperial Valley earthquake using the *Hartzell and Heaton* (1983) kinematic model. Black lines and numbers represent the contour of the slip distribution shown in Figure 3. (**See Color Plate 23.**)

The two fundamental parameters characterizing dynamic fault weakening, breakdown work $W_b$ and slip weakening distance $D_c$, are intrinsically scale dependent (*Ohnaka and Yamashita*, 1989; *Ionescu and Campillo*, 1999; *Cocco and Tinti*, 2008). This means that they cannot be associated with any other physical process controlling dynamic fault weakening at time and length scales smaller than that selected for the macroscopic representation implicit in seismological observations (see *Cocco et al.*, 2006; *Cocco and Tinti*, 2008). Thus, the interpretation of the $D_c$ estimates inferred from seismological data is representative of the macroscopic scale in which the fault zone is shrunk to a virtual mathematical plane of zero thickness. Therefore, it cannot be easily compared with estimates retrieved from laboratory experiments on rock friction and fracture or from those associated with weakening mechanisms occurring at time and length scales smaller than the fault zone thickness.

The results of this study point out that the parameter $T_{acc}$ and peak slip velocity contain the same dynamic information as breakdown stress drop and $D_c$. The inadequate resolution and the limited frequency bandwidth, which characterize inverted kinematic models, reduce the ability to infer the real dynamic traction evolution everywhere on the fault plane. Future attempts to model high-frequency seismic waves are important and will be aided by high-performance computing facilities and high-quality seismic waveforms from borehole seismometers.

## ACKNOWLEDGMENTS

We thank two anonymous reviewers and E. Fukuyama for comments and suggestions that helped us to improve the manuscript.

## REFERENCES

Abercrombie, R. E. and J. R. Rice, (2005), Can observations of earthquake scaling constrain slip weakening?, *Geophys. J. Int.*, **162**, 406-424.

Andrews, D. J., (1976), Rupture propagation with finite stress in antiplane strain, *J. Geophys. Res.*, **81**(20), 3575-3582.

Andrews, D. J., (1999), Test of two methods for faulting in finite-difference calculations, *Bull. Seismol. Soc. Am.*, **89**(4), 931-937.

Andrews, D. J. and Y. Ben-Zion, (1997), Wrinkle-like slip pulse on a fault between different materials, *J. Geophys. Res.*, **102**(B1), 553-571.

Aviles, C. A. and C. H. Scholz, (1987), Fractal analysis applied to characteristic segments of the San Andreas fault, *J. Geophys. Res.*, **92**(B1), 331-344.

Beroza, G. C. and T. Mikumo, (1996), Short slip duration in dynamic rupture in the presence of heterogeneous fault properties, *J. Geophys. Res.*, **101**(B10), 22449-22460.

Beeler, N. M. and T. E. Tullis, (1996), Self-healing slip pulses in dynamic rupture models due to velocity-dependent strength, *Bull. Seismol. Soc. Am.*, **86**(4), 1130-1148.

Beresnev, I. A., (2003), Uncertainties in finite-fault slip inversions: To what extent to believe? (A critical review), *Bull. Seismol. Soc. Am.*, **93**, 2445-2458.

Bizzarri, A., M. Cocco, D. J. Andrews, and E. Boschi, (2001), Solving the dynamic rupture problem with different numerical approaches and constitutive laws, *Geophys. J. Int.*, **144**(3), 656-678.

Boore, D. M., (2005), On pads and filters: Processing strong-motion data, *Bull. Seismol. Soc. Am.*, **76**, 368-369.

Boore, D. M. and J. J. Bommer, (2005), Processing of strong-motion accelerograms: Needs, options and consequences, *Soil Dynam. Earthq. Eng.*, **25**, 93-115.

Bouchon, M., (1997), The state of stress on some faults of the San Andreas system as inferred from near-field strong motion data, *J. Geophys. Res.*, **102**(B6), 11731-11744.

Brown, S. R. and C. H. Scholz, (1985), Broad bandwidth study of the topography of natural rock surfaces, *J. Geophys. Res.*, **90**, 12575-12582.

Burjanek, J., and J. Zaharadnik, (2007), Dynamic stress field of a kinematic earthquake source model with k-squared slip distribution, *Geophys. J. Int.*, **171**(3), 1082-1097.

Cirella, A., A. Piatanesi, E. Tinti, and M. Cocco, (2006), Dynamically consistent source time functions to invert kinematic rupture histories, *Eos, Trans.*, AGU, **87**(52), Fall Meet. Suppl., Abstract, S41B-1323.

Cocco, M., A. Bizzarri, and E. Tinti, (2004), Physical interpretation of the breakdown process using a rate- and state-dependent friction law, *Tectonophys.*, **378**, 241-262.

Cocco, M., P. Spudich, and E. Tinti, (2006), On the mechanical work absorbed on faults during earthquake ruptures, In: *Earthquakes: Radiated Energy and the Physics of Faulting*, edited by Abercrombie, R., A. McGarr., G. Di Toro, and H. Kanamori, *Geophysical Monograph Series*, **170**, 237-254, American Geophysical Union, Washington, D.C., doi:10.1029/170GM24.

Cocco, M. and E. Tinti, (2008), Scale dependence in the dynamics of earthquake propagation: Evidence from seismological and geological observations, *Earth Planet. Sci. Lett.*, **273**(1-2), 123-131.

Cochard, A. and R. Madariaga, (1994), Dynamic faulting under rate-dependent friction, *Pure Appl. Geophys.*, **142**(3-4), 419-445.

Cochard, A. and R. Madariaga, (1996), Complexity of seismicity due to highly rate-dependent friction, *J. Geophys. Res.*, **101**(B11), 25321-25336.

Cochard, A. and J. R. Rice, (2000), Fault rupture between dissimilar materials: Ill-posedness, regularization, and slip-pulse response, *J. Geophys. Res.*, **105**(B11), 25891-25907.

Cohee, B. P. and G. C. Beroza, (1994), A comparison of two methods for earthquake source inversion using strong motion seismograms, *Ann. Geophys.*, **37**, 1515-1538.

Dalguer, L. A., K. Irikura, and J. D. Riera, (2003), Generation of new cracks accompanied by the dynamic shear rupture propagation of the 2000 Tottori (Japan) earthquake, *Bull. Seismol. Soc. Am.*, **93**(5), 2236-2252.

Day, S. M., G. Yu, and D. J. Wald, (1998), Dynamic stress changes during earthquake rupture, *Bull. Seismol. Soc. Am.*, **88**, 512-522.

Delouis, B., D. Giardini, P. Lundgren, and J. Salichon, (2002), Joint inversion of InSar, GPS, teleseismic and strong-motion data for the spatial and temporal distribution of earthquake slip: Application to the 1999 Izmit mainshock, *Bull. Seismol. Soc. Am.*, **92**(1), 278-299.

Dreger, D., E. Tinti, and A. Cirella, (2007), Slip Velocity Parameterization for Broadband Ground Motion Simulation, Abstract for the *Annual Meeting of Seismological Society of America.*

Fukuyama, E. and R. Madariaga, (1998), Rupture dynamics of a planar fault in a 3D elastic medium: Rate- and slip-weakening friction, *Bull. Seismol. Soc. Am.*, **88**(1), 1-17.

Fukuyama, E., T. Mikumo, and K. B. Olsen, (2003), Estimation of the critical slip-weakening distance: Theoretical background, *Bull. Seismol. Soc. Am.*, **93**, 1835-1840.

Fukuyama, E. and T. Mikumo, (2007), Slip-weakening distance estimated at near-fault stations, *Geophys. Res. Lett.*, **34**, L09302, doi:10.1029/2006GL029203.

Guatteri, M. and P. Spudich, (2000), What can strong-motion data tell us about slip-weakening fault-friction laws?, *Bull. Seismol. Soc. Am.*, **90**, 98-116.

Hartzell, S. and T. H. Heaton, (1983), Inversion of strong ground motion and teleseismic waveform data for the fault rupture history of the 1979 Imperial Valley, California, earthquake, *Bull. Seismol. Soc. Am.*, **73**, 1553-1583.

Hartzell, S., P. Liu, C. Mendoza, C. Ji, and K. M. Larson, (2007), Stability and uncertainty of finite-fault slip inversions: Application to the 2004 Parkfield, California, earthquake, *Bull. Seismol. Soc. Am.*, **97**, 1911-1934.

Henry, C., S. Das, and J. H. Woodhouse, (2000), The great March 25, 1998, Antarctic Plate earthquake: Moment tensor and rupture history, *J. Geophys. Res.*, **105**(B7), 16097-16118.

Ide, S. and M. Takeo, (1997), Determination of constitutive relations of fault slip based on seismic wave analysis, *J. Geophys. Res.*, **102**(B12), 27379-27391.

Ionescu, I. R. and M. Campillo, (1999), Influence of the shape of the friction law and fault finiteness on the duration of initiation, *J. Geophys. Res.*, **104**(B2), 3013-3024.

Iwata, T. and H. Sekiguchi (2002), Source process and near-source ground motion during the 2000 Tottori-ken-Seibu earthquake, *Proc. 11th Japan Earthq. Eng. Symp.*, 125-128.

Ji, C., D. J. Wald, and D. V. Helmberger, (2002), Source description of the 1999 Hector Mine, California, earthquake, Part I: Wavelet domain inversion theory and resolution analysis, *Bull. Seismol. Soc. Am.*, **92**(4), 1192-1207.

Marone, C., M. Cocco, E. Richardson, and E Tinti, (2009), The critical slip distance for seismic and aseismic fault zones of finite width, In: *Fault-Zone Properties and Earthquake Rupture Dynamics*, edited by E. Fukuyama, *International Geophysics Series*, **94**, 135-162, Elsevier.

Marone, C. and B. Kilgore, (1993), Scaling of the critical slip distance for seismic faulting with shear strain in fault zones, *Nature*, **362**, 618-621.

Mikumo, T., K. B. Olsen, E. Fukuyama, and Y. Yagi, (2003), Stress-breakdown time and slip-weakening distance inferred from slip-velocity functions on earthquake faults, *Bull. Seismol. Soc. Am.*, **93**(1), 264-282.

Nakamura, H. and T. Miyatake, (2000), An approximate expression of slip velocity time functions for simulation of near-field strong ground motion, *Zisin (J. Seismol. Soc. Jpn.)*, **53**, 1-9 (in Japanese with English abstract).

Nielsen, S. and R. Madariaga, (2003), On the self-healing fracture mode, *Bull. Seismol. Soc. Am.*, **93**(6), 2375-2388.

Ohnaka, M., (2003), A constitutive scaling law and a unified comprehension for frictional slip failure, shear fracture of intact rock, and earthquake rupture, *J. Geophys. Res.*, **108**(B2), 2080, doi:10.1029/2000JB000123.

Ohnaka, M. and T. Yamashita, (1989), A cohesive zone model for dynamic shear faulting based on experimentally inferred constitutive relation and strong motion source parameters. *J. Geophys. Res.*, **94**, 4089-4104.

Palmer, A. C. and J. R. Rice, (1973), The growth of slip surfaces in the progressive failure of over-consolidated clay, *Proc. Roy. Soc. London*, Ser. A, **332**, 527-548.

Perrin, G., J. R. Rice, and G. Zheng, (1995), Self-healing slip pulse on a frictional surface. *J. Mech. Phys. Solids*, **43**, 1461-1495.

Piatanesi, A., E. Tinti, M. Cocco, and E. Fukuyama, (2004), The dependence of traction evolution on the earthquake source time function adopted in kinematic rupture models, *Geophys. Res. Lett.*, **31**, L04609, doi:10.1029/2003GL019225.

Piatanesi, A., A. Cirella, P. Spudich, and M. Cocco, (2007), A global search inversion for earthquake kinematic rupture history: Application to the 2000 western Tottori, Japan earthquake, *J. Geophys. Res.*, **112**, B07314, doi:10.1029/2006JB004821.

Power, W. L., T. E. Tullis, S. R. Brown, G. N. Boitnott, and C. H. Scholz, (1987), Roughness of natural fault surfaces, *Geophys. Res. Lett.*, **14**(1), 29-32.

Pulido, N. and T. Kubo, (2004), Near-fault strong motion complexity of the 2000 Tottori earthquake (Japan) from a broadband source asperity model, *Tectonophys.*, **390**, 177-192.

Rice, J. R., C. G. Sammis, and R. Parsons, (2005), Off-fault secondary failure induced by a dynamic slip-pulse, *Bull. Seismol. Soc. Am.*, **95**(1), 109-134.

Rice, J. R. and M. Cocco, (2007), Seismic fault rheology and earthquake dynamics, In: *Tectonic Faults: Agents of Change on a Dynamic Earth*, edited by M. R. Handy, G. Hirth and N. Hovius (Dahlem Workshop 95, Berlin, January 2005, on *The Dynamics of Fault Zones*), 99-137, MIT Press, Cambridge, MA, USA.

Scholz, C. H., (1988), The critical slip distance for seismic faulting, *Nature*, **336**, 22-29.

Spudich, P. and M. Guatteri, (2004), The effect of bandwidth limitations on the inference of earthquake slip-weakening distance from seismograms, *Bull. Seismol. Soc. Am.*, **94**, 2028-2036

Tinti, E., A. Bizzarri, A. Piatanesi, and M. Cocco, (2004), Estimates of slip weakening distance for different dynamic rupture models, *Geophys. Res. Lett.*, **31**, L02611, doi:10.1029/2003GL018811.

Tinti, E., P. Spudich, and M. Cocco, (2005a), Earthquake fracture energy inferred from kinematic rupture models on extended faults, *J. Geophys. Res.*, **110**, B12303, doi:10.1029/2005JB00364.

Tinti, E., E. Fukuyama, A. Piatanesi and M. Cocco, (2005b), A kinematic source time function compatible with earthquake dynamics, *Bull. Seismol. Soc. Am.*, **95**(4), 1211-1223.

Tinti, E., M. Cocco, E. Fukuyama, and A. Piatanesi, (2008), Dependence of slip weakening distance ($Dc$) on final slip during dynamic rupture of earthquakes, submitted to *Geophys. J. Int.*

Tinti, E., P. Spudich, and M. Cocco, (2008b), Correction to "Earthquake fracture energy inferred from kinematic rupture models on extended faults", *J. Geophys. Res.*, **113**, B07301, doi:10.1029/ 2008JB005829.

Wald, D. J., T. H. Heaton, and K. W. Hudnut, (1996), The slip history of the 1994 Northridge, California, earthquake determined from strong-motion, teleseismic, GPS, and leveling data, *Bull. Seismol. Soc. Am.*, **86** (1, Part B Suppl.), 49-70.

Weertman, J., (1980), Crack tip advance model in fatigue of ductile material, *J. Metals*, **32**(12), 75.

Yasuda, T., Y. Yagi, T. Mikumo, and T. Miyatake, (2005), A comparison between Dc'-values obtained from a dynamic rupture model and waveform inversion, *Geophys. Res. Lett.*, **32**, L14316, doi:10.1029/2005GL023114.

Zheng, G. and J. R. Rice, (1998). Conditions under which velocity-weakening friction allows a self-healing versus a cracklike mode of rupture, *Bull. Seismol. Soc. Am.*, **88**, 1466-1483.

## Chapter 8

# Rupture Dynamics on Bimaterial Faults and Nonlinear Off-Fault Damage

Teruo Yamashita

*Earthquake Research Institute, University of Tokyo, Japan*

We review recent progress in our understanding of the generation mechanism of off-fault damage and rupture dynamics on a bimaterial interface. Although an earthquake fault is traditionally modeled as a planar fault in an isotropic homogeneous medium, it is becoming increasingly clear that such an approach is inadequate for a deep understanding of earthquake rupture dynamics. In fact, field observations indicate the existence of tensile microfractures and shear fault branches that surround earthquake faults; such fractures generally define the damage zone formed near earthquake faults. In spite of its importance, the modeling of damage is not straightforward because microfractures and shear branches are densely distributed in the damage zone and their interactions cannot generally be neglected. However, recent years have brought some progress in the study of this area. It is also known that a geological fault with long slip history is likely to occur on a bimaterial interface. If a fault constitutive friction law dependent on the normal stress is assumed as in a Coulomb friction law, the rupture occurring on a bimaterial interface is enriched by the coupling between slip and normal stress. Although the bimaterial effect on dynamic earthquake rupture is still in debate partly because many factors seem to affect rupture dynamics, active study is now being carried out.

## 1. INTRODUCTION

An earthquake fault is traditionally modeled as a planar fault in an isotropic homogeneous elastic medium. However, a large-displacement fault is known to consist of a hierarchical structure ranging from the scales of fine grains forming a gouge or ultracataclasite zone (generally less than tens of centimeters thick) along the principal plane (Figure 1, top) up to that of the fault system, consisting of discrete fault segments (Figure 1, bottom) or branched fault segments (Figure 1, middle and bottom) (*Ando and Yamashita*, 2007).

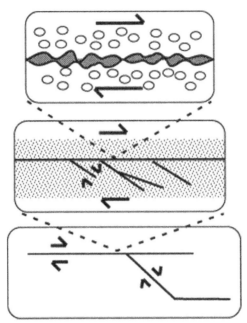

**FIGURE 1**    Schematic illustration of a hierarchical structure of an earthquake fault (from *Ando and Yamashita*, 2007). (bottom) Scale of entire length of the fault system composed of planar segments (macroscopic scale), (middle) length scale of a fault zone involving damage zone (shaded) and branches (mesoscopic-scale), and (top) scale of a frictional surface involving gouge and microcracks (microscopic scale).

A broader zone surrounding the narrow ultracataclasite zone is generally referred to as the damage zone. An inward intensification of microfractures is identified in the damage zone toward the ultracataclasite zone (e.g., *Chester and Logan*, 1987; *Anders and Wiltschko*, 1994). Because the mechanisms of nucleation, propagation, and termination of dynamic earthquake rupture ought to be defined by a physical process in the above entity, the assumption of a planar fault in an isotropic homogeneous medium will not be reasonable for realistic modeling of earthquake rupture.

There have been many field studies documenting the concentration of microfractures near earthquake faults (e.g., *Chester and Logan*, 1987; *Chester et al.*, 1993; *Scholz et al.*, 1993; *Anders and Wiltschko*, 1994; *McGrath and Davison*, 1995; *Vermilye and Scholz*, 1998; *Wilson et al.*, 2003; *Kim et al.*, 2004). Laboratory studies have pointed out the generation of microfractures by shear rupture (*Cox and Scholz*, 1988; *Petit and Barquins*, 1988; *Moore and Lockner*, 1995; *Tenthorey and Cox*, 2006). For example, *Moore and Lockner* (1995) conducted laboratory experiments on shear rupture with granite and found that microfractures are evident around the tips of shear rupture; the microfractures they observed are up to grain size as found in field observations. These microfractures may occur within grains, along grain boundaries, or they may be continuous through two or more grains (e.g., *Anders and Wiltschko*, 1994). In addition, these microfractures have a feature of tensile fracture. *Anders and Wiltschko* (1994) took cores at different distances from a number of faults in their field investigation and found that all microfractures they studied are tensile. The orientations of such microfractures, in general, are indicators of the principal stress distributions at the time the microfractures formed (e.g., *Tapponnier and Brace*, 1976; *Blenkinsop*, 1990) because tensile microfractures will grow in length parallel to the maximum compressive stress. Some researchers believe that microfractures formed during the fault propagation reflect the stress concentrations at propagating fault tips. In fact, fracture mechanics suggest that the fault tip propagates leaving a zone of damaged rock in its wake because of high stress concentration at the fault tip. In this model, the microfracture fabric specifically reflects the fault-tip stress orientation. *Cowie and Scholz* (1992), *Anders and Wiltschko* (1994), *Vermilye and Scholz* (1998), and others actually found, in their field observation of orientations of microfractures formed in the vicinity of exhumed faults, that the microfractures have features reflecting the stress field caused by propagating fault tips.

Another model is also proposed about the generation of microfractures. This model focuses on the progressive accumulation of damage resulting from displacement on an established rough surface (*Scholz*, 1987; *Chester and Chester*, 2000; *Wilson et al.*, 2003). Displacement along a rough surface produces geometric mismatch and local stress concentration. With continued slip, the rock adjacent to a fault surface experiences stress cycling in response to the passing of geometrical irregularities. This cycling leads to continued fracturing and wear and forms the damage zone. *Wilson et al.* (2003) actually found in the investigation of the fracture fabric in the damage zone of the Punchbowl Fault in California that their orientations are most consistent with local damage accumulation from stress cycling associated with slip on a rough fault surface.

Shear branches, whose sizes are much larger than those of the above-mentioned tensile microfractures, are also found widely around faults (see Figure 1, middle); their lengths are known to span a wide spectrum of lengths from scales much smaller than total lengths of faults (e.g., *Tchalenko and Berberian*, 1975; *Sowers et al.*, 1994; *Vermilye and Scholz*, 1998; *Ando and Yamashita*, 2007) to those comparable to them (e.g., *Sowers et al.*, 1994). It is also believed that distribution of microfractures and shear branches evolves with the recurrence of large events. Although many researchers tend to define the damage zone as a zone of densely distributed microfractures (e.g., *Vermilye and Scholz*, 1998), we here define it somewhat broadly taking account of the existence of shear branches. In other words, we define the damage zone as a zone in the vicinity of an earthquake fault where both tensile microfractures and shear branches are distributed densely.

Taking account of the hierarchical fault structure mentioned earlier may promote our understanding of earthquake rupture. It should, however, be noted that we have to detail the concept of spatial scales when the hierarchical fault structure is taken into account in the modeling of earthquake fault (*Ando and Yamashita*, 2007). Although the concept of macroscopic and microscopic length scales is sometimes employed in earthquake source studies, they are mentioned in different ways according to authors. For example, the macroscopic length scale usually refers to the size of samples in laboratory experiments, whereas the microscopic length scale could be used to refer to any objects of smaller sizes such as fault gouge particles, real contact areas, or tensile micro-cracks between grains. For the analysis of natural earthquakes, whereas the macroscopic length scale usually refers to the total fault length, the microscopic length scale could be employed for any objects of smaller sizes such as subfaults assumed in seismic wave inversion analyses or branch faults or fault zone widths observed in the field. It is, however, crucial in the modeling study how microscopic and macroscopic length scales are defined because physical phenomena are ignored if their typical scales are smaller than the assumed microscopic scale. If we consider only the microscopic and macroscopic length scales in the modeling, we may lose physically important properties whose length scales lie between them. In fact, the scale of shear branches mentioned earlier lies between microscopic and macroscopic length scales if microscopic and macroscopic length scales are defined as the length scales of microfractures and the entire fault system. Hence, the consideration of shear branches may bring about a better understanding of the dynamic earthquake rupture.

Another important element that should be taken into account in a realistic modeling of earthquake rupture is that a geological fault with long slip history is likely to bring into contact materials with different elastic properties. *Chester et al.* (1993) and *Chester and Chester* (1998) found in their field observations of exhumed faults that the fault walls are often bordered by materials that are different from each other. Plate boundaries where a number of major earthquakes recur also have a feature of bimaterial interface. For example, the contrast of seismic wave velocities at various locations across the San Andreas Fault in California is known to be about 5% to 30% (e.g., *Shi and Ben-Zion*, 2006). Such contrast may also exist at the boundary between the continental plate and the subducting oceanic plate (e.g., *Kodaira et al.*, 2002). It is well known theoretically that the normal stress remains unchanged on the fault plane if a fault grows on a plane in a homogeneous isotropic medium because of the symmetrical deformation with respect to the fault plane. In contrast, the normal stress change is excited by the propagating fault if it occurs on a bimaterial interface (*Weertman*, 1980). If the fault constitutive friction dependent on the normal stress is assumed as in a Coulomb friction law, the rupture occurring on a bimaterial interface may be largely different from that expected based on a model of rupture embedded in an isotropic homogeneous medium.

All of these considerations suggest that the geometrical complexity of earthquake faults and heterogeneous media should be taken into account in a realistic modeling of earthquake rupture. Theoretical or numerical modelings generally confront many obstacles in such studies, but some progress has been made. We review the progress of such modeling studies here.

## 2. FORMATION OF DAMAGE ZONE DUE TO DYNAMIC FAULT GROWTH

### 2.1. Inference about Orientation and Distribution of Secondary Fractures

Field and laboratory studies have pointed out that microfractures and shear branches are formed in the vicinity of faults. As stated in the Introduction, there exist mainly two views on the mechanism of generation of microfractures. One focuses on the role of stress concentrations at the propagating fault tips, whereas the other focuses on the role of rough fault surfaces. More active research has been made from the former viewpoint, so that we first review studies done from such lines. In fact, many researchers have theoretically inferred the location and orientation of microfractures from their analyses of intense stress perturbation due to the passage of fault tips. Earlier works were based on the quasi-static analysis (e.g., *Reches and Lockner*, 1994; *Vermilye and Scholz*, 1998). For example, *Reches and Lockner* (1994) inferred the orientation and location of tensile microfractures from the distribution of tensile stress caused by shear fault in their quasi-static analysis. *Reches and Dewers* (2005) afterward extended the analysis and calculated the tensile components of stress and strain rates near the dynamically propagating shear fault. They proposed that fault gouge forms by rock pulverization based on their finding of the profound increase of stress intensity and dilation rates with increasing rupture velocity.

A distinct feature observed only in dynamic rupture is the strong growth of off-plane shear stress near the fault tip, relative to those on the primary fault plane, at high fault tip speed (*Rice*, 1980; *Kame and Yamashita*, 1999; *Rice et al.*, 2005). In fact, linear fracture mechanics predict that the maximum hoop shear traction axis shifts from the propagating direction of fault if the fault tip velocity exceeds a certain threshold (*Freund*, 1990; *Kame and Yamashita*, 1999): the hoop traction is defined as the traction on an inclined plane around the fault tip (Figure 2). This implies the occurrence of secondary shear ruptures ahead of the propagating fault tip, which was actually simulated by *Kame and Yamashita* (1999) and will contribute to the formation of shear branches. Dynamic analysis is therefore indispensable for more realistic study of the generation of secondary fractures. *Poliakov et al.* (2002) and *Rice et al.* (2005) examined orientations of potential secondary shear ruptures near the dynamically propagating fault with steady sub-Rayleigh rupture speeds calculating the elastodynamic stress field around the fault and assuming the Coulomb failure criterion and slip weakening friction law. Slip is

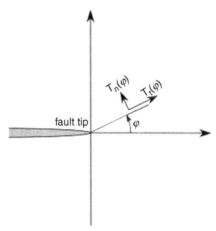

**FIGURE 2**  Hoop shear $T_t(\varphi)$ and hoop normal $T_n(\varphi)$ tractions: they act on an inclined plane emanating from a fault tip.

supposed to occur when the Coulomb failure stress, defined as $\tau + f\sigma_n$, is greater than a certain threshold in the Coulomb failure criterion, where $\tau$ is the shear stress, $\sigma_n$ is the normal stress, and $f$ is the friction coefficient. The normal stress is assumed to be positive for extension in this chapter. They found that important parameters for potential secondary faulting are the rupture velocity as well as the principal directions and ratios components in the prestress field. *Bhat et al.* (2007) extended the work of *Rice et al.* (2005) to the supershear regime. However, all of these authors except *Kame and Yamashita* (1999) did not actually simulate the generation of secondary fractures; they only examined the location and orientation of potential secondary fractures.

It has been pointed out in field observations that faults are irregular surfaces over the entire range, with their surface topography scaling as a generalized fractal (*Power et al.*, 1987; *Scholz*, 1990, pp. 145-152). Based on such observations, some researchers believe that the formation of microfractures is closely related to slip on a rough fault surface as stated in the Introduction. *Chester and Chester* (2000) evaluated the local stress along a wavy frictional fault as a function of surface roughness and others; their stress analysis indicated that microfractures are likely to occur at angles to the fault that are slightly greater than the far-field maximum compressive stress because of the wavy nature of the fault (see also *Wilson et al.*, 2003). Hence, the orientation of microfractures surrounding a fault reflects the geometry of the fault in the model in which the formation of microfractures is attributed to fault surface roughness. However, earthquake faults are generally more complex than just a wavy surface. For example, actual faults are generally segmented and present offsets, kinks, and bifurcations (*Chester and Kirschner*, 2000; *Ando et al.*, 2004; *Kim et al.*, 2004; *Di Toro et al.*, 2005;

*Ando and Yamashita*, 2007). Effects of the geometrical irregularity of faults such as fault bending or branching on dynamic earthquake propagation and seismic wave radiation have been studied actively (*Kame and Yamashita*, 1999; *Aochi et al.*, 2002; *Oglesby and Archuleta*, 2003; *Ando et al.*, 2004; *Madariaga et al.*, 2006; *Ando and Yamashita*, 2007). Although they did not study the generation of microfractures explicitly, intense damage ought to be formed at kinks or offsets during dynamic fault propagation because of high stress concentration there (*Kim et al.*, 2004; *Festa and Vilotte*, 2006a). Numerical analysis will be indispensable for the analysis of a fault with kinks or offsets and damage formation there. However, some care must be taken in the modeling of a fault kink, as *Tada and Yamashita* (1996) have pointed out. They showed, in the context of the boundary integral equation method, that, in the mechanics of 2D curved in-plane shear cracks, a smooth curve, along which the crack orientation changes continuously, and an abrupt kink, across which it changes discontinuously, are not equivalent. In other words, the behavior of fault normal stress at the kink is significantly different in the two approaches.

## 2.2. Modeling of Generation of Tensile Microfractures

In contrast to the studies mentioned in 2.1, *Yamashita* (2000) and *Dalguer et al.* (2003a, 2003b) actually modeled the generation of tensile microfractures that is caused by a dynamically propagating fault. They modeled the damage zone as a zone of densely distributed tensile microfractures. It is generally difficult to model the generation of a large number of microcracks in numerical analyses. *Yamashita* (2000) overcame this difficulty by representing the microfracture distribution by anisotropic properties of the overall elastic coefficients based on the study by *Hudson* (1980). This treatment implies the existence of a large number of dilutely distributed microcracks whose sizes are negligibly small. A tensile microfracture was assumed to be formed if a tensile stress component exceeds a certain threshold locally. He found in his finite difference simulation of 2D in-plane shear rupture that the decrease in the microfracture density is approximated well by a logarithmic function of the distance from the fault plane; microfractures on the dilatational side of the rupture plane are shown to make larger angles to the fault plane than on the compressive side (Figure 3). These are consistent with field observations (*Anders and Wiltschko*, 1994; *Vermilye and Scholz*, 1998). However, there are some shortcomings in the treatment of *Yamashita* (2000): the microfractures are dilutely distributed and the medium in which the fault is embedded is treated as a continuum. *Dalguer et al.* (2003a, 2003b) numerically studied the generation of tensile microfractures on the assumption of a 3D fault using the discrete element method (DEM); DEM has an advantage in that distributed fractures can be introduced with little computational efforts. What is significantly different in the two approaches is that each microfracture is

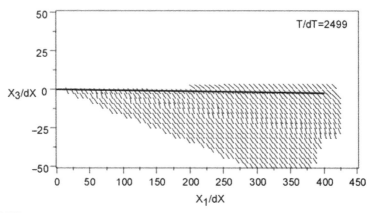

**FIGURE 3**  An example of the distribution of tensile microfractures generated spontaneously by dynamic fault propagation (from *Yamashita*, 2000, with modification). The thick straight line denotes the fault, which lies at $-399 < X_1/dX < 399$, $X_3/dX = 0$ at this instant. Each thin line segment denotes the orientation of a microcrack. Only the right half is illustrated because of symmetry with respect to the origin.

considered individually in the studies of *Dalguer et al.* (2003a, 2003b), whereas distributed microfractures are modeled macroscopically in the study by *Yamashita* (2000). However, the accuracy of the calculation results for densely distributed microfractures is not clear in such a numerical method as DEM; in addition, calculation time and memory requirement will increase significantly if a larger number of densely distributed microfractures are introduced.

In contrast to the approach of *Yamashita* (2000) and *Dalguer et al.* (2003a, 2003b), *Andrews* (2005) modeled the damage zone as a continuous plastic zone in his numerical study on the formation of a damage zone due to dynamic propagation of 2D in-plane shear fault. He assumed that material off the fault is subject to a Coulomb yield criterion and found that energy loss off the fault is much larger than energy loss on the fault specified by the fault constitutive relation. This means that the formation of plastic damage zone affects the energy budget of earthquake faulting (Figure 4). It is also implied that the creation of damage stabilizes the fault growth. We, however, have to note that the magnitude of plastic strain and the extent of plastic zone generated off the fault depend on the assumption of strength of cohesion and others (*Ben-Zion and Shi*, 2005), whose reliable information is indispensable for realistic simulation. Although the damage model of *Andrews* (2005) is quite different from those of *Yamashita* (2000) and *Dalguer et al.* (2003a, 2003b), we find common features in their studies—that is, the damage zone is formed on the extensional side of the fault and its thickness normal to the fault is proportional to the propagation distance of the fault.

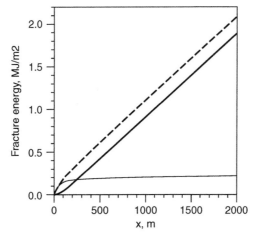

**FIGURE 4**   Energy absorbed near the rupture front. Thin and thick solid curves indicate energy absorbed on and off the fault, respectively. Broken curve indicates total. The abscissa denotes the distance along the fault; the rupture is initiated near $x = 0$ and propagates bilaterally (from *Andrews*, 2005; modified with permission from American Geophysical Union). The off-fault energy change is due to the formation of the plastic zone, whereas the energy change on the fault is related to the fracture energy.

## 2.3. Modeling of Dynamic Generation of Mesoscopic Shear Branches

As mentioned in the Introduction, field investigations have pointed out that shear branches are found widely in the field (e.g., *Tchalenko and Berberian*, 1975; *Vermilye and Scholz*, 1998; *Ando and Yamashita*, 2007) near earthquake faults. It was also shown in quasi-static numerical simulation that secondary shear fractures are generated around faults (*Willson et al.*, 2007). Although it will be unlikely for tensile microfractures to grow to macroscopic ones in the form of tensile rupture, shear branches may dynamically grow into macroscopic ruptures as numerically shown by *Ando and Yamashita* (2007), so that the consideration of such shear branches may be critically important to explaining dynamic earthquake rupture. Shear branches mentioned earlier and in the Introduction will introduce a mesoscopic-scale fault structure in the modeling of earthquake rupture. This mesoscopic-scale structure bridges microscopic- and macroscopic-scale ones if we assume microscopic and macroscopic fault structures as off-fault tensile microfractures and entire fault systems (Figure 1).

*Ando and Yamashita* (2007) pointed out in their theoretical study that the consideration of the mesoscopic shear branches gives us a new perspective of dynamic earthquake rupture. They assumed the nucleation of shear branches at the tip of dynamically propagating 2D in-plane shear fault and their extension into the extensional side of the fault; the maximum hoop shear traction criterion was assumed for their growths (see Figure 2 as to the hoop shear traction). The branches grow behind the propagating fault tip at the high-velocity stage of fault tip velocity because the maximum hoop shear traction occurs behind the fault tip in the slip weakening fault model, which they assumed (see Figure 3 in *Ando and Yamashita*, 2007). It was shown that once the length of the main fault exceeds a certain threshold, a limited number of branches begin unstable growth and their sizes soon become comparable to that of the main fault (Figure 5). This unstable growth will occur mainly because of intense nonlinear interactions between the neighboring branches and stress enhancement at the tips of branches due to the growth of the main fault. This finding indicates the critical importance of considering the meso-scopic-scale fault structure in our effort to understand fault dynamics. It also indicates that there exists no simple linear scaling between small and large events because of the spontaneous transformation of mesoscopic branches into macroscopic ruptures in large events.

## 2.4. Effects of Damage on Earthquake Rupture in a Poroelastic Medium

As pointed out in Sections 2.1 and 2.2, fault propagation is likely to generate a large number of tensile microfractures in the vicinity of the fault, which will be the main source of inelastic porosity in rocks of low porosity. Because the generation of such pores will suck up fluid if the medium is permeated by fluid on and around the fault, earthquake rupture will be affected by the pore generation. Note that it is generally believed that fracture strength and resid-ual frictional strength are dependent on pore-fluid pressure as well as normal stress according to the Coulomb failure criterion. In other words, lower pore-fluid pressure tends to suppress the fault growth.

*Marone et al.* (1990) assumed water-saturated layers of sand sheared between 45° surfaces and measured porosity changes under nominally drained conditions in their laboratory experiment on fault slip. Their experiments indicate a monotonic increase in inelastic porosity with increasing fault slip (see also *Segall and Rice*, 1995). Such increase in porosity may reflect the

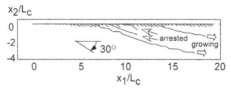

**FIGURE 5** An example of a fault system formed dynamically by the rupture propagation. The main fault, nucleated at $x_1 = 0$, propagates unilaterally along the $x_1$ axis. It is found that two mesoscopic shear branches dynamically grow into macroscopic ruptures (from *Ando and Yamashita*, 2007).

generation of tensile microfractures. The pore creation can reduce the pore fluid pressure because of fluid suction as noted earlier, which tends to suppress the rupture growth. *Garagash and Rudnicki* (2003), *Rice* (2006), *Bizzarri and Cocco* (2006), and *Suzuki and Yamashita* (2007) theoretically studied the effects of inelastic change of porosity on the dynamic slip of a fault embedded in thermoporoelastic media. *Suzuki and Yamashita* (2007) showed in their 1D fault model, assuming a Coulomb friction law, that slip-strengthening behavior emerges when the rate of pore creation is relatively large. This occurs because the inelastic pore creation tends to lower the fluid pressure locally as mentioned earlier and raises the level of residual frictional stress. They also suggested that the slip-strengthening causes pulselike fault slip, which has been revealed in seismic wave inversions of large events (*Heaton*, 1990).

Numerical simulations have also been carried out on the effects of pore creation on seismicity. For example, *Yamashita* (1999) numerically simulated earthquake swarm assuming fluid flow in a narrow porous fault zone and monotonic increase of inelastic porosity with increasing fault slip based on the experiment by *Marone et al.* (1990). The fluid suction due to pore creation was shown to retard the fault growth because of a decrease in the Coulomb failure stress and gives rise to swarmlike activity; the Coulomb failure stress is defined as $\tau + f(\sigma_n + p)$ when the pore-fluid pressure is taken into consideration, where $p$ is the pore-fluid pressure. The pore-fluid pressure is assumed to be positive for compression in this chapter.

The preceding consideration indicates the importance of coupling between the generation of microfractures and pore-fluid pressure in deepening our understanding of earthquake rupture. All the authors mentioned previously in this subsection assumed the generation of microfractures that act as pores only on a fault surface. However, pores should also be created ahead of the propagating fault tip because of stress intensification there. If such pores are taken into account as well as pores on the fault surface, we expect much larger resistance to fault tip growth because fluid suction ahead of the fault tip tends to reduce the Coulomb failure stress there.

## 2.5. Rheology of Damage Zone

When the damage zone is modeled macroscopically, knowledge of its rheology is critical. To quantify the damage evolution macroscopically, several models have been proposed and applied to the brittle deformation of Earth's crust (*Lyakhovsky et al.*, 1997; *Nanjo et al.*, 2005; *Shcherbakov et al.*, 2005; *Ben-Zion and Lyakhovsky*, 2006). Although *Dalguer et al.* (2003a, 2003b) and *Ando and Yamashita* (2007) tried to model the dynamic formation of the damage zone, the concern of *Lyakhovsky et al.* (1997) and others seems to be the quasi-static evolution of damage. Hence, their models cannot directly be applicable to the formation of the damage zone near a dynamically propagating fault. However, their modeling may be helpful, at least partly, for our future modeling of dynamic formation of the damage zone, so that we slightly touch on their studies here.

Many researchers have proposed models for damage rheology with a scalar damage parameter changing from 0 at an undamaged state to 1 at instability (*Lyakhovsky et al.*, 1997; *Nanjo et al.*, 2005) in the category of model referred to as the continuum damage model; the damage parameter will be a measure of microcrack density. Each of the elastic moduli is assumed to be a linear function of the damage parameter in many studies such as *Lyakhovsky et al.* (1997), *Nanjo et al.* (2005), *Shcherbakov et al.* (2005), and *Ben-Zion and Lyakhovsky* (2006), and it decreases as the value of the damage parameter increases. One of the key elements of the theoretical modeling of damage in the continuum damage model seems to lie in the physical development of the evolution equation of the damage parameter (*Lyakhovsky et al.*, 1997); variation of elastic moduli and Poisson's ratio with damage evolution can be expressed in terms of the damage parameter as noted previously. For example, *Lyakhovsky et al.* (1997) proposed an evolution equation of a scalar damage parameter based on thermodynamic principles. We can evaluate the spatiotemporal evolution of damage based on this evolution equation.

There also exist damage models conceptually different from the continuum damage model mentioned earlier. In fact, a model referred to as the discrete fiber-bundle model (*Turcotte et al.*, 2003), which is applicable to the deformation and fracture of composite materials, is also widely employed. It is conceptually assumed in the fiber-bundle model that a composite material is made up of strong fibers embedded in a relatively weak matrix (*Turcotte et al.*, 2003). When stress applied to a fiber-bundle is increased, one of the fibers will fail. The stress from the failed fiber is redistributed on the remaining fibers, and then the weakest fiber among them will fail. The process of failure followed by stress redistribution continues until all fibers fail and no load can be carried. In the fiber-bundle model, the failure statistics of the individual fibers that make up the fiber bundle must be specified to simulate the failure of the composite. In many cases, the statistical distribution of times to failure for the fibers is specified in terms of the stress acting on the fibers (*Coleman*, 1958); this gives a model for the temporal evolution of the failure of fibers. The whole of this process is regarded as the evolution of damage in the fiber-bundle model. *Krajcinovic* (1996) and *Turcotte et al.* (2003) showed, assuming simple models of deformation, that the two models, the continuum damage model and fiber-bundle model, lead to similar results in the limit of a large number of discrete elements.

Although some progress has been made in the quasi-static modeling of damage evolution in the approach mentioned earlier, the development of a damage model valid in the dynamic range is a key issue for the quantitative simulation of dynamic earthquake rupture. The development of such a model is important because the damage formed around a fault affects the fault propagation significantly (*Ando and Yamashita*, 2007).

## 3. FAULT GROWTH ON A BIMATERIAL INTERFACE

### 3.1. Field Observation of Faults

The studies mentioned in Section 2 imply that the repeated occurrence of earthquakes on a planar fault forms asymmetrical distribution of micro-fractures across the fault as actually found for natural faults (*Feng and McEvilly*, 1983; *Eberhart-Phillips and Michael*, 1998; *Dor et al.*, 2006; *Shi and Ben-Zion*, 2006; see also Figure 3). Hence, the repeated occurrence of earthquakes may form a bimaterial structure. In addition, a large number of major earthquakes occur at plate interfaces, which have clear features of bimaterial interfaces that separate mechanically different materials as stated in the Introduction. Hence, the consideration of a fault on a bimaterial inter-face will improve the quality of our understanding of earthquake rupture.

### 3.2. Quasi-Static Features of In-Plane Tensile Crack

Interface cracks between two different elastic media have received consider-able attention in the field of mechanical engineering because of their practical importance (*England*, 1965; *Rice and Sih*, 1965; *Comninou*, 1977; *Rice*, 1988). Although their main concern was the quasi-static nature of in-plane tensile crack, an important discovery was made. In fact, it has been shown that the stress at the tip of the interface crack possesses an oscillatory character of the type $r^{1/2}\sin$ (or cos) of the argument $\varepsilon \log r$ if the tensile stress is released and shear stress is unchanged on the crack, where $r$ is the radial distance from the crack tip and $\varepsilon$ is a function of material constants. This means that the stress changes its sign an infinite number of times as $r$ approaches zero. More-over, the relative displacement also changes its sign an infinite number of times. The preceding finding suggests that the behavior of an in-plane shear crack located on a bimaterial interface is also largely different from that expected from the analysis of a crack embedded in an isotropic homogeneous medium because of the qualitative similarity of the solutions for in-plane shear and tensile cracks. An in-plane shear crack or rupture is referred to as an in-plane crack or rupture for short in the following discussion, because we are interested only in shear cracks.

### 3.3. Theoretical and Numerical Studies of Dynamic Fault Slip

In recent years, seismologists have become increasingly interested in the behavior of dynamic fault propagation on an interface between two different elastic media (Figure 6) because it has been found that mechanical properties are generally different across large-scale faults as mentioned earlier in this section. We have to note that there is a fundamental theoretical distinction between slips along an interface separating similar and dissimilar materials. On a planar fault between materials with identical elastic constants and

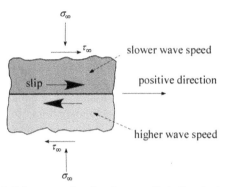

**FIGURE 6** Frictional sliding on an interface between dissimilar elastic materials. The shear modulus is larger in the lower medium. The remotely applied shear and normal stresses are given by $\tau_\infty$ and $\sigma_\infty$.

density, the change in the normal stress is zero because of the symmetry of deformation, so that there is no coupling between slip and normal stress on the fault governed by a friction law dependent on the normal stress. In contrast, slip dynamics on a bimaterial interface is enriched by the coupling between slip and normal stress.

*Weertman* (1980) theoretically studied an in-plane shear dislocation moving along a bimaterial interface at a constant velocity. He showed that the normal stress becomes more extensional at the propagating edge of dislocation with increasing dislocation velocity if the dislocation moves in the direction of slip on the more compliant material (referred to later as the positive direction; see Figure 6). This suggests a possibility of self-sustained propagation of slip pulse on an interface governed by a constant coefficient of friction $f$ independent of slip rate or its history even if the materials are loaded below the friction threshold $\tau_\infty < f\sigma_\infty$ because the normal stress $\sigma$ reduces in the slipping region so as to meet the condition $\tau_\infty = f\sigma$, where $\tau_\infty$ and $\sigma_\infty$ are the remote shear and normal stresses (see also *Rice*, 2001, pp. 15-19). It is also suggested that slip pulse propagates in the positive direction. Such a solution is shown to exist only when the generalized Rayleigh wave exists (*Adams*, 1998). The generalized Rayleigh wave corresponds to wave motion with free slip, $\tau = 0$, but no opening gap at the interface, where $\tau$ is the shear stress. When the generalized Rayleigh wave exists, its speed lies between the Rayleigh wave speeds of the two materials and it is slower than the S wave speed of the more compliant material. The study by *Weertman* (1980) seems to have stimulated seismologists because a mode of fault slip consisting of a slip pulse may explain the lack of observable frictional heat along strike-slip faults in California (*Lachenbruch and Sass*, 1992) and short rise times in earthquake slip histories (*Heaton*, 1990).

Spontaneous dynamic growth of slip on a bimaterial interface was first addressed in 2D finite difference simulations (*Andrews and Ben-Zion*, 1997;

*Harris and Day*, 1997; *Ben-Zion and Andrews*, 1998). *Andrews and Ben-Zion* (1997) performed a calculation of in-plane slip on an interface between elastic materials with 20% contrast in wave speeds. They assumed a classical Coulomb friction law with constant coefficient of friction—that is, the static and dynamic coefficients of friction are equal. The rupture was triggered by elevated localized pore pressure; the pore pressure returns to its ambient value some time after the triggering. As *Weertman* (1980) predicted, they found the self-sustaining narrow pulse of slip velocity propagating with a velocity close to the generalized Rayleigh wave (Figure 7). Propagation of slip is unilateral in the positive direction, where the change of normal traction is tensile. They called this slip pulse a wrinkle-like pulse because particle motion in the more compliant material is analogous to a wrinkle in a carpet. The change of normal traction was shown to become tensile during the slip at the propagating edge of the fault, which can promote slip (see Figure 7). The change of the normal traction is compressive at the other edge of the fault, leading to dynamic strengthening. They also showed that the slip pulse gets narrower and its peak value increases; irregularities in the pulse shape tend to become separate pulses (see Figure 7). *Anooshehpoor and Brune* (1999) also found, in their laboratory experiment with foam rubber, sustained unilateral propagation of slip pulse.

However, in contrast to the finding of *Andrews and Ben-Zion* (1997), *Harris and Day* (1997) and *Andrews and Harris* (2005) obtained a bilateral cracklike solution in their 2D in-plane numerical calculations; asymmetry was most apparent in that slip velocities near the fault tip moving in the positive direction were much larger than those at the opposite direction (Figure 8). Major differences between the modeling of *Harris and Day* (1997) and that of *Andrews and Ben-Zion* (1997) seem to lie in the assumptions of friction law and the nucleation mechanism of rupture. In fact, *Harris and Day* (1997) assumed that the friction coefficient decreases from the static to dynamic coefficient with slip (a slip weakening friction law) and the rupture was triggered by artificially dropping the friction coefficient to its dynamic value over a nucleation zone; on the other hand, the friction coefficient was constant and the rupture was triggered by artificially elevating localized pore-fluid pressure in the study by *Andrews and Ben-Zion* (1997), as mentioned earlier. *Xia et al.* (2005) suggested that the bilateral cracklike growth found in the study by *Harris and Day* (1997) is due to their assumption of slip weakening. They actually observed bilateral rupture growth with asymmetric rupture speed in their laboratory experiment of rupture, in which dynamic rupture was initiated by exploding a wire locally, which qualitatively resembles the nucleation procedure in the simulation by *Andrews and Ben-Zion* (1997). The friction coefficient seems to have dropped at the onset of slip in the experiment by *Xia et al.* (2005) because the bimaterial interface was frictionally held by a far field load. *Shi and Ben-Zion* (2006) numerically simulated in-plane ruptures along a bimaterial interface governed by a slip weakening friction assuming the same procedure of rupture nucleation as assumed by *Andrews and Ben-Zion*

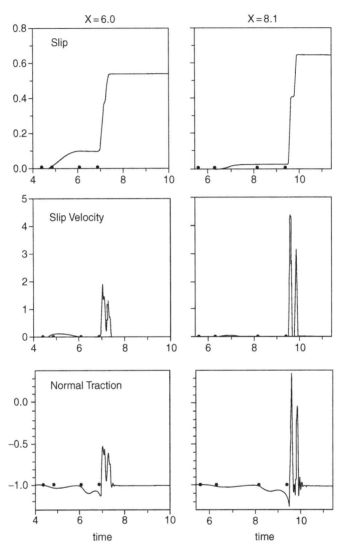

**FIGURE 7** Examples of temporal changes of slip, slip velocity, and normal traction at two locations on the interface (from *Andrews and Ben-Zion*, 1997). The imposed source is located in a zone centered on the origin $x = 0$. Dots mark phase arrivals of faster P, slower P, faster S, and slower S waves from an arbitrarily chosen point within the nucleation zone.

(1997). They obtained a unilaterally propagating slip pulse when the dynamic friction coefficient is larger than $\tau_\infty/\sigma_\infty$ (see also *Andrews and Harris*, 2006), which is roughly consistent with the result obtained by *Andrews and Ben-Zion* (1997). Slip seems to have been triggered in their simulation in spite of such a

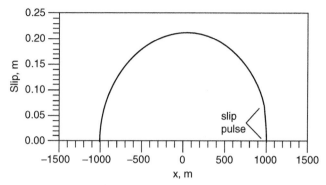

**FIGURE 8**   An example of spatial variation of slip on the interface at a certain instant from the 2D calculation with uniform stress drop (from *Andrews and Harris*, 2005). The steep rise of slip at the right rupture front is a slip pulse.

large friction coefficient because of the strong nucleation procedure; the effect of nucleation conditions of rupture will be discussed in the next paragraph. All of the above considerations suggest that a unilaterally propagating slip pulse can exist if the friction coefficient is constant or the dynamic friction coefficient is larger than $\tau_\infty/\sigma_\infty$ if the slip weakening friction law is assumed. If the friction coefficient drops below $\tau_\infty/\sigma_\infty$ in the slip weakening process, bilateral cracklike rupture will result (*Andrews and Harris*, 2006; *Harris and Day*, 2006).

Nucleation conditions of dynamic rupture are known to influence the style of rupture even for ruptures along an interface between materials with identical elastic constants and density (*Nielsen and Madariaga*, 2003; *Festa and Vilotte*, 2006b; *Shi et al.*, 2008), so that the dependence on nucleation conditions is also expected for ruptures along an interface between dissimilar materials. *Andrews and Harris* (2006) actually pointed out a possibility that a strong dynamic nucleation contributes to the generation of a unilaterally propagating slip pulse significantly on a bimaterial interface. *Cochard and Rice* (2000) also suggested that the difference in the rupture behaviors found by *Andrews and Ben-Zion* (1997) and *Harris and Day* (1997) resulted from the different rupture nucleation procedures as well as the different friction laws. In fact, the rupture nucleation assumed in the study by *Harris and Day* (1997) was slower than in the study by *Andrews and Ben-Zion* (1997). *Rubin and Ampuero* (2007) carried out a detailed numerical study of in-plane ruptures on a bimaterial interface and indicated that cracklike ruptures grow bilaterally as found by *Harris and Day* (1997) if slow rupture nucleation and slip weakening friction are assumed. In contrast, unilaterally propagating slip pulses are generated on a bimaterial interface governed by slip weakening friction with a large friction coefficient exceeding $\tau_\infty/\sigma_\infty$ in cases initiated with strong localized nucleation (*Shi and Ben-Zion*, 2006).

It is, however, becoming clear that the situation is not as simple as mentioned earlier, even if the analysis is restricted to 2D modeling. In fact, in their theoretical study of in-plane ruptures on a bimaterial fault, *Ampuero and Ben-Zion* (2008) showed that the assumed friction law significantly affects the rupture behavior on a bimaterial interface. They showed that with strongly velocity weakening friction, ruptures occur under a wide range of conditions as pulses propagating in the positive direction. They also found a secondary small-scale slip pulse detached from the principal slip pulse after propagation for some distance. *Rubin and Ampuero* (2007) also found, assuming a slip weakening friction law, a secondary small-scale slip pulse spontaneously formed at an instant when a 2D rupture propagating in the positive direction encounters a stress barrier.

Whereas *Andrews and Ben-Zion* (1997) and others studied subshear propagation of slip, the analysis was extended to supershear in later works. *Cochard and Rice* (2000), *Adams* (2001), and *Ranjith and Rice* (2001) found that slip pulses with a velocity near the slower P wave speed can also propagate with a constant friction coefficient in the negative direction on a bimaterial interface. It was shown that the change of normal stress at the tip propagating in the positive direction of slip is compressive in contrast with the case of subshear rupture, and the supershear rupture propagates in the negative direction (*Adams*, 2001). *Xia et al.* (2005) confirmed the existence of supershear rupture propagation in this direction in their laboratory experiment. They found that the rupture growth in the positive direction has stable properties that did not depend on the employed experimental conditions and propagates with a velocity close to the generalized Rayleigh wave; the rupture velocity in the negative direction can be supershear as long as the rupture is allowed to grow to a sufficiently long distance. Supershear rupture was also numerically simulated in the negative direction by *Shi and Ben-Zion* (2006), who assumed slip weakening friction in the 2D finite-difference calculation. However, the supershear rupture was shown to occur only when a vigorous source to trigger the rupture is given. This may correspond to the laboratory observation of *Xia et al.* (2005), who showed that the supershear rupture occurs in the negative direction when driving force for the nucleation of rupture is very high.

It will be more reasonable from a seismological viewpoint to consider both bimaterial effects and damage generation in the modeling of dynamic earthquake rupture because many preexisting faults seem to have a feature of bimaterial interface and microfractures will be generated by dynamic faulting as mentioned before. *Ben-Zion and Shi* (2005) actually carried out numerical simulation of the dynamic growth of 2D in-plane rupture on a fault between different elastic solids in a model that includes dynamic generation of plastic strain off the fault. They found that the slip velocity is stabilized by the damage formation assuming a model of constant friction coefficient.

The irregularity of the slip pulse found in the numerical calculations of *Andrews and Ben-Zion* (1997) suggests some numerical instability (see

Figure 7). *Adams* (1995) actually showed that in-plane steady slip on a bima-terial Coulomb interface has a strong dynamic instability if a constant friction coefficient is assumed. *Ranjith and Rice* (2001) showed a connection between the previously mentioned dynamic instability and the existence of a generalized Rayleigh wave. Hence, careful analysis is required when slip on a bimaterial interface is numerically studied.

## 3.4. Regularization of an Ill-Posed Problem

Following the study by *Andrews and Ben-Zion* (1997), *Cochard and Rice* (2000) investigated the same problem. However, they showed that the prob-lem of slip on a bimaterial interface is ill posed for any value of the friction coefficient when the generalized Rayleigh wave speed is defined. When the generalized Rayleigh wave speed is not defined, the problem remains ill posed for values of the friction coefficient larger than a critical value. This implies that numerical simulations do not converge with grid refinement. *Cochard and Rice* (2000) pointed out that the case studied by *Andrews and Ben-Zion* (1997) falls in the range in which the generalized Rayleigh wave speed is defined. *Cochard and Rice* (2000) actually showed, using a model assumed by *Andrews and Ben-Zion* (1997), in the numerical analysis that the numerical solutions have grid size dependence because of the ill-posedness of the problem.

Some possible regularizations of the problem have been discussed. For example, *Harris and Day* (2005) regularized the problem by making the medium Kelvin-Voigt viscoelastic, which eliminates the exponential growth of short-wavelength perturbations. On the other hand, *Cochard and Rice* (2000) and *Ranjith and Rice* (2001) discussed an experimentally based consti-tutive law that can provide a regularization to the previously mentioned ill-posedness. In fact, *Cochard and Rice* (2000) proposed a law:

$$\dot{\tau}_s = -\frac{|V|+V^*}{L}[\tau_s - f \max(0, -\sigma_\mathrm{n})] \tag{1}$$

which they called a modified version of the simplified Prakash-Clifton law, where $\tau_s$ is the shear strength, the overdot means the time differentiation, $V$ is the slip velocity, $V^*$ is a positive constant, and $L$ is a characteristic slip dis-tance. This law is an example that smoothes into a continuous transition with time or slip the otherwise instantaneous variation of shear strength that would follow from an instantaneous variation in normal stress if the Coulomb fric-tion law was used (*Cochard and Rice*, 2000). *Cochard and Rice* (2000) pointed out that classical slip weakening or rate- and state-dependent constitu-tive law with instantaneous response to change in local normal stress is not sufficient for the study of rupture on a bimaterial interface. However, subsequent analytical work (*Adda-Bedia and Ben Amar*, 2003) showed for steady-state slip pulse of finite size that a physically reasonable solution is not found even for a friction law in which a relaxation time for the response of the shear stress to a sudden variation of normal stress is taken into account; namely, they observed singularity in the stress component normal to the fault, which suggests local fault opening and violates the boundary condition of continuity of the displacement normal to the fault plane. We should, however, note that the observation of normal stress singularity merely suggests the inappropriateness of the assumption of a planar fault. In other words, this may indicate the occurrence of secondary rupture or the formation of damage.

## 3.5. Poroelastic Bimaterial Effects on Fault Slip

*Andrews and Ben-Zion* (1997) considered the effect of pore-fluid pressure only to trigger the dynamic rupture growth. However, *Yamashita* (2007) and *Yamashita and Suzuki* (2009) considered the effect of fluid migration in their studies of fault growth on a bimaterial interface, although their analyses were quasi-static. *Yamashita* (2007) theoretically studied the mechanism of afterslip assuming a 2D in-plane fault on an interface between dissimilar poroelastic media assuming contrasts in the Biot-Willis coefficient and drained and undrained Poisson's ratios; the Biot-Willis coefficient is given by $1 - K_s/K$, where $K$ is the drained bulk modulus and $K_s$ is the bulk modulus of a solid constituent. He found that the positive feedback between the fluid pressure raised at the extending fault tip and evolving fault slip promotes the unilateral quasi-static fault tip extension. Such feedback occurs because of coupling between slip and normal stress for faulting on a bimaterial interface as mentioned before. *Yamashita and Suzuki* (2009) theoretically investigated the mechanism of afterslip assuming a fault on an interface between two poroelastic media of different diffusivity. They found that the diffusivity contrast is more effective for the afterslip than the contrasts in Poisson's ratios and the Biot-Willis parameter. The extension distance of fault is also found to be largest when one of the media is impermeable. We may gain an insight into the generation mechanism of slow slip phenomena discovered recently (e.g., *Ide et al.*, 2007) from such studies, because many researchers now tend to believe the involvement of high-pressure pore fluid in the generation of slow slip phenomena (e.g., *Ito et al.*, 2007).

Some progress is also being made in our understanding of dynamic rupture on an interface between dissimilar poroelastic media (*Rudnicki and Rice*, 2006; *Vredevoogd et al.*, 2007), although it still seems to be in a germinal stage. *Rudnicki and Rice* (2006) took into account poroelasticity in the vicinity of a fault and showed that the sign and magnitude of the change of the effective normal stress depend not only on the elastic dissimilarity of the materials bounding the fault but also on the differences of the poroelastic properties of the materials very near the fault. *Vredevoogd et al.* (2007) studied the behavior of a 1D fault slip on an interface between materials of different permeability and found the highest pressure buildup off the fault. This will again suggest a possibility of the generation of off-fault secondary rupture or damage as also implied in the analysis of *Adda-Bedia and Ben Amar* (2003) as mentioned in the preceding subsection. In other words, these studies suggest that a planar fault is unstable if it is located on a bimaterial interface.

## 3.6. How Much Are Earthquake Ruptures Influenced by Bimaterial Effects?

How much earthquake ruptures are influenced by bimaterial effects has been under debate. Two major issues concern (1) the importance of the wrinkle-like slip pulse (*Andrews and Harris*, 2005, 2006; *Ben-Zion*, 2006b) and (2) the preferred direction of rupture (*Harris and Day*, 2005, 2006; *Ben-Zion*, 2006a). However, these issues are closely related because cracklike ruptures generally propagate bilaterally. As discussed in Section 3.3, it is becoming clear through active studies on 2D in-plane ruptures that bilaterally propagating cracklike ruptures are observed if we assume slow rupture nucleation and slip weakening friction. As mentioned, the asymmetry is apparent in the slip velocities near the two tips in a cracklike rupture (*Andrews and Harris*, 2005; see Figure 8). Hence, the wrinkle-like slip pulse will not be important as long as we assume slow rupture nucleation and slip weakening friction in 2D analyses.

More realistic modeling is being done. For example, 3D effects of heterogeneities in stress and strength are taken into account in some studies. *Andrews and Harris* (2005) found in their 3D simulation that the wrinkle-like slip pulse is only a small component of the calculated solution; they assumed slip weakening and stress drop whose distribution is heterogeneous on the fault. They found that rupture propagation is determined primarily by the spatial distribution of potential stress drop. This suggests that the wrinkle-like slip pulse found in the study by *Andrews and Ben-Zion* (1997) is not important in actual earthquake rupture dynamics. *Brietzke et al.* (2009) also carried out a 3D numerical simulation of ruptures on a bimaterial interface in which heterogeneous stress drop and slip weakening friction law were assumed as in *Andrews and Harris* (2005). However, they found a range of model parameters for which the rupture is transformed from cracklike into pulselike slip during its propagation, which contrasts much with the simulation result of *Andrews and Harris* (2005). Further study will be required concerning what causes such differences in the results in spite of apparently similar modelings.

A recent issue will be about the secondary small-scale pulse discussed in Section 3.3 and mentioned in the preceding paragraph. The size of the secondary pulse is comparable or smaller than the size of the process zone of the parent rupture front (*Ampuero and Ben-Zion*, 2008), so that numerical analysis of high resolution is required for the study of the secondary slip pulse. In addition, small-scale material heterogeneity may affect its generation. Although the size of the secondary pulse is small, it may affect earthquake generation. As we will describe at the end of this subsection, *Rubin and Ampuero* (2007) suggested that asymmetrical distribution of microearthquake aftershocks found at bimaterial segments of the San Andreas Fault in California is largely due to the stress perturbation resulting from this secondary slip pulse.

The preceding consideration implies that more systematic study is required to explain the generation mechanism of slip pulse. We will have to consider a large body of theoretical results and observations before proffering a conclusion on the behavior of earthquake rupture because many factors seem to affect dynamic rupture on a bimaterial interface as actually noted by *Andrews and Harris* (2005), *Harris and Day* (2005), *Ben-Zion* (2006a, 2006b), and *Brietzke et al.* (2007). What is becoming increasingly clear is that not only the contrast in the elastic constants but also the assumed friction law, nucleation mechanism of rupture, 3D effects, and stress and strength heterogeneities significantly affect the behavior of dynamic fault slip on a bimaterial interface. More detailed and systematic studies are especially required about 3D effects and stress and strength heterogeneities. Even if a large-scale slip pulse propagating in the positive direction is excited, its propagation may be suppressed by the presence of strong heterogeneities; however, it is not clear how large stress and strength heterogeneities exist on real faults (*Ampuero and Ben-Zion*, 2008). We need knowledge about how such heterogeneities develop with the recurrence of earthquakes. We should note that these heterogeneities will be affected significantly by the formation of damage. The generation of a small-scale secondary pulse may especially be sensitive to the detailed structure of damage.

Some findings were made from seismological and geological observations that may be related to bimaterial effects on earthquake ruptures (*Rubin and Gillard*, 2000; *Rubin*, 2002; *Dor et al.*, 2006; *Rubin and Ampuero*, 2007). *Rubin and Gillard* (2000) and *Rubin* (2002) showed from seismological observations the asymmetrical distribution of microearthquake aftershocks along the central San Andreas Fault in California; this region has lower-velocity rocks to the northwest at seismogenic depths. *Rubin and Ampuero* (2007) suggested, based on their simulation study, that such asymmetrical distribution is due to the secondary slip pulse generated when the rupture tip extending in the positive direction encounters a stress barrier. They implied that the number of aftershocks is smaller in the positive direction of the fault because this stress pulse smoothes the stress field and reduces the static stress change beyond the rupture tip. *Dor et al.* (2006) reported that near-surface damage occurs predominantly on one side of major strike-slip faults in California. They proposed that this results from unidirectional rupture growth in the positive direction. Note that damage will be formed on two sides of a fault if the fault propagates bilaterally, as illustrated in Figure 3.

## 3.7. Macroscopic Parameter Affected by the Existence of Fault at Bimaterial Interface

Although the seismic moment remains one of the few fundamental macroscopic properties of earthquake sources that can be reliably estimated, it has been shown theoretically that the moment density associated with a specified slip distribution on a bimaterial interface is fundamentally ambiguous (*Heaton and Heaton*, 1989; *Ben-Zion*, 2001; *Ampuero and Dahlen*, 2005). This ambiguity arises because the scalar moment of an earthquake is defined as $M_o = \mu \langle \Delta u \rangle A$, where $\mu$ is the rigidity in the vicinity of source, $\langle \Delta u \rangle$ is the average slip, and $A$ is the fault area (*Aki*, 1966). In the case of a bimaterial interface with a discontinuity in the rigidity, there is no obvious choice for the fault rigidity $\mu$, so the earthquake moment $M_o$ is not well defined (*Ampuero and Dahlen*, 2005). This bimaterial ambiguity is an argument for abandoning the moment density representation of an earthquake source and replacing it with a potency density representation (*Heaton and Heaton*, 1989; *Ben-Zion*, 2001; *Ampuero and Dahlen*, 2005). The potency density tensor $p_{kl} = 1/2\,\Delta u\,(n_k e_l + n_l e_k)$ depends only on the slip $\Delta u_k = \Delta u e_k$ and is independent of the discontinuous elastic stiffness, where $n_k$ denotes the unit normal to the fault.

## 4. CONCLUDING REMARKS

As noted in the Introduction, it is not reasonable to assume a planar fault embedded in an isotropic homogeneous medium in the reasonable modeling of dynamic earthquake rupture. It is well known in field observations of exhumed faults that actual faults are generally segmented and present offsets, kinks, and bifurcations. In addition, a damage zone is generally found around earthquake faults, and it is considered to evolve spatiotemporarily with the recurrence of earthquakes. Also, plate boundaries and other major fault zone structures are known to have clear features of bimaterial interfaces that separate mechanically different materials. We have reviewed recent progress in our understanding of the effects of bimaterial structure and damage formation on earthquake rupture and showed that these effects are significant for improving our understanding of earthquake rupture. Although it is true that considerable progress has been achieved, we have to admit that we have not arrived at a deep understanding of these effects. For example, if the damage zone is treated macroscopically as in *Yamashita* (2000) and *Andrews* (2005), the individual intense interaction between microcracks formed near the principal fault cannot be taken into account. If this interaction intensifies with increasing crack density, a limited number of microcracks may grow into a macroscopic one as typically simulated by *Ando and Yamashita* (2007). However, it is still difficult, in numerical calculation, to assume a large number of microcracks and their interactions and to simulate their spatiotemporal change.

## ACKNOWLEDGMENTS

Discussion with J.-P. Ampuero and his comments on the manuscript helped to improve this chapter. I appreciate R. A. Harris and an anonymous reviewer for reviewing the manuscript and providing helpful comments. I also thank Gilbert Brietzke for sending me a copy of his poster presented at the American Geophysical Union 2007 fall meeting.

## REFERENCES

Adams, G. G., (1995), Self-excited oscillations of two elastic half-spaces sliding with a constant coefficient of friction, *J. Appl. Mech.*, **62**, 867-872.

Adams, G. G., (1998), Steady sliding of two elastic half-spaces with friction reduction due to interface stick-slip, *J. Appl. Mech.*, **65**, 470-475.

Adams, G. G., (2001), An intersonic slip pulse at a frictional interface between dissimilar materials, *J. Appl. Mech*, **68**, 81-86.

Adda-Bedia, M. and M. Ben Amar, (2003), Self-sustained slip pulses of finite size between dissimilar materials, *J. Mech. Phys. Solids*, **51**, 1849-1861.

Aki, K., (1966), Generation and propagation of G waves from the Niigata earthquake of June 16, 1964, 2. Estimation of earthquake moment, released energy, and strain-drop from the G wave spectrum, *Bull. Earthq. Res. Inst.*, **44**, 23-88.

Ampuero, J.-P. and F. A. Dahlen, (2005), Ambiguity of the moment tensor, *Bull. Seismol. Soc. Amer.*, **95**, 390-400, doi:10.1785/0120040103.

Ampuero, J.-P. and Y. Ben-Zion, (2008), Cracks, pulses and macroscopic asymmetry of dynamic rupture on a bimaterial interface with velocity-weakening friction, *Geophys. J. Int.*, **173**, 674-692.

Anders, M. H. and D. V. Wiltschko, (1994), Microfracturing, paleostress and the growth of faults, *J. Struct. Geol.*, **16**, 795-815.

Ando, R., T. Tada, and T. Yamashita, (2004), Dynamic evolution of a fault system through interactions between fault segments, *J. Geophys. Res.*, **109**, B05303, doi:10.1029/2003JB002665.

Ando, R. and T. Yamashita, (2007), Effects of mesoscopic-scale fault structure on dynamic earthquake ruptures: Dynamic formation of geometrical complexity of earthquake faults, *J. Geophys. Res.*, **112**, B09303, doi:10.1029/2006JB004612.

Andrews, D. J., (2005), Rupture dynamics with energy loss outside the slip zone, *J. Geophys. Res.*, **110**, B01307, doi:10.1029/2004JB003191.

Andrews, D. J. and Y. Ben-Zion, (1997), Wrinkle-like slip pulse on a fault between different materials, *J. Geophys. Res.*, **102**, 553-571.

Andrews, D. J. and R. A. Harris, (2005), The wrinkle-like slip pulse is not important in earthquake dynamics, *Geophys. Res. Lett.*, **32**, L23303, doi:10.1029/2005GL023996.

Andrews, D. J. and R. A. Harris, (2006), Reply to comment by Y. Ben-Zion on "The wrinkle-like slip pulse is not important in earthquake dynamics," *Geophys. Res. Lett.*, **33**, L06311, doi:10.1029/2006GL02573.

Anooshehpoor, A. and J. N. Brune, (1999), Wrinkle-like Weertman pulse at the interface between two blocks of foam rubber with different velocities, *Geophys. Res. Lett.*, **26**, 2025-2028.

Aochi, H., E. Fukuyama, and R. Madariaga, (2002), Effect of normal stress during rupture propagation along nonplanar faults, *J. Geophys. Res.*, **107**, doi:10.1029/20001JB000500.

Ben-Zion, Y., (2001), On quantification of the earthquake source, *Seism. Res. Lett.*, **72**, 151-152.

Ben-Zion, Y., (2006a), Comment on "Material contrast does not predict earthquake rupture propagation" by R. A. Harris and S. M. Day, *Geophys. Res. Lett.*, **33**, L13310, doi:10.1029/2005GL025652.

Ben-Zion, Y., (2006b), Comment on "The wrinkle-like slip pulse is not important in earth-quake dynamics" by D. J. Andrews and R. A. Harris, *Geophys. Res. Lett.*, **33**, L06310, doi:10.1029/2005GL025372.

Ben-Zion, Y. and D. J. Andrews, (1998), Properties and implications of dynamic rupture along a material interface, *Bull. Seismol. Soc. Am.*, **88**, 1085-1094.

Ben-Zion, Y. and Z. Shi, (2005), Dynamic rupture on a material interface with spontaneous generation of plastic strain in the bulk, *Earth Planet. Sci. Lett.*, **236**, 486-496.

Ben-Zion, Y. and V. Lyakhovsky, (2006), Analysis of aftershocks in a lithospheric model with seismogenic zone governed by damage rheology, *Geophys. J. Int.*, **165**, 197-210.

Bhat, H. S., R. Dmowska, G. C. P. King, Y. Klinger, and J. R. Rice, (2007), Off-fault damage patterns due to supershear ruptures with application to the 2001 Mw 8.1 Kokoxili (Kunlun) Tibet earthquake, *J. Geophys. Res.*, **112**, B06301, doi:10.1029/2006JB004425.

Bizzarri, A. and M. Cocco, (2006), A thermal pressurization model for the spontaneous dynamic rupture propagation on a three-dimensional fault: 2. Traction evolution and dynamic parameters, *J. Geophys. Res.*, **111**, B05304, doi:10.1029/2005JB003864.

Blenkinsop, T. G., (1990), Correlation of paleotectonic fracture and microfracture orientations in cores with seismic anisotropy at Cajon Pass drill hole, southern California, *J. Geophys. Res.*, **95**, 11143-11150.

Brietzke, G. B., A. Cochard, and H. Igel, (2007), Dynamic rupture along bimaterial interfaces in 3D, *Geophys. Res. Lett.*, **34**, L11305, doi:10.1029/2007GL029908.

Brietzke, G. B., A. Cochard, and H. Igel, (2009), Importance of bimaterial interfaces for earthquake 1 dynamics and strong ground motion, submitted to *Geophys. J. Int.*

Chester, F. M. and J. M. Logan, (1987), Composite planar fabric of gouge from the Punchbowl fault, California, *J. Struct. Geology*, **9**, 621-634.

Chester, F. M., J. P. Evans, and R. L. Biegel, (1993), Internal structure and weakening mechanisms of the San-Andreas fault, *J. Geophys. Res.*, **98**(B1), 771-786.

Chester, F. M. and J. S. Chester, (1998), Ultracataclasite structure and friction processes of the Punchbowl fault, San Andreas System, California, *Tectonophys.*, **295**(1-2), 199-221.

Chester, F. M. and J. S. Chester, (2000), Stress and deformation along wavy frictional faults, *J. Geophys. Res.*, **105**, 23421-23430.

Chester, F. M. and D. L. Kirschner, (2000), Geochemical investigation of fluid involvement in exhumed faults of the San Andreas System, In: *National Earthquake Hazards Reduction Program, Annual Project Summary*, **41**, U.S. Geological Survey.

Cochard, A. and J. R. Rice, (2000), Fault rupture between dissimilar materials: Ill-posedness, regularization, and slip-pulse response, *J. Geophys. Res.*, **105**(B11), 25891-25907.

Coleman, B. D., (1958), Statistics and time dependence of mechanical breakdown in fibers, *J. Appl. Phys.*, **29**, 968-983.

Cowie, P. A. and C. H. Scholz, (1992), Physical explanation for the displacement-length relationship of faults using a post-yield fracture mechanics model, *J. Struct. Geology*, **14**, 1133-1148.

Cox, S. J. D. and C. H. Scholz, (1988), Rupture initiation in shear fracture of rocks: An experimental study, *J. Geophys. Res.*, **93**, 3307-3320.

Comninou, M., (1977), The interface crack, *J. Appl. Mech.*, **44**, 631-636.

Dalguer L. A., K. Irikura, and J. D. Riera, (2003a), Generation of new cracks accompanied by the dynamic shear rupture propagation of the 2000 Tottori (Japan) earthquake, *Bull. Seismol. Soc. Am.*, **93**(5), 2236-2252.

Dalguer L. A., K. Irikura, and J. D. Riera, (2003b), Simulation of tensile crack generation by three-dimensional dynamic shear rupture propagation during an earthquake, *J. Geophys. Res.*, **108**, 2144, doi:10.1029/2001JB001738.

Di Toro, G., S. Nielsen, and G. Pennacchioni, (2005), Earthquake rupture dynamics frozen in exhumed ancient faults, *Nature*, **436**, 1009-1012.

Dor, O., T. K. Rockwell, and Y. Ben-Zion, (2006), Geological observations of damage asymmetry in the structure of the San Jacinto, San Andreas and Punchbowl faults in southern California: A possible indicator for preferred rupture propagation direction, *Pure Appl. Geophys.*, **163**, 301-349.

Eberhart-Phillips, D. and A. J. Michael, (1998), Seismotectonics of the Loma Prieta, California, region determined from three-dimensional Vp, Vp/Vs, and seismicity, *J. Geophys. Res.*, **103**, 21099-21120.

England. A. H., (1965), A crack between dissimilar media, *J. Appl. Mech.*, **32**, 400-402.

Feng, R. and T. V. McEvilly, (1983), Interpretation of seismic reflection profiling data for the structure of the San Andreas fault zone, *Bull. Seismol. Soc. Am.*, **73**, 1701-1720.

Festa, G. and J.-P. Vilotte, (2006a), Dynamic rupture propagation and radiation along kinked faults, *Geophys. Res. Abst.*, **8**, 06228.

Festa, G. and J.-P. Vilotte, (2006b), Influence of the rupture initiation on the intersonic transition: Crack-like versus pulse-like modes, *Geophys. Res. Lett.*, **33**, L15320, doi:10.1029/2006GL026378.

Freund, L. B., (1990), *Dynamic Fracture Mechanics*, Cambridge University Press, Cambridge.

Garagash, D. I. and J. W. Rudnicki, (2003), Shear heating of a fluid-saturated slip-weakening dilatants fault zone: 1. Limiting regimes, *J. Geophys. Res.*, **108**, 2121, doi:10.1029/2001JB001653.

Harris, R. A. and S. M. Day, (1997), Effects of a low velocity zone on a dynamic rupture, *Bull. Seismol. Soc. Am.*, **87**, 1267-1280.

Harris, R. A. and S. M. Day, (2005), Material contrast does not predict earthquake rupture propagation direction, *Geophys. Res. Lett.*, **32**, L23301, doi:10.1029/2005GL023941.

Harris, R. A. and S. M. Day, (2006), Reply to comment by Y. Ben-Zion on "Material contrast does not predict earthquake propagation direction," *Geophys. Res. Lett.*, **33**, L13311, doi:10.1029/2006GL026811.

Heaton, T. H. and R. E. Heaton, (1989), Static deformations from point forces and force couples located in welded elastic Poissonian half-spaces: Implications for seismic moment tensors, *Bull. Seismol. Soc. Am.*, **79**, 813-841.

Heaton, T. H., (1990), Evidence for and implications of self-healing pulses of slip in earthquake rupture, *Phys. Earth Planet. Interi.*, **64**, 1-20.

Hudson, J. A., (1980), Overall properties of a cracked solid, *Math. Proc. Camb. Phil. Soc.*, **88**, 371-384.

Ide, S., G. C. Beroza, D. R. Shelly, and T. Uchide, (2007), A scaling law for slow earthquakes, *Nature*, **477**, 76-79.

Ito, Y., K. Obara, K. Shiomi, S. Sekine, and H. Hirose, (2007), Slow earthquakes coincident with episodic tremors and slow slip events, *Science*, **315**, 503-506.

Kame, N. and T. Yamashita, (1999), Simulation of the spontaneous growth of a dynamic crack without constraints on the crack tip path, *Geophys. J. Int.*, **139**(2), 345-358.

Kim, Y.-S., D. C. P. Peacock, and D. J. Sanderson, (2004), Fault damage zones, *J. Struct. Geol.*, **26**, 503-517.

Kodaira, S., E. Kurashimo, J.-O. Park, T. Takahashi, A. Nakanishi, S. Miura, T. Iwasaki, N. Hirata, K. Ito, and Y. Kaneda, (2002), Structural factors controlling the rupture process of a megathrust earthquake at the Nankai trough seismogenic zone, *Geophys. J. Int.*, **149**, 815-835.

Krajcinovic, D., (1996), *Damage Mechanics*, Elsevier, Amsterdam.

Lachenbruch, A. H. and J. H. Sass, (1992), Heat flow from Cajon Pass, fault strength and tectonic implications, *J. Geophys. Res.*, **97**, 4995-5030.

Lyakhovsky, V., Y. Ben-Zion, and A. Agnon, (1997), Distributed damage, faulting, and friction, *J. Geophys. Res.*, **102**, 27635-27649.

Madariaga, R., J.-P. Ampuero, and M. Adda-Bedia, (2006), Seismic radiation from simple models of earthquakes, In: *Earthquakes: Radiated Energy and the Physics of Faulting*, edited by Abercrombie, R., A. McGarr, G. Di Toro, and H. Kanamori, *Geophysical Monograph Series*, **170**, 223-236, American Geophysical Union, Washington, D.C.

Marone, C., C. B. Raleigh, and C. H. Scholz, (1990), Frictional behavior and constitutive modeling of simulated fault gouge, *J. Geophys. Res.*, **95**(B5), 7007-7025.

McGrath A. G. and I. Davison, (1995), Damage zone geometry around fault tips, *J. Struct. Geol.*, **17**, 1011-1024.

Moore, D. E. and D. A. Lockner, (1995), The role of microcracking in shear-fracture propagation in granite, *J. Struct. Geol.*, **17**, 95-114.

Nanjo, K. Z., D. L. Turcotte, and R. Shcherbakov, (2005), A model of damage mechanics for the deformation of the continental crust, *J. Geophys. Res.*, **110**, B07403, doi:10.1029/2004JB003438.

Nielsen, S. and R. Madariaga, (2003), On the self-healing fracture mode, *Bull. Seismol. Soc. Am.*, **93**(6), 2375-2388.

Oglesby, D. D. and R. J. Archuleta, (2003), The three-dimensional dynamics of a nonplanar thrust fault, *Bull. Seismol. Soc. Am.*, **93**, 2222-2235.

Petit, J.-P. and M. Barquins, (1988), Can natural faults propagate under mode II conditions?, *Tectonics*, **7**, 1243-1256.

Poliakov, A. N. B., R. Dmowska, and J. R. Rice, (2002), Dynamic shear rupture interactions with fault bends and off-axis secondary faulting, *J. Geophys. Res.*, **107**(B11), 2295, doi:10.1029/2001JB000572.

Power, W. L., T. E. Tullis, S. R. Brown, G. N. Boitnott, and C. H. Scholz, (1987), Roughness of natural fault surfaces, *Geophys. Res. Lett.*, **14**(1), 29-32.

Ranjith, K. and J. R. Rice, (2001), Slip dynamics at an interface between dissimilar materials, *J. Mech. Phys. Solids*, **49**, 341-361.

Reches, Z. and D. A. Lockner, (1994), Nucleation and growth of faults in brittle rocks, *J. Geophys. Res.*, **99**, 18159-18173.

Reches, Z. and T. A. Dewers, (2005), Gouge formation by dynamic pulverization during earthquake rupture, *Earth Planet. Sci. Lett.*, **235**, 361-374.

Rice, J. R., (1980), The mechanics of earthquake rupture, In: *Physics of the Earth's Interior*, Proc. International School of Physics "Enrico Fermi," edited by Dziewonski, A. M. and E. Boschi, Italian Physical Society, Bologna, and North Holland Publishing Co., Amsterdam, 555-649.

Rice, J. R., (1988), Elastic fracture mechanics concepts for interfacial cracks, *J. Appl. Mech.*, **110**, 98-103.

Rice, J. R., (2001), New perspectives in crack and fault dynamics, In: *Mechanics for a New Millennium* (Proceedings of the 20th International Congress of Theoretical and Applied Mechanics, 27 Aug-2 Sept 2000, Chicago), edited by Aref H. and J. W. Phillips, Kluwer Academic Publishers, 1-23.

Rice, J. R., (2006), Heating and weakening of faults during earthquake slip, *J. Geophys. Res.*, **111**, B05311, doi:10.1029/2005JB004006.

Rice, J. R. and G. C. Sih, (1965), Plane problems of cracks in dissimilar media, *J. Appl. Mech.*, **32**, 418-423.

Rice, J. R., C. G. Sammis, and R. Parsons, (2005), Off-fault secondary failure induced by a dynamic slip-pulse, *Bull. Seismol. Soc. Am.*, **95**(1), 109-134.

Rubin, A. M., (2002), Aftershocks of microearthquakes as probes of the mechanics of rupture, *J. Geophys. Res.*, **107**, 2142, doi:10.1029/2001JB000496.

Rubin, A. M. and D. Gillard, (2000), Aftershock asymmetry/rupture directivity among central San Andreas fault microearthquakes, *J. Geophys. Res.*, **105**, 19095-19109.

Rubin, A. M. and J.-P. Ampuero, (2007), Aftershock asymmetry on a bimaterial interface, *J. Geophys. Res.*, **112**, B05307, doi:10.1029/2006JB004337.

Rudnicki, J. W. and J. R. Rice, (2006), Effective normal stress alteration due to pre pressure changes induced by dynamic slip propagation on a plane between dissimilar materials, *J. Geophys. Res.*, **111**, B10308, doi:10.1029/2006JB004396.

Scholz, C. H., (1987), Wear and gouge formation in brittle faulting, *Geology*, **15**, 493-495.

Scholz, C. H., (1990), *The mechanics of earthquakes and faulting*, Cambridge Univ. Press, Cambridge.

Scholz, C. H., N. H. Dawers, J. Z. Yu, and M. H. Anders, (1993), Fault growth and fault scaling laws—preliminary-results, *J. Geophys. Res.*, **98**, 21951-21961.

Segall, P. and J. R. Rice, (1995), Dilatancy, compaction, and slip instability of a fluid infiltrated fault, *J. Geophys. Res.*, **100**(B11), 22155-22171.

Shcherbakov, R., D. L. Turcotte, and J. B. Rundle, (2005), Aftershock statistics, *Pure Appl. Geophys.*, **162**, 1051-1076, doi:10.1007/s00024-004-2661-8.

Shi, Z. and Y. Ben-Zion, (2006), Dynamic rupture on a bimaterial interface governed by slip-weakening friction, *Geophys. J. Int.*, **165**, 469-484.

Shi, Z., Y. Ben-Zion, and A. Needleman, (2008), Properties of dynamic rupture and energy partition in a two-dimensional elastic solid with a frictional interface, *J. Mech. Phys. Solids*, **56**, 5-24, doi:10.1016/j.jmps.2007.04.006.

Sowers, J. M., J. R. Unruh, W. R. Lettis, and T. D. Rubin, (1994), Relationship of the Kickapoo Fault to the Johnson Valley and Homestead Valley Faults, San-Bernardino County, California, *Bull. Seismol. Soc. Am.*, **84**, 528-536.

Suzuki, T. and T. Yamashita, (2007), Understanding of slip-weakening and strengthening in a single framework of modeling and its seismological implications, *Geophys. Res. Lett.*, **34**, L13303, doi:10.1029/2007GL030260.

Tada, T. and T. Yamashita, (1996), The paradox of smooth and abrupt bends in two-dimensional in-plane shear-crack mechanics, *Geophys. J. Int.*, **127**, 795-800.

Tapponnier, A. P. and W. F. Brace, (1976), Development of stress-induced microcracks in Westerly granite, *Int. J. Rock. Mech. Min. Sci.*, **13**, 103-112.

Tchalenko, J. S. and M. Berberian, (1975), Dasht-E Bayaz fault, Iran—Earthquake and earlier related structures in bed rock, *Geol. Soc. Am. Bull.*, **86**, 703-709.

Tenthorey, E. and S. F. Cox, (2006), Cohesive strengthening of fault zones during the interseismic period: An experimental study, *J. Geophys. Res.*, **111**, B09202, doi:10.1029/2005JB004122.

Turcotte, D. L., W. I. Newman, and R. Shcherbakov, (2003), Micro and macroscopic models of rock fracture, *Geophys. J. Int.*, **152**, 718-728.

Vermilye, J. M. and C. H. Scholz, (1998), The process zone: A microstructural view of fault growth, *J. Geophys. Res.*, **103**, 12223-12237.

Vredevoogd, M. A., D. D. Oglesby, and S. K. Park, (2007), Fluid pressurization due to frictional heating on a fault at a permeability contrast, *Geophys. Res. Lett.*, **34**, L18304, doi:10.1029/2007GL030754.

Weertman, J., (1980), Unstable slippage across a fault that separates elastic media of different elastic constants, *J. Geophys. Res.*, **85**, 1455-1461.

Willson, J. P., J. L. Rebecca, and Z. K. Shiptonx, (2007), Simulating spatial and temporal evolution of multiple wing cracks around faults in crystalline basement rocks, *J. Geophys. Res.*, **112**, B08408, doi:10.1029/2006JB004815.

Wilson, J. E., J. S. Chester, and F. M. Chester, (2003), Microfracture analysis of fault growth and wear processes, Punchbowl fault, San Andreas system, California, *J. Struct. Geology*, **25**, 1855-1873.

Xia, K., A. J. Rosakis, H. Kanamori, and J. R. Rice, (2005), Laboratory earthquakes along inhomogeneous faults: Directionality and supershear, *Science*, **308**, 681-684.

Yamashita, T., (1999), Pore creation due to fault slip in a fluid-permeated fault zone and its effects on seismicity: Generation mechanism of earthquake swarm, *Pure Appl. Geophys.*, **155**, 625-647.

Yamashita, T., (2000), Generation of microcracks by dynamic shear rupture and its effects on rupture growth and elastic wave radiation, *Geophys. J. Int.*, **143**, 395-406.

Yamashita, T., (2007), Postseismic quasi-static fault slip due to pore pressure change on a bimaterial interface, *J. Geophys. Res.*, **112**, B05304, doi:10.1029/2006JB004667.

Yamashita, T. and T. Suzuki, (2009), Quasi-static fault slip on an interface between poroelastic media with different hydraulic diffusivity: A generation mechanism of afterslip, *J. Geophys. Res.*, in press.

## Chapter 9

# Boundary Integral Equation Method for Earthquake Rupture Dynamics

Taku Tada

*Department of Architecture, Faculty of Engineering, Tokyo University of Science, Tokyo, Japan*
*Geological Survey of Japan, National Institute of Advanced Industrial Science and Technology,*
*Tsukuba, Japan*

The boundary integral equation method (BIEM) is widely used in numerical modeling studies that deal with problems of earthquake rupture dynamics. This chapter reviews basic concepts and procedures used in the method and presents fundamental equations for a number of typical cases, both 2D and 3D. A special focus is given on the popular approach that makes use of (1) the time-domain representation, (2) regularization of hypersingularities through integration by parts, (3) the collocation method, and (4) the piecewise-constant approximation for the slip rate function.

## 1. INTRODUCTION

The *boundary integral equation method* (BIEM) is a class of numerical modeling methods that is widely used to solve problems of physical sciences and engineering (e.g., *Beskos*, 1987, 1997; *Bouchon and Sánchez-Sesma*, 2007). It has found broad application in the numerical simulation of fault dynamics, and the results have provided a wealth of insights into the physics of earthquake rupture. The BIEM is now considered to be one of the most viable tools of numerical modeling in this perspective, alongside other popularly used categories of methods, such as the finite difference method (FDM) and the finite element method (FEM).

In the BIEM, partial differential equations, defined in the domain of a medium, are converted into integral equations defined on the boundaries of the said domain. This lowers the dimension of the problem by one. For example, an elastodynamic problem, defined in a 2D-medium space, is replaced by an integral equation problem defined on its boundaries, which is 1D, whereas a problem in a 3D space is replaced by a 2D integral equation problem. The integral equations are then solved numerically by dividing the boundaries, on which they are defined, into a mesh of discrete elements of finite size. With emphasis on this aspect, the BIEM is also often called the *boundary element method* (BEM).

This chapter reviews the most central concepts and procedures of the BIEM as applied to the study of fault rupture dynamics. We intend it to be a convenient reference for all users and potential users of the BIEM, because reviews available in research articles often focus on specific aspects of the whole procedure, and comprehensive tutorials on the entire perspective are not always abundant in the literature. Note that this chapter is specifically concerned with seismological applications. Although a huge number of studies have addressed mathematically equivalent problems in other fields of research such as materials science and mechanical engineering, we do not attempt to make an exhaustive cross-disciplinary review of all available publications.

The term *crack* is a synonym for *fault* that is fairly often found in the literature dealing with earthquake rupture dynamics. Although the two terms are not always distinguished rigorously, talking of a *crack* tends to emphasize the stress concentration near the rupture tip, whereas the word *fault* is preferred when the emphasis is placed on understanding the physics of slip under friction. We mostly stick to the term *fault* throughout this chapter.

## 2. BASIC EQUATIONS

### 2.1. General Description

Earthquake faults, on which rupture occurs, behave as boundaries to the surrounding medium, because the displacement is discontinuous across them. (In the following, we use the term *slip* to refer to the amount of displacement discontinuity across a fault.) BIEMs in earthquake rupture dynamics applications therefore begin with writing down basic integral equations of elastodynamics in such a way that they are defined on the fault surface alone.

The representation theorem of elasticity (e.g., *Aki and Richards*, 2002) states that the displacement field in an unbounded elastic medium is expressed by a spatiotemporal convolution of the distribution of slip on the fault surface, which is embedded therein, with appropriate integration kernels:

$$u_k(\mathbf{x}, t) = \int_\Gamma dS(\xi) \int_0^t d\tau \Delta u_i(\xi, \tau) c_{ijpq} n_j(\xi) \frac{\partial}{\partial \xi_q} G_{kp}(\mathbf{x}, t - \tau; \xi, 0), \quad (1)$$

where $u_k(\mathbf{x}, t)$ is the displacement in the $k$th direction at receiver location $\mathbf{x}$ and time $t$, $\Gamma$ is the fault surface, $\xi$ is the source point lying on $\Gamma$, $\Delta u_i(\xi, \tau)$ is the slip across the fault in the $i$th direction at location $\xi$ and time $\tau$, $c_{ijpq}$ is the elastic constants, and $\mathbf{n}(\xi)$ is the unit vector that is normal to the fault surface

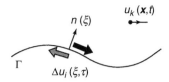

**FIGURE 1**   Nomenclature of symbols for the displacement, slip, and fault-normal unit vector.

at location $\xi$ (Figure 1). We designate one side of $\Gamma$ as the positive side and the other as the negative side. The slip $\Delta u_i(\xi, \tau)$ is defined as the relative displacement on the positive side of $\Gamma$ with reference to the negative side. The normal vector $n(\xi)$ should be so defined that it points from the negative to the positive side of $\Gamma$. Here and in all of the following equations, summation over repeated indices is implied.

The function $G_{kp}(\mathbf{x}, t - \tau; \xi, 0)$, in the integrand of Equation 1, represents the displacement *Green's function*, which is defined as the displacement that should arise in the $k$th direction at receiver location $\mathbf{x}$ and time $t - \tau$ in response to a unit force applied in the $p$th direction at source location $\xi$ and time 0. The use of the difference $t - \tau$ in the argument of $G_{kp}$ indicates that the Green's function depends on time only through the amount of delay between the input time $\tau$ and the output time $t$:

$$G_{kp}(\mathbf{x}, t; \xi, \tau) = G_{kp}(\mathbf{x}, t - \tau; \xi, 0). \tag{2}$$

If the medium considered is unbounded and homogeneous, a similar translatability also holds with respect to space. In other words, the Green's function depends on the input and the output spatial coordinates only through their vectorial difference:

$$G_{kp}(\mathbf{x}, t - \tau; \xi, 0) = G_{kp}(\mathbf{x} - \xi, t - \tau; 0, 0). \tag{3}$$

Equations 2 and 3 warrant the following reciprocity properties for derivatives of the Green's function:

$$\frac{\partial}{\partial \tau} G_{ij}(\mathbf{x}, t - \tau; \xi, 0) = -\frac{\partial}{\partial t} G_{ij}(\mathbf{x}, t - \tau; \xi, 0), \tag{4}$$

$$\frac{\partial}{\partial \xi_k} G_{ij}(\mathbf{x}, t - \tau; \xi, 0) = -\frac{\partial}{\partial x_k} G_{ij}(\mathbf{x}, t - \tau; \xi, 0). \tag{5}$$

By virtue of Equation 5, Equation 1 is rewritten as

$$u_k(\mathbf{x}, t) = -\int_{\Gamma} dS(\xi) \int_0^t d\tau \Delta u_i(\xi, \tau) c_{ijpq} n_j(\xi) \frac{\partial}{\partial x_q} G_{kp}(\mathbf{x}, t - \tau; \xi, 0), \tag{6}$$

and the stress components $\sigma_{kl}(\mathbf{x}, t)$, at any arbitrary location $\mathbf{x}$ and time $t$, are obtained as

$$\sigma_{kl}(\mathbf{x}, t) = \sigma_{kl}^0 + c_{klrs} \frac{\partial}{\partial x_s} u_r(\mathbf{x}, t)$$

$$= \sigma_{kl}^0 - \int_\Gamma dS(\xi) \int_0^t d\tau \Delta u_i(\xi, \tau) c_{ijpq} c_{klrs} n_j(\xi) \tag{7}$$

$$\frac{\partial^2}{\partial x_q \partial x_s} G_{rp}(\mathbf{x}, t - \tau; \xi, 0),$$

on condition that $\Gamma$ and $\Delta u_i(\xi, \tau)$ are smooth enough. The term $\sigma_{kl}{}^0$ accounts for a field of initial, homogeneously applied stress that may be present.

Equation 7 should hold in the limiting case where the receiver location $x$ approaches the fault surface $\Gamma$. The *shear traction* $T(x, t)$ on $\Gamma$, namely the stress component that acts on the fault surface in the direction of a unit vector $t(x)$ that is tangential to the fault at location $x$, is therefore given by

$$T(\mathbf{x}, t) = n_k(\mathbf{x}) t_l(\mathbf{x}) \sigma_{kl}(\mathbf{x}, t)$$

$$= T^0(\mathbf{x}) - \int_\Gamma dS(\xi) \int_0^t d\tau \Delta u_i(\xi, \tau) c_{ijpq} c_{klrs} n_k(\mathbf{x}) t_l(\mathbf{x}) n_j(\xi) \tag{8}$$

$$\frac{\partial^2}{\partial x_q \partial x_s} G_{rp}(\mathbf{x}, t - \tau; \xi, 0) \quad (\mathbf{x} \in \Gamma),$$

where

$$T^0(\mathbf{x}) = n_k(\mathbf{x}) t_l(\mathbf{x}) \sigma_{kl}^0 \tag{9}$$

accounts for the initial traction arising from the presence of the initial stress. Equation 8 gives the fundamental relationship that should hold between the spatiotemporal distributions of traction and of slip on a fault that is embedded in an unbounded and homogeneous medium.

The most typical problem to be encountered in fault dynamics modeling studies is one of solving for the unknown slip function $\Delta u(x, t)$ (the subscript has been dropped for simplicity) when a boundary condition is given in terms of the traction $T(x, t)$. The simplest boundary condition states that all broken patches on the fault are traction-free:

$$T(\mathbf{x}, t) = 0 \quad (\mathbf{x} \in \Gamma(t)), \tag{10}$$

where $\Gamma(t)$ denotes the part of the fault $\Gamma$ that is slipping at time $t$. (The geometry of $\Gamma(t)$ may evolve with $t$—in other words, it may grow, move, or shrink with time.) More realistic numerical models usually assume that traction and slip on the broken parts of the fault are mutually related via some *friction law* (or *constitutive law*), such as a triangular slip weakening law

$$T(\mathbf{x}, t) = \begin{cases} T_c(1 - \Delta u(\mathbf{x}, t)/D_c) & \Delta u(\mathbf{x}, t) < D_c \\ 0 & \Delta u(\mathbf{x}, t) \geq D_c \end{cases} \quad (\mathbf{x} \in \Gamma(t)), \tag{11}$$

(*Ida*, 1972) or any other law that depends on the slip or on its time derivative, the slip rate. All such varieties of friction laws, including the simple traction-free assumption (Equation 10), can be described in a general, conceptual manner as

$$T(\mathbf{x}, t) = F(\Delta u(\mathbf{x}, t), \Delta \dot{u}(\mathbf{x}, t)) \quad (\mathbf{x} \in \Gamma(t)), \tag{12}$$

where $F$ is an appropriate functional that differs from one friction law to another.

The friction law may also possibly depend on the fault-normal traction $N(x, t)$, as in the case of the classical friction law involving the coefficients of static and kinetic friction or on some state variables. Such potential dependence, however, has been dropped here for the sake of simplicity.

BIEMs in earthquake rupture dynamics applications basically consist of solving the boundary integral equation (Equation 8) under a boundary condition given by Equation 12, plus an appropriate initial condition, which is most often that the medium is at rest at time $t = 0$. If you take a different angle of view, this may also be looked on as solving the set of two equations, Equations 8 and 12, simultaneously for the two unknown functions, namely the traction $T(x, t)$ and the slip $\Delta u(x, t)$.

Once the slip function $\Delta u(x, t)$ is known everywhere on the fault surface, displacement and stress at any arbitrary location in the rest of the medium can be easily obtained by simple forward calculation using Equations 6 and 7.

## 2.2. Planar Fault of Two-Dimensional Nature

The general formulation, described in the foregoing section, may appear too abstract and conceptual for many readers. To improve understanding, we illustrate some concrete, relatively plain forms the boundary integral equation (Equation 8) takes when the problem deals with a planar shear fault of 2D nature that is embedded in an unbounded, homogeneous, and isotropic medium. Let us recall that, in an isotropic medium, the elastic constants can be written as

$$c_{ijpq} = \lambda \delta_{ij} \, \delta_{pq} + \mu(\delta_{ip} \, \delta_{jq} + \delta_{iq} \, \delta_{jp}), \tag{13}$$

where $\lambda$ and $\mu$ are the Lamé constants and $\delta$ denotes the Kronecker delta.

We first consider an *antiplane shear* (*mode III*) fault, lying on the $x_1 x_3$-plane ($x_2 = 0$) and slipping in the $x_3$-direction, in a 2D problem setting that does not depend on the $x_3$-coordinate (Figure 2a). We regard the side of positive $x_2$ as the fault plane's positive side. Because $\Delta u_3(\xi, \tau)$ is the only nonzero component of slip and because $n_2(\xi) = 1$ is the only nonzero component of the fault-normal unit vector, we have, by virtue of Equation 6, that

$$u_k(\mathbf{x}, t) = -\int_\Gamma dS(\xi) \int_0^t d\tau \Delta u_3(\xi, \tau) c_{32pq} \frac{\partial}{\partial x_q} G_{kp}(\mathbf{x}, t - \tau; \xi, 0). \tag{14}$$

The only nonzero components of $c_{32pq}$ are $c_{3223}$ and $c_{3232}$, both equaling $\mu$. Because the antiplane problem is independent of $x_3$, however, the term

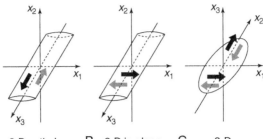

A    2-D anti-plane   B   2-D in-plane   C     3-D

**FIGURE 2**   Planar shear fault problems. (a) 2D antiplane case. (b) 2D in-plane case. (c) 3D case.

containing $c_{3223}$ should vanish, because it also contains the derivative of $G_{kp}$ with respect to $x_3$. Equation 14 therefore reduces to

$$u_k(\mathbf{x},t) = -\mu \int_\Gamma dS(\xi) \int_0^t d\tau \Delta u_3(\xi,\tau) \frac{\partial}{\partial x_2} G_{k3}(\mathbf{x}, t-\tau; \xi, 0). \qquad (15)$$

The shear traction on the fault, or the component of stress that acts in the $+x_3$-direction on the positive side of the fault plane, is given by

$$T(\mathbf{x},t) = \sigma_{23}(\mathbf{x},t)$$
$$= \sigma_{23}^0 + \mu \left[ \frac{\partial}{\partial x_2} u_3(\mathbf{x},t) + \frac{\partial}{\partial x_3} u_2(\mathbf{x},t) \right] \quad (\mathbf{x} \in \Gamma). \qquad (16)$$

The $x_3$-independence of the antiplane problem again reduces this to

$$T(\mathbf{x},t) = \sigma_{23}^0 + \mu \frac{\partial}{\partial x_2} u_3(\mathbf{x},t)$$
$$= T^0 - \mu^2 \int_\Gamma dS(\xi) \int_0^t d\tau \Delta u_3(\xi,\tau) \frac{\partial^2}{\partial x_2^2} G_{33}(\mathbf{x}, t-\tau; \xi, 0) \quad (\mathbf{x} \in \Gamma).$$
$$(17)$$

Equation 17 gives the concrete form of Equation 8 as specialized to the case of the planar, antiplane shear fault in question.

We next consider an *in-plane shear (mode II)* fault, lying on the $x_1 x_3$-plane and slipping in the $x_1$-direction (Figure 2b), in an $x_3$-independent 2D problem setting as was dealt with in Figure 2a. Because $\Delta u_1(\xi, \tau)$ and $n_2(\xi) = 1$ are the only nonzero components of the corresponding vectors, we have

$$u_k(\mathbf{x},t) = -\int_\Gamma dS(\xi) \int_0^t d\tau \Delta u_1(\xi,\tau) c_{12pq} \frac{\partial}{\partial x_q} G_{kp}(\mathbf{x}, t-\tau; \xi, 0). \qquad (18)$$

The only nonzero components of $c_{12pq}$ are $c_{1212}$ and $c_{1221}$, both equaling $\mu$, so that

$$u_k(\mathbf{x}, t) = -\mu \int_\Gamma dS(\xi) \int_0^t d\tau \Delta u_1(\xi, \tau)$$

$$\left[ \frac{\partial}{\partial x_2} G_{k1}(\mathbf{x}, t - \tau; \xi, 0) + \frac{\partial}{\partial x_1} G_{k2}(\mathbf{x}, t - \tau; \xi, 0) \right]. \tag{19}$$

The shear traction, or the stress component acting in the $+x_1$-direction on the positive ($+x_2$-) side of the fault plane, is therefore given by

$$T(\mathbf{x}, t) = \sigma_{21}(\mathbf{x}, t) = \sigma_{21}^0 + \mu \left[ \frac{\partial}{\partial x_2} u_1(\mathbf{x}, t) + \frac{\partial}{\partial x_1} u_2(\mathbf{x}, t) \right]$$

$$= T^0 - \mu^2 \int_\Gamma dS(\xi) \int_0^t d\tau \Delta u_1(\xi, \tau)$$

$$\left[ \frac{\partial^2}{\partial x_2^2} G_{11} + \frac{\partial^2}{\partial x_1 \partial x_2}(G_{12} + G_{21}) + \frac{\partial^2}{\partial x_1^2} G_{22} \right] \tag{20}$$

$$= T^0 - \mu^2 \int_\Gamma dS(\xi) \int_0^t d\tau \Delta u_1(\xi, \tau)$$

$$\left( \frac{\partial^2}{\partial x_2^2} G_{11} + 2 \frac{\partial^2}{\partial x_1 \partial x_2} G_{12} + \frac{\partial^2}{\partial x_1^2} G_{22} \right) \quad (\mathbf{x} \in \Gamma),$$

where we used the symmetry $G_{ij} = G_{ji}$ that holds for the Green's functions in an unbounded, homogeneous, and isotropic medium (see Equation 21). Equation 20 is the desired, in-plane shear case counterpart of Equation 17.

## 2.3. Three- and Two-Dimensional Green's Functions

Despite the 2D nature of the problem, the medium is still treated as 3D in Equations 17 and 20, which means that the integral over $\Gamma$ should be regarded as a surface integral over the $x_1 x_3$-plane. The functions $G_{33}, G_{11}, G_{12}$, and $G_{22}$, appearing on the right-hand sides of Equations 17 and 20, should therefore be the Green's functions of 3D elastodynamics. They are given, in the general form, by

$$G_{ij}(\mathbf{x}, t - \tau; \xi, 0) = \frac{1}{4\pi\mu r} \left[ (\delta_{ij} - 3\gamma_i\gamma_j)I(t - \tau, r) \right.$$

$$+ (\delta_{ij} - \gamma_i\gamma_j)\delta(t - \tau - r/c_T) \tag{21}$$

$$\left. + p^2\gamma_i\gamma_j\delta(t - \tau - r/c_L) \right],$$

where

$$r = |\mathbf{x} - \xi| \tag{22}$$

denotes the distance between the source (input) and receiver (output) locations,

$$\gamma_i = (x_i - \xi_i)/r \tag{23}$$

the direction cosine of the source-to-receiver vector, $\delta(\cdot)$ the delta function, $c_T$ the S-wave velocity, $c_L$ the P-wave velocity,

$$p = c_T/c_L, \tag{24}$$

$$
\begin{aligned}
I(t - \tau, r) &= \int_1^p dv \cdot v \delta(t - \tau - vr/c_T) = -c_T^2 \int_{1/c_L}^{1/c_T} d\lambda \cdot \lambda \delta(t - \tau - \lambda r) \\
&= -\frac{c_T^2(t - \tau)}{r^2} \Big[ -H(t - \tau - r/c_T) + H(t - \tau - r/c_L) \Big],
\end{aligned} \tag{25}
$$

and $H(\cdot)$ denotes the Heaviside step function (e.g., *Achenbach*, 1973, Equation 3.96; *Eringen and Şuhubi*, 1975, Equation 5.10.30; *Zhang*, 1991; *Fukuyama and Madariaga*, 1998; *Tada et al.*, 2000; *Aki and Richards*, 2002, Equation 10.2).

It is helpful to note, for the sake of later developments in Section 3.3, that these 3D Green's functions can also be written in an implicit, but systematic manner as

$$
\begin{aligned}
G_{11} &= \frac{\partial^2}{\partial x_1^2} J + G_T \quad G_{23} = \frac{\partial^2}{\partial x_2 \partial x_3} J \\
G_{22} &= \frac{\partial^2}{\partial x_2^2} J + G_T \quad G_{31} = \frac{\partial^2}{\partial x_3 \partial x_1} J, \\
G_{33} &= \frac{\partial^2}{\partial x_3^2} J + G_T \quad G_{12} = \frac{\partial^2}{\partial x_1 \partial x_2} J
\end{aligned} \tag{26}
$$

where

$$
J = \frac{c_T^2}{4\pi\mu r} \Bigg[ \int_0^{t-\tau-r/c_L} ds \cdot s \delta(t - \tau - r/c_L - s) \\
- \int_0^{t-\tau-r/c_T} ds \cdot s \delta(t - \tau - r/c_T - s) \Bigg] \tag{27}
$$

and

$$
G_T = \frac{1}{4\pi\mu r} \delta(t - \tau - r/c_T) \tag{28}
$$

(*Achenbach*, 1973, Equation 3.94; *Tada et al.*, 2000).

If you replace the surface integral over $\Gamma$ with a line integral along the $x_1$-coordinate alone, and instead "push" into the definition of the Green's functions the process of integrating over the $x_3$-coordinate, Equations 17 and 20 become

$$
T(\mathbf{x}, t) = T^0 - \mu^2 \int_\Gamma d\xi_1 \int_0^t d\tau \Delta u_3(\xi_1, \tau) \frac{\partial^2}{\partial x_2^2} G_{33}^{(2D)}(\mathbf{x}, t - \tau; \xi, 0) \quad (\mathbf{x} \in \Gamma) \tag{29}
$$

$$T(\mathbf{x}, t) = T^0 - \mu^2 \int_\Gamma d\xi_1 \int_0^t d\tau \, \Delta u_1(\xi_1, \tau)$$

$$\left( \frac{\partial^2}{\partial x_2^2} G_{11}^{(2D)} + 2 \frac{\partial^2}{\partial x_1 \partial x_2} G_{12}^{(2D)} + \frac{\partial^2}{\partial x_1^2} G_{22}^{(2D)} \right) \quad (\mathbf{x} \in \Gamma), \tag{30}$$

where $G_{ij}^{(2D)}$ denotes the 2D counterpart of $G_{ij}$ and is given by

$$G_{33}^{(2D)}(\mathbf{x}, t - \tau; \xi, 0) = \frac{1}{2\pi\mu} \frac{H(t - \tau - r/c_T)}{\sqrt{(t - \tau)^2 - (r/c_T)^2}} \tag{31}$$

$$G_{\alpha\beta}^{(2D)}(\mathbf{x}, t - \tau; \xi, 0) = \frac{1}{2\pi\mu} \frac{c_T^2}{r^2} \left\{ \gamma_\alpha \gamma_\beta \left[ 2(t - \tau)^2 - (r/c_L)^2 \right] \right.$$

$$\left. - \delta_{\alpha\beta} \left[ (t - \tau)^2 - (r/c_L)^2 \right] \right\} \frac{H(t - \tau - r/c_L)}{\sqrt{(t - \tau)^2 - (r/c_L)^2}}$$

$$- \frac{1}{2\pi\mu} \frac{c_T^2}{r^2} \left\{ \gamma_\alpha \gamma_\beta \left[ 2(t - \tau)^2 - (r/c_T)^2 \right] - \delta_{\alpha\beta}(t - \tau)^2 \right\}$$

$$\frac{H(t - \tau - r/c_T)}{\sqrt{(t - \tau)^2 - (r/c_T)^2}} \quad (\alpha, \beta = 1, 2)$$

$$\tag{32}$$

(e.g., *Achenbach*, 1973, Equations 3.128 and 3.132; *Eringen and Şuhubi*, 1975, Equation 5.12.11; *Zhang*, 1991; *Tada and Yamashita*, 1997).

## 2.4. Planar Fault of Three-Dimensional Nature

Some additional complexity arises when the problem goes 3D. Let us illustrate this with the simplest case of a planar shear fault that lies on the $x_1x_2$-plane in an unbounded, homogeneous, and isotropic medium. The fault plane is now taken to be normal to the $x_3$-coordinate, unlike the aforementioned 2D case where it was normal to the $x_2$-axis (Figure 2c). We designate the side of positive $x_3$ as the positive side of the fault plane.

Now that the slip vector has two degrees of freedom, we have, for the displacement in the medium,

$$u_k(\mathbf{x}, t) = -\mu \int_\Gamma dS(\xi) \int_0^t d\tau \left[ \Delta u_1(\xi, \tau) \left( \frac{\partial}{\partial x_3} G_{k1} + \frac{\partial}{\partial x_1} G_{k3} \right) \right.$$

$$\left. + \Delta u_2(\xi, \tau) \left( \frac{\partial}{\partial x_3} G_{k2} + \frac{\partial}{\partial x_2} G_{k3} \right) \right]. \tag{33}$$

The shear traction also has two degrees of freedom. The component of traction that acts in the $+x_1$-direction on the positive side of the fault plane, which we denote by $T_1(\boldsymbol{x}, t)$, is given by

$$
\begin{aligned}
T_1(\mathbf{x},t) &= \sigma_{31}(\mathbf{x},t) = \sigma_{31}^0 + \mu\left[\frac{\partial}{\partial x_3}u_1(\mathbf{x},t) + \frac{\partial}{\partial x_1}u_3(\mathbf{x},t)\right] \\
&= T_1^0 - \mu^2 \int_\Gamma dS(\xi) \int_0^t d\tau\left[\Delta u_1(\xi,\tau)\left(\frac{\partial^2}{\partial x_3^2}G_{11} + 2\frac{\partial^2}{\partial x_1\partial x_3}G_{13} + \frac{\partial^2}{\partial x_1^2}G_{33}\right)\right. \\
&\qquad \left. +\Delta u_2(\xi,\tau)\left(\frac{\partial^2}{\partial x_3^2}G_{12} + \frac{\partial^2}{\partial x_2\partial x_3}G_{13} + \frac{\partial^2}{\partial x_1\partial x_3}G_{23} + \frac{\partial^2}{\partial x_1\partial x_2}G_{33}\right)\right] \\
&\quad (\mathbf{x}\in\Gamma).
\end{aligned}
$$

(34)

By virtue of the symmetry between the $x_1$- and $x_2$-coordinates, the expression for the traction component $T_2(\boldsymbol{x}, t)$, which acts in the $+x_2$-direction, is easily obtained by interchanging the indices 1 and 2 in Equation 34.

In the most general case, we have four unknown functions to be solved for, namely the traction components $T_1(\boldsymbol{x}, t)$ and $T_2(\boldsymbol{x}, t)$, and the slip components $\Delta u_1(\boldsymbol{x}, t)$ and $\Delta u_2(\boldsymbol{x}, t)$. They are constrained by four relations, namely the boundary integral equation (Equation 34), its $T_2(\boldsymbol{x}, t)$ counterpart, plus two friction laws that relate the two-degree-of-freedom traction with the two-degree-of-freedom slip, its time derivative, and potentially also some state variables.

## 3. REGULARIZATION

### 3.1. Hypersingularities in the Integration Kernels

The boundary integral equations, derived in Section 2, are *hypersingular*; in other words, their *integration kernels* (integrand components apart from the slip function[s]) are not integrable in the Cauchy principal value sense (e.g., *Burridge*, 1969; *Zhang*, 1991, 2002; *Koller et al.*, 1992; *Cochard and Madariaga*, 1994; *Tada and Yamashita*, 1997; *Fukuyama and Madariaga*, 1998; *Frangi and Novati*, 1999; *Kame and Yamashita*, 1999b; *Tada et al.*, 2000; *Zhang and Chen*, 2006a).

In fact, the 3D Green's function $G_{ij}$ of elastodynamics, given by Equation 21, possesses, when integrated over time $t$, a singularity that diverges as $1/r$ when $r \to 0$. [The term $I(t - \tau, r)$, defined by Equation 25, produces a constant when integrated over $t$.] Accordingly, second-order spatial derivatives of the Green's function, such as $\partial^2 G_{33}/\partial x_2{}^2$ that appears in the integrand of Equation

17, diverge as $1/r^3$. Singularities on the order of at most $1/r^2$ can be integrable on surface $\Gamma$ at $r = 0$ ($x = \xi$) in the sense of Cauchy principal values, but those on the order of $1/r^3$ are not.

A similar argument holds for the 2D case as well. The 2D Green's functions, given by Equations 31 and 32, are regular at $r = 0$, as the $1/r^2$ singularities in the first and second halves of the right-hand side of Equation 32 cancel each other out. Accordingly, their second-order spatial derivatives, appearing in the integrands in Equations 29 and 30, have $1/r^2$ singularities. These are hypersingularities, because only up to $1/r$ singularities can be integrable along a single coordinate axis in the Cauchy principal value sense.

The hypersingular nature of the boundary integral equations poses no small difficulty at the later stage of numerical implementation, when the integration kernels have to be integrated over finite spatiotemporal domains (Section 4.2). Although it is not strictly impossible to squarely evaluate the hypersingular integrals in the sense of Hadamard finite part integrals (e.g., *Kame and Yamashita*, 1999b; *Zhang*, 2002), one approach that has popularly been taken in fault dynamics modeling studies is to use the integration-by-parts technique to *regularize* them or to rewrite them in equivalent forms that are at most integrable in the sense of Cauchy principal values. In the following sections, we illustrate this procedure by referring to the cases of 2D and 3D planar faults that were described earlier.

## 3.2. Planar Two-Dimensional Antiplane Fault

In Equation 29, the representation of traction on a planar 2D antiplane shear fault involves a hypersingular kernel $\partial^2 G_{33}/\partial x_2{}^2$ that is integrated with respect to the $\xi_1$-coordinate. This specific form is not immediately suitable to integration by parts, because the integrand involves no differentiation with respect to $\xi_1$. To make the integration by parts possible, we make use of the equation of motion

$$\frac{1}{c_T^2}\frac{\partial^2}{\partial t^2}G_{33}^{(2D)}(\mathbf{x},t;\xi,\tau) = \left(\frac{\partial^2}{\partial x_1^2}+\frac{\partial^2}{\partial x_2^2}\right)G_{33}^{(2D)}(\mathbf{x},t-\tau;\xi,0), \qquad (35)$$

and rewrite Equation 29 as

$$T(\mathbf{x},t) = T^0 - \mu^2\int_\Gamma d\xi_1 \int_0^t d\tau\, \Delta u_3(\xi_1,\tau)$$
$$\left(-\frac{\partial^2}{\partial x_1^2}+\frac{1}{c_T^2}\frac{\partial^2}{\partial t^2}\right)G_{33}^{(2D)}(\mathbf{x},t-\tau;\xi,0) \quad (\mathbf{x}\in\Gamma). \qquad (36)$$

Using the Green's functions' reciprocity properties stated in Equations 4 and 5, and carrying out integration by parts, we obtain

$$T(\mathbf{x}, t) = T^0 - \mu^2 \int_\Gamma d\xi_1 \int_0^t d\tau \; \Delta u_3(\xi_1, \tau)$$

$$\left( \frac{\partial^2}{\partial \xi_1 \partial x_1} + \frac{1}{c_T^2} \frac{\partial^2}{\partial \tau^2} \right) G_{33}^{(2D)}(\mathbf{x}, t - \tau; \xi, 0)$$

$$= T^0 + \mu^2 \int_\Gamma d\xi_1 \int_0^t d\tau \frac{\partial}{\partial \xi_1} \Delta u_3(\xi_1, \tau) \frac{\partial}{\partial x_1} G_{33}^{(2D)}(\mathbf{x}, t - \tau; \xi, 0)$$

$$- \frac{\mu^2}{c_T^2} \int_\Gamma d\xi_1 \int_0^t d\tau \frac{\partial^2}{\partial \tau^2} \Delta u_3(\xi_1, \tau) G_{33}^{(2D)}(\mathbf{x}, t - \tau; \xi, 0) \quad (\mathbf{x} \in \Gamma).$$

(37)

In Equation 37, the hypersingularity at $r = 0$ has been removed, but the first integral term on the right-hand side possesses another hypersingularity at the wavefront, which is on the order of $((t - \tau)^2 - (r/c_T)^2)^{-3/2}$ (*Zhang*, 1991; *Tada and Yamashita*, 1997). This can be regularized by applying yet another step of integration by parts, in the wake of *Cochard and Madariaga* (1994), by noticing

$$\frac{\partial}{\partial x_1} G_{33}^{(2D)}(\mathbf{x}, t - \tau; \xi, 0) = \frac{1}{2\pi\mu} \frac{\gamma_1}{r} \frac{\partial}{\partial \tau} \frac{t - \tau}{\sqrt{(t - \tau)^2 - (r/c_T)^2}} H(t - \tau - r/c_T).$$

(38)

Considering that $r = |x_1 - \xi_1|$ and $\gamma_1/r = 1/(x_1 - \xi_1)$ when both $x$ and $\xi$ lie on the $x_1$-axis, we have

$$T(\mathbf{x}, t) = T^0 - \frac{\mu}{2\pi} \int_\Gamma d\xi_1 \int_0^t d\tau \frac{\partial}{\partial \xi_1} \Delta \dot{u}_3(\xi_1, \tau)$$

$$\frac{t - \tau}{x_1 - \xi_1} \frac{H(t - \tau - |x_1 - \xi_1|/c_T)}{\sqrt{(t - \tau)^2 - \left((x_1 - \xi_1)/c_T\right)^2}}$$

$$- \frac{\mu}{2\pi c_T^2} \int_\Gamma d\xi_1 \int_0^t d\tau \frac{\partial}{\partial \tau} \Delta \dot{u}_3(\xi_1, \tau)$$

$$\frac{H(t - \tau - |x_1 - \xi_1|/c_T)}{\sqrt{(t - \tau)^2 - \left((x_1 - \xi_1)/c_T\right)^2}} \quad (\mathbf{x} \in \Gamma),$$

(39)

where $\Delta \dot{u}_3(\xi_1, \tau)$, with a superscript dot, denotes the time derivative of the slip $\Delta u_3(\xi_1, \tau)$ with respect to $\tau$.

## 3.3. Planar Three-Dimensional Fault

Regularization of hypersingularities becomes a much more difficult affair when the problem goes 3D. However, a relatively simple method to cope with this case was presented by *Tada et al.* (2000), and we outline it here.

In view of the symmetry relationship of Equation 26, and the equation of motion

$$\frac{1}{c_T^2} \frac{\partial^2}{\partial t^2} G_T = \left( \frac{\partial^2}{\partial x_1^2} + \frac{\partial^2}{\partial x_2^2} + \frac{\partial^2}{\partial x_3^2} \right) G_T \tag{40}$$

that holds for the term $G_T$ defined by Equation 28, terms in the integrand of Equation 34 that contain differentiation with respect to $x_3$ can be rewritten as

$$\frac{\partial^2}{\partial x_3^2} G_{11} = \frac{\partial^4}{\partial x_1^2 \partial x_3^2} J + \frac{\partial^2}{\partial x_3^2} G_T$$

$$= \frac{\partial^2}{\partial x_1^2} (G_{33} - G_T) - \left( \frac{\partial^2}{\partial x_1^2} + \frac{\partial^2}{\partial x_2^2} \right) G_T + \frac{1}{c_T^2} \frac{\partial^2}{\partial t^2} G_T \tag{41}$$

$$\frac{\partial^2}{\partial x_1 \partial x_3} G_{13} = \frac{\partial^4}{\partial x_1^2 \partial x_3^2} J = \frac{\partial^2}{\partial x_1^2} (G_{33} - G_T) \tag{42}$$

$$\frac{\partial^2}{\partial x_3^2} G_{12} = \frac{\partial^2}{\partial x_2 \partial x_3} G_{13} = \frac{\partial^2}{\partial x_1 \partial x_3} G_{23}$$

$$= \frac{\partial^4}{\partial x_1 \partial x_2 \partial x_3^2} J = \frac{\partial^2}{\partial x_1 \partial x_2} (G_{33} - G_T). \tag{43}$$

Substituting Equations 41 to 43 into Equation 34 and carrying out integration by parts, we have

$$T_1(\mathbf{x}, t) = T_1^0 - \mu^2 \int_\Gamma dS(\xi) \int_0^t d\tau \left\{ \Delta u_1(\xi, \tau) \left[ \frac{\partial^2}{\partial x_1^2} (4G_{33} - 3G_T) \right. \right.$$

$$\left. - \left( \frac{\partial^2}{\partial x_1^2} + \frac{\partial^2}{\partial x_2^2} \right) G_T + \frac{1}{c_T^2} \frac{\partial^2}{\partial t^2} G_T \right] + \Delta u_2(\xi, \tau) \frac{\partial^2}{\partial x_1 \partial x_2} (4G_{33} - 3G_T) \right\}$$

$$= T_1^0 - \mu^2 \int_\Gamma dS(\xi) \int_0^t d\tau \left[ \left( \frac{\partial}{\partial \xi_1} \Delta u_1(\xi, \tau) + \frac{\partial}{\partial \xi_2} \Delta u_2(\xi, \tau) \right) \right.$$

$$\frac{\partial}{\partial x_1} (4G_{33} - 3G_T) - \left( \frac{\partial}{\partial \xi_1} \Delta u_1(\xi, \tau) \frac{\partial}{\partial x_1} + \frac{\partial}{\partial \xi_2} \Delta u_1(\xi, \tau) \frac{\partial}{\partial x_2} \right) G_T$$

$$\left. + \frac{1}{c_T^2} \frac{\partial^2}{\partial \tau^2} \Delta u_1(\xi, \tau) G_T \right] \quad (\mathbf{x} \in \Gamma). \tag{44}$$

A parallel expression for the traction component $T_2(\mathbf{x}, t)$ can be obtained by interchanging the indices 1 and 2 in Equation 44. Using the summation convention and the Greek subscripts $\alpha$ and $\beta$, which run over the indices 1 and 2, the pair of these boundary integral equations can be summarized in a relatively compact form as

$$T_\alpha(\mathbf{x}, t) = T_\alpha^0 - \mu^2 \int_\Gamma dS(\xi) \int_0^t d\tau \left[ \frac{\partial}{\partial \xi_\beta} \Delta u_\beta(\xi, \tau) \frac{\partial}{\partial x_\alpha} (4G_{33} - 3G_T) \right.$$
$$\left. - \frac{\partial}{\partial \xi_\beta} \Delta u_\alpha(\xi, \tau) \frac{\partial}{\partial x_\beta} G_T + \frac{1}{c_T^2} \Delta \ddot{u}_\alpha(\xi, \tau) G_T \right] \quad (\mathbf{x} \in \Gamma).$$

(45)

Considering that $\gamma_3 = 0$ when both $\mathbf{x}$ and $\xi$ lie on the $x_1 x_2$-plane, we have the following expressions for the first-order spatial derivatives of the Green's functions appearing in the integrand of Equation 45:

$$\frac{\partial}{\partial x_\alpha} (4G_{33} - 3G_T) = \frac{\gamma_\alpha}{4\pi\mu r^2} \left[ -12I(t - \tau, r) - 5\delta(t - \tau - r/c_T) \right.$$
$$\left. + 4p^2 \delta(t - \tau - r/c_L) - \frac{r}{c_T} \dot{\delta}(t - \tau - r/c_T) \right]$$

(46)

$$\frac{\partial}{\partial x_\beta} G_T = \frac{\gamma_\beta}{4\pi\mu r^2} \left[ -\delta(t - \tau - r/c_T) - \frac{r}{c_T} \dot{\delta}(t - \tau - r/c_T) \right].$$

(47)

Substituting these into Equation 45, carrying out integration by parts, and also making use of the identity

$$\int_\Gamma dS(\xi) \frac{1}{r} \Delta \ddot{u}_i(\xi, t - r/c) = 2\pi c \Delta \dot{u}_i(\mathbf{x}, t) - c \int_\Gamma dS(\xi) \frac{\gamma_\alpha}{r} \frac{\partial}{\partial \xi_\alpha} \Delta \dot{u}_i(\xi, t - r/c)$$

(48)

which holds when both $\mathbf{x}$ and $\xi$ lie on the $x_1 x_2$-plane (see *Tada et al.*, 2000, Appendix B, for proof), we obtain

$$T_\alpha(\mathbf{x}, t) = T_\alpha^0 + \frac{3\mu}{\pi} \int_\Gamma dS(\xi) \frac{\gamma_\alpha}{r^2} \int_1^p dv \cdot v \frac{\partial}{\partial \xi_\beta} \Delta u_\beta(\xi, t - vr/c_T)$$
$$+ \frac{5\mu}{4\pi} \int_\Gamma dS(\xi) \frac{\gamma_\alpha}{r^2} \frac{\partial}{\partial \xi_\beta} \Delta u_\beta(\xi, t - r/c_T)$$
$$- \frac{\mu}{\pi} p^2 \int_\Gamma dS(\xi) \frac{\gamma_\alpha}{r^2} \frac{\partial}{\partial \xi_\beta} \Delta u_\beta(\xi, t - r/c_L)$$
$$+ \frac{\mu}{4\pi c_T} \int_\Gamma dS(\xi) \frac{\gamma_\alpha}{r} \frac{\partial}{\partial \xi_\beta} \Delta \dot{u}_\beta(\xi, t - r/c_T)$$
$$- \frac{\mu}{4\pi} \int_\Gamma dS(\xi) \frac{\gamma_\beta}{r^2} \frac{\partial}{\partial \xi_\beta} \Delta u_\alpha(\xi, t - r/c_T)$$
$$- \frac{\mu}{2c_T} \Delta \ddot{u}_\alpha(\mathbf{x}, t) \quad (\mathbf{x} \in \Gamma).$$

(49)

In Equations 48 and 49, the partial derivative with respect to $\xi_\beta$ is supposed to affect only the first argument $\xi$ of the slip/slip rate function; it does not affect the source-receiver distance $r$ that appears in the second argument.

Equation 49, first derived by *Fukuyama and Madariaga* (1998), is the regularized boundary integral equation that defines the relationship between the traction and slip on a planar fault of 3D nature.

## 3.4. Planar Two-Dimensional In-Plane Fault

To derive a regularized version of the boundary integral equation for the planar 2D in-plane fault, it is more convenient to start from the 3D theory, described in Section 3.3, than to squarely face Equation 30 and find out a way to regularize it. Interchanging the indices 2 and 3 in Equation 44, dropping the $x_3$-dependence of the problem (equating to zero all derivatives with respect to $x_3$ or $\xi_3$), and also equating the antiplane slip component $\Delta u_3(\xi, \tau)$ to zero, we have

$$
\begin{aligned}
T(\mathbf{x}, t) &= T^0 - 4\mu^2 \int_\Gamma dS(\xi) \int_0^t d\tau \frac{\partial}{\partial \xi_1} \Delta u_1(\xi, \tau) \frac{\partial}{\partial x_1} (G_{22} - G_T) \\
&\quad - \frac{\mu^2}{c_T^2} \int_\Gamma dS(\xi) \int_0^t d\tau \frac{\partial^2}{\partial \tau^2} \Delta u_1(\xi, \tau) G_T \\
&= T^0 - 4\mu^2 \int_\Gamma d\xi_1 \int_0^t d\tau \frac{\partial}{\partial \xi_1} \Delta u_1(\xi_1, \tau) \frac{\partial}{\partial x_1} \left( G_{22}^{(2D)} - G_{33}^{(2D)} \right) \\
&\quad - \frac{\mu^2}{c_T^2} \int_\Gamma d\xi_1 \int_0^t d\tau \frac{\partial^2}{\partial \tau^2} \Delta u_1(\xi_1, \tau) G_{33}^{(2D)} \quad (\mathbf{x} \in \Gamma),
\end{aligned}
\tag{50}
$$

because $G_T$ becomes $G_{33}^{(2D)}$ when integrated over the $\xi_3$-coordinate.

The in-plane case counterpart of Equation 38, or the identity that helps to conduct an additional step of integration by parts on Equation 50, is

$$
\frac{\partial}{\partial x_1} \left( G_{22}^{(2D)} - G_{33}^{(2D)} \right) = \frac{c_T^2}{2\pi\mu} \frac{\gamma_1}{r} \frac{\partial}{\partial \tau} (t - \tau)
$$

$$
\left\{ \left[ \frac{4\gamma_2^2 - 1}{r^2} \sqrt{(t-\tau)^2 - (r/c_L)^2} \right. \right.
$$

$$
\left. + \frac{\gamma_2^2}{c_L^2} \frac{1}{\sqrt{(t-\tau)^2 - (r/c_L)^2}} \right] H(t - \tau - r/c_L)
\tag{51}
$$

$$
- \left[ \frac{4\gamma_2^2 - 1}{r^2} \sqrt{(t-\tau)^2 - (r/c_T)^2} \right.
$$

$$
\left. \left. + \frac{\gamma_2^2}{c_T^2} \frac{1}{\sqrt{(t-\tau)^2 - (r/c_T)^2}} \right] H(t - \tau - r/c_T) \right\}.
$$

Considering that $r = |x_1 - \xi_1|$, $\gamma_1/r = 1/(x_1 - \xi_1)$ and $\gamma_2 = 0$ when both $x$ and $\xi$ lie on the $x_1$-axis, we finally have

$$
T(\mathbf{x}, t) = T^0 - \frac{2\mu c_T^2}{\pi} \int_\Gamma d\xi_1 \int_0^t d\tau \frac{\partial}{\partial \xi_1} \Delta \dot{u}_1(\xi_1, \tau) \frac{t - \tau}{(x_1 - \xi_1)^3}
$$

$$
\times \left[ \sqrt{(t-\tau)^2 - \left((x_1 - \xi_1)/c_L\right)^2} \, H(t - \tau - |x_1 - \xi_1|/c_L) \right.
$$

$$
\left. - \sqrt{(t-\tau)^2 - \left((x_1 - \xi_1)/c_T\right)^2} \, H(t - \tau - |x_1 - \xi_1|/c_T) \right] \quad (52)
$$

$$
- \frac{\mu}{2\pi c_T^2} \int_\Gamma d\xi_1 \int_0^t d\tau \frac{\partial}{\partial \tau} \Delta \dot{u}_1(\xi_1, \tau)
$$

$$
\frac{H(t - \tau - |x_1 - \xi_1|/c_T)}{\sqrt{(t-\tau)^2 - \left((x_1 - \xi_1)/c_T\right)^2}} \quad (\mathbf{x} \in \Gamma).
$$

An equivalent of Equation 52 was first published by *Tada and Yamashita* (1997).

## 3.5. Isolating the Instantaneous Response Term

The topic of this section is not directly related to the regularization of hypersingular integrals, but it can provide a useful insight for the physical interpretation of the boundary integral equations.

Equation 49, which defines the relationship between the traction at location $x$ and time $t$, on the one hand, and the entire past history of slip/slip rate distribution on the planar 3D fault, on the other hand, contains, on its right-hand side, a term that equals $-\mu/(2c_T)$ times the slip rate at location $x$ and time $t$. This represents the *instantaneous response*—slip, when it occurs on a fault, has an instantaneous influence on the traction at the very location where it has occurred.

The instantaneous response does not appear explicitly in Equations 39 and 52, which deal with planar 2D faults of an antiplane and in-plane nature, respectively. To derive equivalent expressions in which instantaneous response terms are isolated, we make use of the 2D counterpart of Equation 48:

$$
\int_\Gamma d\xi_1 \int_0^t d\tau \frac{H(t - \tau - |x_1 - \xi_1|/c)}{\sqrt{(t-\tau)^2 - \left((x_1 - \xi_1)/c\right)^2}} \Delta \ddot{u}_i(\xi_1, \tau)
$$

$$
= \pi c \Delta \dot{u}_i(x_1, t) - \int_\Gamma d\xi_1 \int_0^t d\tau \frac{x_1 - \xi_1}{t - \tau} \quad (53)
$$

$$
\frac{H(t - \tau - |x_1 - \xi_1|/c)}{\sqrt{(t-\tau)^2 - \left((x_1 - \xi_1)/c\right)^2}} \frac{\partial}{\partial \xi_1} \Delta \dot{u}_i(\xi_1, \tau),
$$

which is obtained by integrating Equation 48 along the $x_2$-axis. Substitution of Equation 53 into Equation 39 yields

$$T(\mathbf{x},t) = T^0 - \frac{\mu}{2c_T}\Delta\dot{u}_3(x_1,t) - \frac{\mu}{2\pi}\int_\Gamma d\xi_1 \int_0^t d\tau \frac{\partial}{\partial\xi_1}\Delta\dot{u}_3(\xi_1,\tau)$$

$$\frac{\sqrt{(t-\tau)^2 - \left((x_1-\xi_1)/c_T\right)^2}}{(t-\tau)(x_1-\xi_1)}H(t-\tau-|x_1-\xi_1|/c_T) \quad (\mathbf{x}\in\Gamma).$$

(54)

Equation 54 was first derived by *Cochard and Madariaga* (1994). Likewise, substitution of Equation 53 into Equation 52 yields

$$T(\mathbf{x},t) = T^0 - \frac{\mu}{2c_T}\Delta\dot{u}_1(x_1,t)$$

$$- \frac{2\mu c_T^2}{\pi}\int_\Gamma d\xi_1 \int_0^t d\tau \frac{\partial}{\partial\xi_1}\Delta\dot{u}_1(\xi_1,\tau)\frac{t-\tau}{(x_1-\xi_1)^3}$$

$$\sqrt{(t-\tau)^2 - ((x_1-\xi_1)/c_L)^2}H(t-\tau-|x_1-\xi_1|/c_L)$$

$$+ \frac{\mu c_T^2}{2\pi}\int_\Gamma d\xi_1 \int_0^t d\tau \frac{\partial}{\partial\xi_1}\Delta\dot{u}_1(\xi_1,\tau)$$

(55)

$$\frac{\left[2(t-\tau)^2 - ((x_1-\xi_1)/c_T)^2\right]^2}{(t-\tau)(x_1-\xi_1)^3}$$

$$\frac{H(t-\tau-|x_1-\xi_1|/c_T)}{\sqrt{(t-\tau)^2 - ((x_1-\xi_1)/c_T)^2}} \quad (\mathbf{x}\in\Gamma).$$

## 4. SPATIOTEMPORAL DISCRETIZATION

### 4.1. Boundary Elements and Time Steps

Once the boundary integral equation(s) and the boundary condition(s) have been formulated, the next step to take is to discretize the model space boundaries (fault surface in our case) on which they are defined. In a dynamic problem, the progression of time must also be discretized.

When the model space is 2D and the model fault 1D (as in Equations 39/54 and 52/55), the fault should be divided into elements of finite length. When the model space is 3D and the fault 2D (as in Equation 49), the elements should, accordingly, be 2D.

When flat faults are dealt with, linear or planar elements suffice, but otherwise, the use of curved fault elements should be considered as a possible option. In many cases, however, fault surfaces are represented in an approximate way by meshes of linear or planar elements even when they are curved. Rectangles and triangles are the fairly common choices of such planar mesh elements in 3D problems.

The size of the mesh elements may be either homogeneous all over the fault, or variable according to different degrees of precision that are needed in different parts of the fault.

Time is also divided into steps of finite duration. The time step intervals are usually taken to be homogeneous over the whole model time, but again, the use of variable time step intervals should not be ruled out as a possible option.

## 4.2. Discretizing the Equations

When the fault surface and the flow of time have both been discretized, so have to be the unknown functions and the integral equations. We illustrate this with the example of the widely used *collocation method*, combined with what is known as the *piecewise-constant approximation* for the slip rate function (e.g., *Zhang*, 1991, 2002; *Zhang and Gross*, 1993; *Cochard and Madariaga*, 1994; *Seelig and Gross*, 1997; *Tada and Yamashita*, 1997; *Fukuyama and Madariaga*, 1998; *Kame and Yamashita*, 1999b; *Aochi et al.*, 2000b; *Tada and Madariaga*, 2001). In the following, we denote the total number of mesh elements by $N_X$, the total number of time steps by $N_T$, the surface of the $n$th mesh element by $\Gamma_n$, and the endpoint of the $q$th time step by $t_q$ (the interval covered by the $q$th time step is $t_{q-1} \leq t \leq t_q$).

Let us generalize all foregoing representations of the regularized boundary integral equations of fault dynamics and rewrite them in a unified, conceptual manner as

$$T(\mathbf{x}, t) = T^0(\mathbf{x}) + \int_\Gamma dS(\xi) \int_0^t d\tau \, \hat{K}(\mathbf{x}, t - \tau; \xi, 0) \Delta \dot{u}(\xi, \tau) \quad (\mathbf{x} \in \Gamma), \quad (56)$$

where $\hat{K}$ denotes the generalized integration kernel, which may include any number of differential operators that act on the slip rate function $\Delta \dot{u}(\mathbf{x}, t)$.

The unknown slip rate is then approximated by a function that has a constant value within each of the $N_X \times N_T$ discrete spatiotemporal elements but can be discontinuous across their borders (Figure 3). In other words, we approximate the slip rate function by a linear combination of $N_X \times N_T$ base functions, each of which takes a constant value of unity within one out of the $N_X \times N_T$ discrete spatiotemporal elements, but equals zero outside that specific element:

$$\Delta \dot{u}(\mathbf{x}, t) = \sum_{n=1}^{N_X} \sum_{q=1}^{N_T} V_{nq} \, H(\mathbf{x} \in \Gamma_n) \left[ H(t - t_{q-1}) - H(t - t_q) \right]. \quad (57)$$

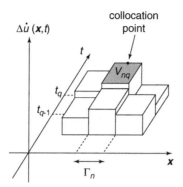

**FIGURE 3**   Piecewise-constant approximation for the slip rate function.

In Equation 57, $V_{nq}$ denotes the discretized slip rate, and $H(x \in \Gamma_n)$ denotes a function that equals unity when $x$ lies within $\Gamma_n$ and zero otherwise. The function $H(t - t_{q-1}) - H(t - t_q)$, the difference of two Heaviside step functions, equals unity when $t_{q-1} < t < t_q$ and zero otherwise. If slip does not occur on mesh element $n$ during time step $q$, a zero value should be assigned to $V_{nq}$.

Substituting Equation 57 into Equation 56, we obtain

$$T(\mathbf{x}, t) = T^0(\mathbf{x}) + \sum_{n=1}^{N_X} \sum_{q=1}^{N_T} V_{nq} \int_{\Gamma} dS(\xi) \int_0^t d\tau \hat{K}(\mathbf{x}, t - \tau; \xi, 0) \tag{58}$$

$$H(\xi \in \Gamma_n)[H(\tau - t_{q-1}) - H(\tau - t_q)] \quad (\mathbf{x} \in \Gamma).$$

In Equation 58, we have neither replaced the domain $\Gamma$ of surface integration with $\Gamma_n$, nor the interval $0 \leq \tau \leq t$ of temporal integration with $t_{q-1} \leq \tau \leq t_q$. This is because the kernel $\hat{K}$ may include differential operators that act on the $H(\cdot)$ functions.

We would like to determine the $N_X \times N_T$ unknown values of the discrete slip rate $V_{nq}$ so that Equation 58 is satisfied at $N_X \times N_T$ combinations of the location $x$ and time $t$, called *collocation points*, each one of them representing a different spatiotemporal element (Figure 3). Consideration of symmetry mandates that the $n$th spatial element $\Gamma_n$ should be collocated at its center of gravity, $x_n$. (Under the piecewise-constant approximation, the collocation point should lie in the interior of $\Gamma_n$ at any rate, because the regularization procedure relies on the premise that the first-order spatial derivative of the slip function is continuous, at least at locations where that function is evaluated [*Koller et al.*, 1992; *Tada and Yamashita*, 1997].) The flow of time, however, is not symmetrical, so the $q$th time step, $t_{q-1} \leq t \leq t_q$, need not necessarily be collocated at its midpoint. It is in fact a customary practice to collocate the $q$th time step at its endpoint $t_q$ (e.g., *Aochi et al.*, 2000b; *Cochard and Madariaga*, 1994; *Fukuyama and Madariaga*, 1998), although the possibility of other choices should not be ruled out (e.g., *Tada and Madariaga*, 2001).

Setting $x = x_m$ and $t = t_p$ in Equation 58, we have

$$T_{mp} = T_m^0 + \sum_{n=1}^{N_X} \sum_{q=1}^{p} V_{nq} K_{mp/nq}, \tag{59}$$

where

$$T_{mp} = T(\mathbf{x}_m, t_p), \quad T_m^0 = T^0(\mathbf{x}_m) \tag{60}$$

denote the discretized traction values. The summation over the time step $q$ is bounded by $p$, because the traction during time step $p$ is subject to no influence of slip that takes place in the future. The discretized integration kernel,

$$K_{mp/nq} = \int_\Gamma dS(\xi) \int_0^t d\tau \, \hat{K}(\mathbf{x}_m, t_p - \tau; \xi, 0) H(\xi \in \Gamma_n) \tag{61}$$
$$\left[ H(\tau - t_{q-1}) - H(\tau - t_q) \right],$$

may be regarded as a response function, representing the influence that a unit slip rate $V_{nq}$ on element $(n, q)$ should have on the traction $T_{mp}$ at the collocation point on element $(m, p)$. Concrete procedures to calculate $K_{mp/nq}$ shall be illustrated for a few typical cases in Section 5.

If the time steps have homogeneous intervals $\Delta t$, $K_{mp/nq}$ depends on the time indices only through their mutual difference, $p - q$:

$$K_{mp/nq} = K_{m,p-q/n0}, \tag{62}$$

whereupon Equation 59 can be rewritten as

$$T_{mp} = T_m^0 + \sum_{n=1}^{N_X} \sum_{q=1}^{p} V_{nq} K_{m,p-q/n0}. \tag{63}$$

Similar translatability does not, however, generally hold for the space indices $m$ and $n$, because the boundary elements are not necessarily laid out in space or numbered in a regular manner.

Meanwhile, Equation 12 for the boundary condition can be discretized to

$$T_{mp} = F(D_{mp}, V_{mp}) \quad \left( \mathbf{x}_m \in \Gamma(t_p) \right), \tag{64}$$

where the discretized displacement, $D_{mp} = \Delta u(x_m, t_p)$, is given by a weighted sum of past slip rate values, and equals

$$D_{mp} = \sum_{q=1}^{p} V_{mq} \Delta t \tag{65}$$

if all time step intervals are equal. The problem of solving the boundary integral Equation 56, under the boundary condition of Equation 12, has thus been reduced to the problem of defining Equations 63, 64, and 65 for all combinations of element indices $(m, p)$ on which slip is occurring and solving them simultaneously for three classes of unknown quantities, namely $T_{mp}$, $V_{mp}$, and $D_{mp}$.

## 4.3. Implicit Time-Marching Scheme

In dealing with problems of fault ruptures that evolve with time, it is customary to solve Equation 63 in a *time-marching scheme*, starting from time step one and moving successively forward to the next time step when all unknowns have been solved for in the previous time step. Let us illustrate this by separating, on the right-hand side of Equation 63, terms concerning the current time step under inspection, which is indexed $p$, and terms representing the accumulated effects of all slip that occurred before:

$$T_{mp} = T_m^0 + \sum_{n=1}^{N_X} V_{np} K_{m0/n0} + \sum_{n=1}^{N_X} \sum_{q=1}^{p-1} V_{nq} K_{m,p-q/n0}. \tag{66}$$

Separating terms in the same manner in Equation 65 and substituting the latter into Equation 64, we get

$$T_{mp} = F(V_{mp}\Delta t + D_{m,p-1}, V_{mp}) \quad \left(\mathbf{x}_m \in \Gamma(t_p)\right). \tag{67}$$

Equating the right-hand sides of Equations 66 and 67 and transposing terms, we obtain

$$\sum_{n=1}^{N_X} V_{np} K_{m0/n0} - F(V_{mp}\Delta t + D_{m,p-1}, V_{mp})$$
$$= -T_m^0 - \sum_{n=1}^{N_X} \sum_{q=1}^{p-1} V_{nq} K_{m,p-q/n0} \quad \left(\mathbf{x}_m \in \Gamma(t_p)\right). \tag{68}$$

Equation 68 is rendered in such a way that all quantities on the right-hand side are known at time step $p$. This defines a set of simultaneous equations that have to be solved for the unknown slip rate values $V_{mp}$, corresponding to all mesh elements $m$ where slip is occurring during time step $p$. The coefficients $K_{m0/n0}$ represent interactions among different mesh elements. In general, individual $V_{mp}$ cannot be obtained by simple forward calculation, but should instead be solved for with costly inverse matrix techniques (*implicit scheme*) (e.g., *Koller et al.*, 1992; *Zhang and Gross*, 1993; *Siebrits and Crouch*, 1994).

## 4.4. Courant-Friedrichs-Lewy Condition and the Explicit Time-Marching Scheme

When the time step interval $\Delta t$ is taken to be sufficiently small with respect to the mesh size, all interaction terms vanish, and the simultaneous Equation 68 decouples into a set of mutually independent equations, each of which involves only a single $V_{mp}$ (e.g., *Peirce and Siebrits*, 1996, 1997; *Seelig and Gross*, 1997; *Tada and Madariaga*, 2001; *Bhat et al.*, 2004). This significantly alleviates computational costs, as it obviates the need to solve bulky inverse matrix problems.

Let us denote by $\Delta x$ the minimum distance between the collocation point of a certain mesh element and the surface occupied by the nearest element. If, for example, the mesh consists of square elements of side length $L$ that are laid out seamlessly on a flat or nearly flat fault plane, $\Delta x$ equals $L/2$, whereas for a mesh of equilateral triangles of side length $L$ laid out seamlessly on a flat or nearly flat fault plane, $\Delta x$ equals $L/(2\sqrt{3})$. Let us also denote by $c$ the velocity of the fastest waves involved, namely, the S-wave velocity $c_T$ in the 2D antiplane problem, and the P-wave velocity $c_L$ in the 2D in-plane or the 3D problem.

If $\Delta t$ is so small that

$$c\Delta t/\Delta x \le 1, \tag{69}$$

slip that takes place on a certain mesh element $n$ during a given time step, say $t_{q-1} \le t \le t_q = t_{q-1} + \Delta t$, is unable to influence the collocation point of any other mesh element $m$ by time $t_q$, where that same time step is collocated (Figure 4). This means that all members of $K_{m0/n0}$ should vanish except when $m = n$. The only class of nonzero members, $K_{m0/m0}$, represents the instantaneous response of the traction on mesh element $m$ to a unit slip rate that takes place on that same element during the current time step, and equals $-\mu/(2c_T)$ (Section 3.5). Equation 68 is therefore simplified to

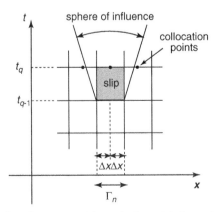

**FIGURE 4**   Slip that takes place on a certain mesh element during a given time step does not, when Equation 69 is satisfied, influence the collocation point of any other mesh element corresponding to the same time step.

$$-\frac{\mu}{2c_T}V_{mp} - F(V_{mp}\Delta t + D_{m,p-1}, V_{mp})$$

$$= -T_m^0 - \sum_{n=1}^{N_X}\sum_{q=1}^{p-1} V_{nq}K_{m,p-q/n0} \quad \left(\mathbf{x}_m \in \Gamma(t_p)\right), \tag{70}$$

which involves only a single unknown $V_{mp}$ on the left-hand side.

The quotient on the left-hand side of Equation 69, $c\Delta t/\Delta x$, is called the *Courant-Friedrichs-Lewy (CFL) number*, and is known to play an important role in a wide range of time-marching numerical schemes. Equation 69 itself is called the *CFL condition*. As will be explained in Section 7, the CFL number is also closely related to the numerical stability of the scheme.

In the special case of the traction-free boundary condition (Equation 10), in particular, Equation 70 reduces to

$$V_{mp} = \frac{2c_T}{\mu}\left[T_m^0 + \sum_{n=1}^{N_X}\sum_{q=1}^{p-1} V_{nq}K_{m,p-q/n0}\right] \quad \left(\mathbf{x}_m \in \Gamma(t_p)\right), \tag{71}$$

which means that any $V_{mp}$ can be solved for by simple forward calculation or by summing up the effects of all slip that occurred in the past (*explicit scheme*).

The triangular slip weakening friction law, described by Equation 11, also allows for an explicit scheme of solution. In fact, setting

$$F(V_{mp}\Delta t + D_{m,p-1}, V_{mp}) = T_c[1 - (V_{mp}\Delta t + D_{m,p-1})/D_c] \atop (V_{mp}\Delta t + D_{m,p-1} < D_c) \tag{72}$$

in Equation 70 and solving for $V_{mp}$, we obtain

$$V_{mp} = \frac{T_m^0 + \sum_{n=1}^{N_X}\sum_{q=1}^{p-1} V_{nq}K_{m,p-q/n0} - T_c(1 - D_{m,p-1}/D_c)}{\mu/(2c_T) - (T_c/D_c)\,\Delta t} \tag{73}$$

$$\left(V_{mp}\Delta t + D_{m,p-1} < D_c, \mathbf{x}_m \in \Gamma(t_p)\right),$$

for all slipping parts of the fault that lie in the slip weakening regime. Equation 71 should be used instead, however, once the slip on the element $(m, p)$ has exceeded the critical slip distance $D_c$.

## 5. EVALUATING DISCRETE INTEGRATION KERNELS

### 5.1. Planar Two-Dimensional Antiplane Fault

The final step we have to go through before numerical modeling becomes feasible is to evaluate the discrete integration kernels $K_{mp/nq}$, which were discussed in Section 4.2. These are defined by Equation 61 in the specific case of the collocation method combined with a piecewise-constant approximation, although other parallel expressions should be used instead if the method of discretization used is different.

Let us return to the simplest case of the 2D antiplane shear fault lying on the $x_1x_3$-plane (Figure 2a) and illustrate the concrete procedure to carry out the integral in Equation 61. Because the mesh element $\Gamma_m$ reduces in this case to a finite interval on the $x_1$-coordinate axis, say $x_1^{m-1/2} < x_1 < x_1^{m+1/2}$ centered on $x_1^m$, Equation 61 becomes

$$
K_{mp/nq} = \int_\Gamma d\xi_1 \int_0^t d\tau \, \hat{K}(x_1^m, t_p - \tau; \xi_1, 0) \Big[ H(\xi_1 - x_1^{n-1/2}) \\
- H(\xi_1 - x_1^{n+1/2}) \Big] [H(\tau - t_{q-1}) - H(\tau - t_q)],
\tag{74}
$$

where

$$
\hat{K}(x_1, t - \tau; \xi_1, 0) = -\frac{\mu}{2\pi} \frac{H(t - \tau - |x_1 - \xi_1|/c_T)}{\sqrt{(t - \tau)^2 - \left((x_1 - \xi_1)/c_T\right)^2}} \\
\left( \frac{t - \tau}{x_1 - \xi_1} \frac{\partial}{\partial \xi_1} + \frac{1}{c_T^2} \frac{\partial}{\partial \tau} \right).
\tag{75}
$$

Here and in the following, mesh element numbers such as $m$ and $n$ are written in superscripts instead of subscripts, wherever necessary, to avoid collision with coordinate axis indices such as 1 and 2.

It is, in fact, convenient to divide Equation 74 into four different parts, each corresponding to a hypothetical homogeneous unit slip rate on a "quadrant" on the spatiotemporal plane that continues to infinite farness and to perpetual future (Figure 5). Referring to the spatial translatability of the Green's function (Equation 3), we have

$$
K_{mp/nq} = \int_{-\infty}^\infty d\xi_1 \int_{-\infty}^\infty d\tau \, \hat{K}(x_1^m, t_p - \tau; \xi_1, 0) \Big[ H(\xi_1 - x_1^{n-1/2}) \\
H(\tau - t_{q-1}) - H(\xi_1 - x_1^{n-1/2}) H(\tau - t_q) \\
-H(\xi_1 - x_1^{n+1/2}) H(\tau - t_{q-1}) + H(\xi_1 - x_1^{n+1/2}) H(\tau - t_q) \Big] \\
= \int_{-\infty}^\infty d\xi_1 \int_{-\infty}^\infty d\tau \, \hat{K}(x_1^m - x_1^{n-1/2}, t_p - t_{q-1} - \tau; \xi_1, 0) \, H(\xi_1) H(\tau) \\
- \int_{-\infty}^\infty d\xi_1 \int_{-\infty}^\infty d\tau \, \hat{K}(x_1^m - x_1^{n-1/2}, t_p - t_q - \tau; \xi_1, 0) \, H(\xi_1) H(\tau) \\
- \int_{-\infty}^\infty d\xi_1 \int_{-\infty}^\infty d\tau \, \hat{K}(x_1^m - x_1^{n+1/2}, t_p - t_{q-1} - \tau; \xi_1, 0) \, H(\xi_1) H(\tau) \\
+ \int_{-\infty}^\infty d\xi_1 \int_{-\infty}^\infty d\tau \, \hat{K}(x_1^m - x_1^{n+1/2}, t_p - t_q - \tau; \xi_1, 0) \, H(\xi_1) H(\tau),
\tag{76}
$$

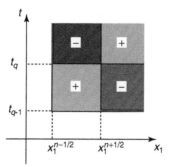

**FIGURE 5**   A homogeneous slip rate, taking place within a finite length interval and a finite time window alone, is equivalent to sums and differences of four homogeneous slip rates, each continuing to infinite farness and to perpetual future.

whereupon the evaluation of $K_{mp/nq}$ reduces to the problem of calculating four integrals of exactly parallel forms, and taking their sums and differences:

$$
K_{mp/nq} = L\left(x_1^m - x_1^{n-1/2}, t_p - t_{q-1}\right) - L\left(x_1^m - x_1^{n-1/2}, t_p - t_q\right) \\
- L\left(x_1^m - x_1^{n+1/2}, t_p - t_{q-1}\right) + L\left(x_1^m - x_1^{n+1/2}, t_p - t_q\right)
\tag{77}
$$

$$
L(x_1, t) = \int_{-\infty}^{\infty} d\xi_1 \int_{-\infty}^{\infty} d\tau \, \hat{K}(x_1, t - \tau; \xi_1, 0) \, H(\xi_1)H(\tau).
\tag{78}
$$

The function $L(x_1, t)$ represents the traction response, at an arbitrary location $x_1$ on the fault and at time $t$, due to a hypothetical homogeneous unit slip rate that starts to take place at time $t = 0$ on the entire positive part $x_1 > 0$ of the fault line and continues forever afterward. It is calculated as

$$
\begin{aligned}
L(x_1, t) = & -\frac{\mu}{2\pi} \int_{-\infty}^{\infty} d\xi_1 \int_{-\infty}^{\infty} d\tau \frac{H(t - \tau - |x_1 - \xi_1|/c_T)}{\sqrt{(t - \tau)^2 - \left((x_1 - \xi_1)/c_T\right)^2}} \\
& \left[\frac{t - \tau}{x_1 - \xi_1} \delta(\xi_1)H(\tau) + \frac{1}{c_T^2} H(\xi_1)\delta(\tau)\right] \\
= & -\frac{\mu}{2\pi} \int_0^{\infty} d\tau \frac{t - \tau}{x_1} \frac{H(t - \tau - |x_1|/c_T)}{\sqrt{(t - \tau)^2 - (x_1/c_T)^2}} \\
& -\frac{\mu}{2\pi c_T^2} \int_0^{\infty} d\xi_1 \frac{H(t - |x_1 - \xi_1|/c_T)}{\sqrt{t^2 - \left((x_1 - \xi_1)/c_T\right)^2}} \\
= & -\frac{\mu}{2\pi x_1} \int_{-\infty}^t d\tau \frac{\tau}{\sqrt{\tau^2 - (x_1/c_T)^2}} H(\tau - |x_1|/c_T) \\
& -\frac{\mu}{2\pi c_T^2} \int_{-\infty}^{x_1} d\xi_1 \frac{H(t - |\xi_1|/c_T)}{\sqrt{t^2 - (\xi_1/c_T)^2}}.
\end{aligned}
\tag{79}
$$

The first integral term on the right-hand side of Equation 79 should be evaluated on a $\tau$ interval that satisfies both $\tau \leq t$ and $\tau \geq |x_1|/c_T$. Such an interval does not exist if $t \leq |x_1|/c_T$, whereas otherwise, it is defined by $|x_1|/c_T \leq \tau \leq t$. In the second term, the range of integration is given by

$$
\begin{array}{ll}
\text{None} & \text{if either } t < 0 \text{ or } x_1 \leq -c_T t \\
-c_T t \leq \xi_1 \leq x_1 & \text{if both } t > 0 \text{ and } -c_T t \leq x_1 \leq c_T t \\
-c_T t \leq \xi_1 \leq c_T t & \text{if both } t > 0 \text{ and } c_T t \leq x_1
\end{array}
$$

(Figure 6). Therefore,

$$
\begin{aligned}
L(x_1, t) = & -\frac{\mu}{2\pi x_1} H(t - |x_1|/c_T) \int_{|x_1|/c_T}^{t} \frac{\tau \, d\tau}{\sqrt{\tau^2 - (x_1/c_T)^2}} \\
& -\frac{\mu}{2\pi c_T^2} H(t - |x_1|/c_T) \int_{-c_T t}^{x_1} \frac{d\xi_1}{\sqrt{t^2 - (\xi_1/c_T)^2}} \\
& -\frac{\mu}{2\pi c_T^2} H(x_1) \, H(t)[1 - H(t - |x_1|/c_T)] \\
& \int_{-c_T t}^{c_T t} \frac{d\xi_1}{\sqrt{t^2 - (\xi_1/c_T)^2}} \\
= & -\frac{\mu}{2\pi x_1} H(t - |x_1|/c_T)\sqrt{t^2 - (x_1/c_T)^2} \\
& +\frac{\mu}{2\pi c_T} H(t - |x_1|/c_T)\left(\arccos\frac{x_1}{c_T t} - \pi\right) \\
& -\frac{\mu}{2c_T} H(x_1) \, H(t)[1 - H(t - |x_1|/c_T)].
\end{aligned}
$$

(80)

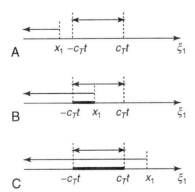

**FIGURE 6**   Range of integration for the second term on the right-hand side of Equation 79.

Noting the property

$$
\arccos\left(x_1/(c_T t)\right) =
\begin{cases}
\arccos\left(\,|x_1|/(c_T t)\right) & \text{if } x_1 \geq 0 \\
\pi - \arccos\left(\,|x_1|/(c_T t)\right) & \text{if } x_1 < 0
\end{cases}
\tag{81}
$$

that holds for the arccosine function, Equation 80 finally reduces to

$$
L(x_1, t) = -\frac{\mu}{2\pi\,|x_1|}\,\mathrm{sgn}(x_1)H(t-|x_1|/c_T)\sqrt{t^2 - (x_1/c_T)^2}
$$

$$
+\frac{\mu}{2\pi c_T}H(t-|x_1|/c_T)\left[\mathrm{sgn}(x_1)\arccos\frac{|x_1|}{c_T t} - \pi H(x_1)\right]
$$

$$
-\frac{\mu}{2c_T}H(x_1)H(t)[1 - H(t-|x_1|/c_T)]
\tag{82}
$$

$$
= -\frac{\mu}{2c_T}H(x_1)H(t) - \frac{\mu}{2\pi c_T}\left[\sqrt{(c_T t/x_1)^2 - 1} - \arccos\frac{|x_1|}{c_T t}\right]
$$
$$
\mathrm{sgn}(x_1)H(t-|x_1|/c_T),
$$

where $\mathrm{sgn}(x_1)$ denotes the signature function that equals 1 when $x_1 > 0$ and equals $-1$ when $x_1 < 0$. The first term on the right-hand side represents the instantaneous response that we discussed in Section 3.5.

An equivalent to Equation 82 was first derived by *Cochard and Madariaga* (1994). The discrete integration kernels, $K_{mp/nq}$, can be evaluated by substituting Equation 82 into Equation 77. A BIEM scheme, using this category of integration kernels, was employed in a later simulation study by *Cochard and Madariaga* (1996).

## 5.2. Planar Two-Dimensional In-Plane Fault

In the case of the 2D in-plane shear fault lying on the $x_1 x_3$-plane (Figure 2b), the integration kernel operator $\hat{K}$ should be taken as

$$
\hat{K}(x_1, t-\tau; \xi_1, 0) = -\frac{2\mu c_T^2}{\pi}\frac{t-\tau}{(x_1 - \xi_1)^3}\left[\sqrt{(t-\tau)^2 - ((x_1 - \xi_1)/c_L)^2}\right.
$$

$$
H(t-\tau-|x_1 - \xi_1|/c_L)
$$

$$
- \sqrt{(t-\tau)^2 - ((x_1 - \xi_1)/c_T)^2}
$$

$$
\left. H(t-\tau-|x_1 - \xi_1|/c_T)\right]\frac{\partial}{\partial \xi_1}
\tag{83}
$$

$$
-\frac{\mu}{2\pi c_T^2}\frac{H(t-\tau-|x_1 - \xi_1|/c_T)}{\sqrt{(t-\tau)^2 - ((x_1 - \xi_1)/c_T)^2}}\frac{\partial}{\partial \tau}.
$$

The in-plane counterpart of Equation 79 is therefore given by

$$
L(x_1, t) = -\frac{2\mu c_T^2}{\pi} \int_0^\infty d\tau \frac{t-\tau}{x_1^3} \left[ \sqrt{(t-\tau)^2 - (x_1/c_L)^2} H(t-\tau - |x_1|/c_L) \right.
$$
$$
\left. - \sqrt{(t-\tau)^2 - (x_1/c_T)^2} H(t-\tau - |x_1|/c_T) \right]
$$
$$
- \frac{\mu}{2\pi c_T^2} \int_0^\infty d\xi_1 \frac{H(t - |x_1 - \xi_1|/c_T)}{\sqrt{t^2 - \left((x_1 - \xi_1)/c_T\right)^2}}
$$
$$
= -\frac{2\mu c_T^2}{\pi x_1^3} \int_{-\infty}^t d\tau\, \tau \left[ \sqrt{\tau^2 - (x_1/c_L)^2} H(\tau - |x_1|/c_L) \right.
$$
$$
\left. - \sqrt{\tau^2 - (x_1/c_T)^2} H(\tau - |x_1|/c_T) \right]
$$
$$
- \frac{\mu}{2\pi c_T^2} \int_{-\infty}^{x_1} d\xi_1 \frac{H(t - |\xi_1|/c_T)}{\sqrt{t^2 - (\xi_1/c_T)^2}}. \tag{84}
$$

Following the same train of logic as in Section 5.1, we obtain

$$
L(x_1, t) = -\frac{2\mu c_T^2}{3\pi x_1^3} \left\{ H(t - |x_1|/c_L) \left[ t^2 - (x_1/c_L)^2 \right]^{3/2} \right.
$$
$$
\left. -H(t - |x_1|/c_T) \left[ t^2 - (x_1/c_T)^2 \right]^{3/2} \right\}
$$
$$
+ \frac{\mu}{2\pi c_T} H(t - |x_1|/c_T) \left( \arccos\frac{x_1}{c_T t} - \pi \right)
$$
$$
- \frac{\mu}{2c_T} H(x_1)H(t) \left[ 1 - H(t - |x_1|/c_T) \right] \tag{85}
$$
$$
= -\frac{\mu}{2c_T} H(x_1)H(t) - \frac{\mu}{2\pi c_T}\frac{4}{3} p^3 \left[ (c_L t/x_1)^2 - 1 \right]^{3/2}
$$
$$
\operatorname{sgn}(x_1) H(t - |x_1|/c_L)
$$
$$
+ \frac{\mu}{2\pi c_T} \left\{ \frac{4}{3} \left[ (c_T t/x_1)^2 - 1 \right]^{3/2} + \arccos\frac{|x_1|}{c_T t} \right\}
$$
$$
\operatorname{sgn}(x_1) H(t - |x_1|/c_T),
$$

where the reader is referred to Equation 24 for the definition of $p$. An equivalent to Equation 85 was first published by *Tada and Madariaga* (2001). Substituting Equation 85 into Equation 77 yields the discrete integration kernels $K_{mp/nq}$. A BIEM scheme using this category of integration kernels was employed in a simulation study by *Kame and Uchida* (2008).

## 5.3. Planar Three-Dimensional Fault

Let us go on to discuss the case of the 3D fault on the $x_1x_2$-plane (Figure 2c). Because in this case both the traction and slip have two degrees of freedom, it is appropriate to rewrite Equation 56 as

$$T_\alpha(\mathbf{x}, t) = T_\alpha^0(\mathbf{x}) + \int_\Gamma dS(\xi) \int_0^t d\tau \hat{K}_{\alpha/\beta}(\mathbf{x}, t - \tau; \xi, 0)\Delta \dot{u}_\beta(\xi, \tau) \quad (\mathbf{x} \in \Gamma),$$

(86)

where the Greek subscripts are again meant to run over 1 and 2, and summation over repeated subscripts is implied. The operator $\hat{K}_{\alpha/\beta}$ is given by

$$\begin{aligned}
\hat{K}_{\alpha/\beta}(\mathbf{x}, t - \tau; \xi, 0) = {}& \frac{3\mu}{\pi} \frac{\gamma_\alpha}{r^2} \int_1^p dv \cdot v \, H(t - \tau - vr/c_T) \frac{\partial}{\partial \xi_\beta} \\
& + \frac{5\mu}{4\pi} \frac{\gamma_\alpha}{r^2} H(t - \tau - r/c_T) \frac{\partial}{\partial \xi_\beta} \\
& - \frac{\mu}{\pi} p^2 \frac{\gamma_\alpha}{r^2} H(t - \tau - r/c_L) \frac{\partial}{\partial \xi_\beta} \\
& + \frac{\mu}{4\pi c_T} \frac{\gamma_\alpha}{r} \delta(t - \tau - r/c_T) \frac{\partial}{\partial \xi_\beta} \\
& - \frac{\mu}{4\pi} \frac{\gamma_\gamma}{r^2} H(t - \tau - r/c_T)\delta_{\alpha\beta} \frac{\partial}{\partial \xi_\gamma} \\
& - \frac{\mu}{2c_T} \delta_{\alpha\beta}\delta(\mathbf{x} - \xi)\delta(t - \tau).
\end{aligned}$$

(87)

If we choose to divide the fault into a set of rectangles with their sides parallel to the coordinate axes, it is convenient to rewrite Equation 61 as

$$\begin{aligned}
K_{\alpha klp/\beta mnq} = {}& \int_\Gamma dS(\xi) \int_0^t d\tau \, \hat{K}_{\alpha/\beta}\left(x_1^k, x_2^l, t_p - \tau; \xi_1, \xi_2, 0\right) \\
& \left[H\left(\xi_1 - x_1^{m-1/2}\right) - H\left(\xi_1 - x_1^{m+1/2}\right)\right] \\
& \times \left[H\left(\xi_2 - x_2^{n-1/2}\right) - H\left(\xi_2 - x_2^{n+1/2}\right)\right] \\
& \left[H\left(\tau - t_{q-1}\right) - H\left(\tau - t_q\right)\right],
\end{aligned}$$

(88)

where a pair of two symbols, $kl$ or $mn$, is now used, instead of just one, to designate a single mesh element.

Following the same train of logic as in Section 5.1, we can divide Equation 88 into eight different parts, each corresponding to a hypothetical homogeneous unit slip rate in an "octant" of the $x$-$t$ parameter space, which is spread over a quadrant on the $x_1x_2$-plane and lasts to eternal future:

$$
\begin{aligned}
K_{\alpha klp/\beta mnq} = {} & \int_{\Gamma} dS(\xi) \int_{0}^{t} d\tau \, \hat{K}_{\alpha/\beta}\left(x_1^k, x_2^l, t_p - \tau; \xi_1, \xi_2, 0\right) \\
& \times \Big[ H\left(\xi_1 - x_1^{m-1/2}\right) H\left(\xi_2 - x_2^{n-1/2}\right) H\left(\tau - t_{q-1}\right) \\
& - H\left(\xi_1 - x_1^{m-1/2}\right) H\left(\xi_2 - x_2^{n-1/2}\right) H\left(\tau - t_q\right) \\
& - H\left(\xi_1 - x_1^{m-1/2}\right) H\left(\xi_2 - x_2^{n+1/2}\right) H\left(\tau - t_{q-1}\right) \\
& + H\left(\xi_1 - x_1^{m-1/2}\right) H\left(\xi_2 - x_2^{n+1/2}\right) H\left(\tau - t_q\right) \\
& - H\left(\xi_1 - x_1^{m+1/2}\right) H\left(\xi_2 - x_2^{n-1/2}\right) H\left(\tau - t_{q-1}\right) \\
& + H\left(\xi_1 - x_1^{m+1/2}\right) H\left(\xi_2 - x_2^{n-1/2}\right) H\left(\tau - t_q\right) \\
& + H\left(\xi_1 - x_1^{m+1/2}\right) H\left(\xi_2 - x_2^{n+1/2}\right) H\left(\tau - t_{q-1}\right) \\
& - H\left(\xi_1 - x_1^{m+1/2}\right) H\left(\xi_2 - x_2^{n+1/2}\right) H\left(\tau - t_q\right) \Big] \qquad (89) \\
= {} & L_{\alpha/\beta}\left(x_1^k - x_1^{m-1/2}, x_2^l - x_2^{n-1/2}, t_p - t_{q-1}\right) \\
& - L_{\alpha/\beta}\left(x_1^k - x_1^{m-1/2}, x_2^l - x_2^{n-1/2}, t_p - t_q\right) \\
& - L_{\alpha/\beta}\left(x_1^k - x_1^{m-1/2}, x_2^l - x_2^{n+1/2}, t_p - t_{q-1}\right) \\
& + L_{\alpha/\beta}\left(x_1^k - x_1^{m-1/2}, x_2^l - x_2^{n+1/2}, t_p - t_q\right) \\
& - L_{\alpha/\beta}\left(x_1^k - x_1^{m+1/2}, x_2^l - x_2^{n-1/2}, t_p - t_{q-1}\right) \\
& + L_{\alpha/\beta}\left(x_1^k - x_1^{m+1/2}, x_2^l - x_2^{n-1/2}, t_p - t_q\right) \\
& + L_{\alpha/\beta}\left(x_1^k - x_1^{m+1/2}, x_2^l - x_2^{n+1/2}, t_p - t_{q-1}\right) \\
& - L_{\alpha/\beta}\left(x_1^k - x_1^{m+1/2}, x_2^l - x_2^{n+1/2}, t_p - t_q\right)
\end{aligned}
$$

$$
L_{\alpha/\beta}(x_1, x_2, t) = \int_{-\infty}^{\infty} d\xi_1 \int_{-\infty}^{\infty} d\xi_2 \int_{-\infty}^{\infty} d\tau \, \hat{K}_{\alpha/\beta}(x_1, x_2, t - \tau; \xi_1, \xi_2, 0) \, H(\xi_1) H(\xi_2) H(\tau). \qquad (90)
$$

The function $L_{\alpha/\beta}(x_1, x_2, t)$ represents the $x_\alpha$-component of the traction response, at an arbitrary location $(x_1, x_2)$ on the fault and at time $t$, due to a hypothetical homogeneous unit slip rate in the $x_\beta$-direction that begins to take place at time $t = 0$ on the entire first quadrant $x_1, x_2 > 0$ of the fault plane and continues forever afterward. After some algebra, it is calculated as

$$
\begin{aligned}
L_{1/1}(x_1, x_2, t) = {} & -\frac{\mu}{2c_T} H(x_1) H(x_2) H(t) \\
& - \frac{\mu c_T^2}{3\pi x_1^3} \left[ g_1(x_1, x_2, c_L) - g_1(x_1, x_2, c_T) \right] \\
& - \frac{\mu}{4\pi x_2} g_2(x_2, x_1, c_T) + \frac{\mu}{4\pi c_T} \left[ g_3(x_1, x_2, c_T) \right. \\
& \left. + g_3(x_2, x_1, c_T) \right]
\end{aligned} \qquad (91)
$$

$$L_{2/1}(x_1, x_2, t) = +\frac{\mu c_T^2}{6\pi x_r^3}\left(t^3 - \frac{3x_r^2}{c_L^2}t + \frac{2x_r^3}{c_L^3}\right)H(t - x_r/c_L)$$

$$-\frac{\mu c_T^2}{6\pi x_r^3}\left(t^3 - \frac{3x_r^2}{2c_T^2}t + \frac{x_r^3}{2c_T^3}\right)H(t - x_r/c_T),\qquad (92)$$

where

$$g_1(x_1, x_2, c) = \left[t^2 - (x_1/c)^2\right]^{3/2}\left[2H(x_2)\,H(t - |x_1|/c)\right.$$
$$- \mathrm{sgn}(x_2)H(t - x_r/c)\Big]$$
$$+ \left[\frac{x_2(2x_r^2 + x_1^2)}{2x_r^3}t^3 - \frac{3x_1^2 x_2}{2c^2 x_r}t\right]H(t - x_r/c) \qquad (93)$$

$$g_2(x_1, x_2, c) = \sqrt{t^2 - (x_1/c)^2}[2H(x_2)\,H(t - |x_1|/c) - \mathrm{sgn}(x_2)H(t - x_r/c)]$$
$$+ \frac{x_2}{x_r}t\,H(t - x_r/c) \qquad (94)$$

$$g_3(x_1, x_2, c) = \arctan\frac{\sqrt{t^2 - (x_1/c)^2}}{x_1/c}[2H(x_2)\,H(t - |x_1|/c)$$
$$- \mathrm{sgn}(x_2)H(t - x_r/c)] + \arctan\frac{x_2}{x_1}H(t - x_r/c) \qquad (95)$$

and

$$x_r = \sqrt{x_1^2 + x_2^2}. \qquad (96)$$

The function $L_{2/2}$ is symmetric in form to $L_{1/1}$, whereas $L_{1/2}$ is equal to $L_{2/1}$.

Equivalents to Equations 91 and 92 were first derived by *Fukuyama and Madariaga* (1998) and were later rewritten in a simpler form by *Tada* (2005). The discrete integration kernels $K_{\alpha klp/\beta mnq}$ can be obtained by substituting Equations 91 and 92 into Equation 89. A BIEM scheme, using this category of integration kernels, was employed in later simulation studies by *Fukuyama and Madariaga* (2000), *Fukuyama and Olsen* (2002), *Fukuyama et al.* (2002, 2003), and *Aochi and Ide* (2004).

## 5.4. Interface with the Two-Dimensional Theory

The consistency between the above-described 3D and 2D formulations can be checked out in a simple manner as we outline next.

Suppose that the entire positive-$x_1$ part of the $x_1x_2$-plane is subject to a homogeneous unit slip rate in the $x_2$-direction, starting at time 0 and lasting forever afterward. Such a situation is equivalent to the sum of a similar slip that takes place on the first quadrant $x_1, x_2 > 0$ and another such slip that takes place on the fourth quadrant $x_1 > 0$, $x_2 < 0$. By rotating the coordinate axes, it is easy to see that the $+x_2$-component traction at location $(x_1, x_2)$ due to a slip in the $+x_2$-direction on the fourth quadrant is equivalent to the $-x_1$-component traction at location $(-x_2, x_1)$ due to a slip in the $-x_1$-direction on the first quadrant (Figure 7). The total $x_2$-component traction response at $(x_1, x_2)$ due to the slip on the said half-plane should therefore be given by

$$L_{2/2}(x_1, x_2, t) + L_{1/1}(-x_2, x_1, t). \tag{97}$$

This gives an expression that is equal to the right-hand side of Equation 82, the corresponding integration kernel derived from the 2D antiplane theory.

Likewise, the $x_1$-component traction response at location $(x_1, x_2)$ due to an everlasting homogeneous unit slip rate in the $x_1$-direction on the entire positive-$x_1$ part of the $x_1x_2$-plane should be given by

$$L_{1/1}(x_1, x_2, t) + L_{2/2}(-x_2, x_1, t). \tag{98}$$

This gives an expression that coincides with the right-hand side of Equation 85, the corresponding integration kernel derived from the 2D in-plane theory.

## 6. DEALING WITH NONPLANAR FAULTS

### 6.1. Overview

In the foregoing sections, we dealt with planar faults alone to illustrate how the BIEM can be adapted to concrete problems of fault dynamics. The use of the BIEM, in fact, can be extended to faults of any arbitrary geometry. In this section, we give a brief review of how this can be done and point out a number of major differences with the planar-fault theory.

A difficulty arises when you try to regularize Equation 8. What we want to do is to take one order of derivative off the Green's function $G_{rp}(\mathbf{x}, t - \tau; \xi, 0)$ and, through integration by parts, move that one order of derivative onto the slip function $\Delta u_i(\xi, \tau)$. In Equation 8, the derivative of the Green's function is defined in terms of the *global Cartesian coordinate system*, with its axis

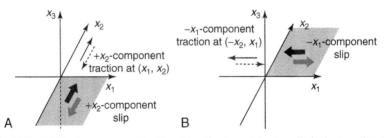

**FIGURE 7** (a) The $+x_2$-component traction at location $(x_1, x_2)$ due to a slip in the $+x_2$-direction on the fourth quadrant. (b) The $-x_1$-component traction at location $(-x_2, x_1)$ due to a slip in the $-x_1$-direction on the first quadrant, which is equivalent to (a).

directions ($x_1$, $x_2$, and $x_3$) fixed over the whole model space. The slip function $\Delta u_i$, by contrast, is differentiable only along directions that are locally tangential to the fault surface, which may change orientation from place to place and does not always agree with the fixed directions of the global coordinate axes. Given this discrepancy, it is not necessarily obvious how the integration by parts can be conducted.

*Tada and Yamashita* (1997) and *Tada et al.* (2000) got over this question by resorting to the concept of a *local Cartesian coordinate system*, say with axes called $x_n$, $x_s$, and $x_t$, which is defined only on the surface of the fault and rotates in accordance with its curvature. One of the three orthogonal axes, $x_n$, is always held normal to the fault surface, whereas the other two, $x_s$ and $x_t$, are always tangential (Figure 8). Derivatives with respect to $x_1$, $x_2$, and $x_3$, on the right-hand side of Equation 8, are rewritten, by way of simple coordinate rotation, into expressions that involve derivatives with respect to $x_n$, $x_s$, and $x_t$. Terms that involve derivatives with respect to $x_n$ are again rewritten, by way of formulae similar to Equation 35 or Equations 41 to 43, into terms that involve derivatives with respect to $x_s$ and $x_t$. Integration by parts then takes away one order of derivative with respect to $x_s$ or $x_t$ off each term in the integration kernel and reattaches that to the slip function.

Simply stated, this is the gist of what *Tada and Yamashita* (1997) and *Tada et al.* (2000) did to derive regularized boundary integral equations that relate dynamic slip and traction on nonplanar faults. The final expressions they derived are lengthy and cumbersome.

In practice, however, one does not necessarily have to rely on those bulky expressions to numerically model the dynamics of nonplanar faults with the BIEM. A major simplification results when one approximates a nonplanar fault with a mesh of small, planar fault elements. The dynamic interactions

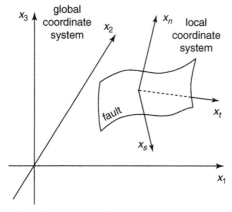

**FIGURE 8**   Global and local Cartesian coordinate systems.

among different parts of a single nonplanar fault then reduce to dynamic inter-actions among different fault elements, each of them planar in geometry.

The discrete integration kernels $K_{mp/nq}$ for planar faults, which we have described in Section 5, cannot, however, be directly brought to use in the numerical modeling of nonplanar faults. They represented the traction response, which should arise on the $m$th part of a planar fault, due to slip that takes place on the $n$th part of the same planar fault (Figure 9a). What we need in the present case, by contrast, is the traction response on one planar fault element, which is numbered $m$, due to slip that takes place on another planar fault element, which is numbered $n$. Unlike in the case of Section 5, the loca-tion on fault element $m$, at which we want to evaluate the traction, does not necessarily lie on the plane of fault element $n$. In addition, the surface of fault element $m$, on which shear traction is defined, is not necessarily oriented par-allel to fault element $n$ (Figure 9b).

How, then, can the response functions $K_{mp/nq}$ be evaluated in the present case? This shall be the subject of the next section.

## 6.2. Evaluating Discrete Integration Kernels

In Section 5, the global Cartesian coordinate system was defined in such a way that the fault in question lies on the $x_1x_2$- or the $x_1x_3$-plane. In the nonpla-nar fault problem, however, there is generally no obvious way to define the global coordinate system, which means that any choice can be fine. Instead, we make use of a local Cartesian coordinate system, which is defined differ-ently for every fault element $n$ so that the latter is parallel to two out of the three coordinate axes.

From here on, we have slightly different stories for the 2D and 3D situa-tions. In the 2D setting, the antiplane $x_3$-axis does not need to be redefined. The system of local coordinates $(x_1^n, x_2^n, x_3)$ should be simply so defined that fault element $n$ is parallel to the $x_1^n$-axis. In the 3D setting, all three coordi-nate axes must generally be redefined. The system of local coordinates $(x_1^n, x_2^n, x_3^n)$ should be so defined that fault element $n$ is parallel to the

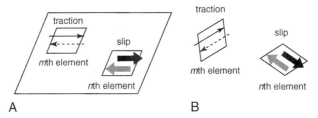

**FIGURE 9**    (a) Traction response on the $m$th part of a planar fault due to slip that takes place on the $n$th part of the same planar fault. (b) Traction response on planar fault element $m$ due to slip that takes place on planar fault element $n$. The two elements do not necessarily lie on the same plane or have identical orientations.

$x_1{}^n x_2{}^n$-plane. In either case, the global and local coordinate systems should be mutually interchangeable by way of a simple rotational transform, plus an optional lateral translation.

For the moment, we restrict ourselves to discussing the 3D situation, because how the story can be adapted to the 2D case should be fairly easy to envisage by way of analogy. Suppose we want to calculate $K_{\alpha mp/\beta nq}$, the shear traction in the $x_\alpha{}^m$-direction on fault element $m$ at time step $p$ in response to slip in the $x_\beta{}^n$-direction on fault element $n$ during time step $q$. Let $e_1{}^n$, $e_2{}^n$, and $e_3{}^n$ be the unit vectors, written in terms of the global coordinate system, which denote the orientations of the $x_1{}^n$-, $x_2{}^n$-, and $x_3{}^n$-axes, respectively (Figure 10).

For every possible combination of fault elements $m$ and $n$, and also for every possible difference $p - q$ in the time step indices, we calculate the traction response function $K_{\alpha mp/\beta nq}$ in the following, five-step manner.

*Step 1.* Calculate how the geometry and location of fault element $n$ should be described when seen in the local $x_1{}^n x_2{}^n x_3{}^n$-coordinate system. When fault element $n$ has a polygonal shape, for example, the coordinates of its apices in the $x_1{}^n x_2{}^n x_3{}^n$-coordinate system must be identified.

*Step 2.* Calculate what coordinates the collocation point on fault element $m$ should have when seen in the $x_1{}^n x_2{}^n x_3{}^n$-coordinate system. Let us denote this by a position vector, $x^{m/n} = (x_1{}^{m/n}, x_2{}^{m/n}, x_3{}^{m/n})$.

*Step 3.* Calculate how the unit tangential and normal vectors on fault element $m$, namely $e_\alpha{}^m$ and $e_3{}^m$ (Figure 10), should each be written when seen in the $x_1{}^n x_2{}^n x_3{}^n$-coordinate system. Let us denote these by vectors $e_\alpha{}^{m/n}$ and $e_3{}^{m/n}$, respectively. These can be obtained fairly straightforwardly by taking vector products:

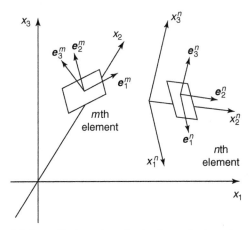

**FIGURE 10**  The global coordinate system, the local coordinate system pertaining to fault element $n$, and unit vectors tangential and normal to fault elements $m$ and $n$.

$$e_\alpha^{m/n} = \begin{pmatrix} e_\alpha^m \bullet e_1^n \\ e_\alpha^m \bullet e_2^n \\ e_\alpha^m \bullet e_3^n \end{pmatrix}, e_3^{m/n} = \begin{pmatrix} e_3^m \bullet e_1^n \\ e_3^m \bullet e_2^n \\ e_3^m \bullet e_3^n \end{pmatrix}. \tag{99}$$

*Step 4.* Calculate the tensorial stress $\sigma^{mp/\beta nq}$, in the framework of the $x_1^n x_2^n x_3^n$-coordinate system, that should arise at location $x^{m/n}$ and time $t_p$ in response to a homogeneous unit slip rate that takes place in the $x_\beta^n$-direction on fault element $n$ during time step $q$. The geometry and location of fault element $n$ in the $x_1^n x_2^n x_3^n$-coordinate system, an essential piece of information here, has already been obtained in step 1. Because fault element $n$ is parallel to the $x_1^n x_2^n$-plane, the calculation of $\sigma^{mp/\beta nq}$ should involve a string of mathematical procedures that are somewhat more complicated than, but nevertheless basically similar to, what we described in Section 5.

*Step 5.* The function $K_{\alpha mp/\beta nq}$, or the magnitude of shear traction acting in the $x_\alpha^m$-direction on fault element $m$, is obtained by taking the product of (1) the unit normal vector to fault element $m$ transposed, (2) the stress tensor at location $x^{m/n}$, and (3) the unit tangential vector to fault element $m$, namely

$$K_{\alpha mp/\beta nq} = \left( e_3^{m/n} \right)^T \left( \sigma^{mp/\beta nq} \right) \left( e_\alpha^{m/n} \right) = \left( \sigma^{mp/\beta nq} \right)_{ij} \left( e_3^{m/n} \right)_i \left( e_\alpha^{m/n} \right)_j. \tag{100}$$

When there are $N_X$ fault elements and $N_T$ time steps in total, step 1 must be followed $N_X$ times, steps 2 and 3 $N_X^2$ times each, and steps 4 and 5 must each be conducted $N_X^2 \times N_T$ times.

## 6.3. Inventory of Available Stress Response Functions

In the aforementioned five-step procedure, step 4, where the stress response tensor $\sigma^{mp/\beta nq}$ has to be calculated, represents the most demanding part. In a number of typical cases, however, analytical expressions of $\sigma^{mp/\beta nq}$ are available.

### 6.3.1. Linear Fault Element in a Two-Dimensional Medium (Figure 11a)

*Tada and Madariaga* (2001) calculated all components of the stress tensor that should arise, at any location and time in a 2D homogeneous medium, in response to a homogeneous unit slip rate which takes place (1) either in the antiplane ($x_3$-) or the in-plane ($x_1$-) direction, (2) on the entire positive part of the $x_1$-axis, and (3) starting at time $t = 0$ and lasting forever afterward. Let us call the antiplane and in-plane such stress response tensors $T^3(x_1, x_2, t)$ and $T^1(x_1, x_2, t)$, respectively.

By following the same logic as in Equations 76 to 78, the stress response tensors $\sigma^{mp/3nq}$ and $\sigma^{mp/1nq}$ at location $x^{m/n} = (x_1^{m/n}, x_2^{m/n})$ and time $t_p$, corresponding to a similar slip that is localized on a linear fault element $n$

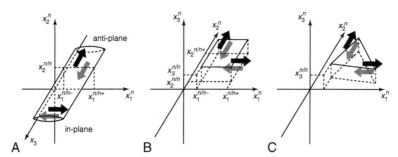

**FIGURE 11**   Categories of situations for which analytical expressions are available for the stress response tensor. (a) Linear fault element in a 2D medium. (b) Rectangular fault element in a 3D medium. (c) Triangular fault element in a 3D medium.

and a finite time step $q$, can be obtained by taking sums and differences of four similar expressions of $T^3$ and $T^1$. If the location of fault element $n$ is given by $x_1^{n/n-} < x_1^n < x_1^{n/n+}$ and $x_2^n = x_2^{n/n}$ in the local $x_1^n x_2^n x_3$-coordinate system and if time step $q$ covers the interval $t_{q-1} < t < t_q$, we have

$$
\begin{aligned}
\sigma^{mp/3nq} = \ & T^3 \left( x_1^{m/n} - x_1^{n/n-}, x_2^{m/n} - x_2^{n/n}, t_p - t_{q-1} \right) \\
& - T^3 \left( x_1^{m/n} - x_1^{n/n-}, x_2^{m/n} - x_2^{n/n}, t_p - t_q \right) \\
& - T^3 \left( x_1^{m/n} - x_1^{n/n+}, x_2^{m/n} - x_2^{n/n}, t_p - t_{q-1} \right) \\
& + T^3 \left( x_1^{m/n} - x_1^{n/n+}, x_2^{m/n} - x_2^{n/n}, t_p - t_q \right).
\end{aligned}
\tag{101}
$$

A parallel relationship holds for $\sigma^{mp/1nq}$ and $T^1$.

### 6.3.2. Rectangular Fault Element in a Three-Dimensional Medium (Figure 11b)

*Aochi et al.* (2000b) derived expressions for all stress components that should arise, at any location and time in a 3D homogeneous medium, in response to a homogeneous unit slip rate which takes place (1) either in the $x_1$- or the $x_2$-direction, (2) on the entire first quadrant $x_1$, $x_2 > 0$ of the $x_1x_2$-plane, and (3) starting at time $t = 0$ and continuing forever after. *Tada* (2005, 2006) later rewrote their expressions in a simpler form. Let $T^\beta(x_1, x_2, x_3, t)$ denote such a stress response tensor corresponding to slip in the $x_\beta$-direction.

In exactly the same way as in the flow of Equations 88 to 90, the stress response tensor, corresponding to a similar slip that is localized on a rectangular fault element $n$ and a finite time step $q$, can be obtained by taking sums and differences of eight similar expressions of $T^\beta$. If the location of fault element $n$ is given by $x_1^{n/n-} < x_1^n < x_1^{n/n+}$, $x_2^{n/n-} < x_2^n < x_2^{n/n+}$, and $x_3^n = x_3^{n/n}$ in the local $x_1^n x_2^n x_3^n$-coordinate system, and if time step $q$ covers the interval $t_{q-1} < t < t_q$, the corresponding stress response tensor $\sigma^{mp/\beta nq}$ at location $x^{m/n} = (x_1^{m/n}, x_2^{m/n}, x_3^{m/n})$ and time $t_p$ is given by

$$\boldsymbol{\sigma}^{mp/\beta nq} = \boldsymbol{T}^{\beta}\left(x_1^{m/n} - x_1^{n/n-}, x_2^{m/n} - x_2^{n/n-}, x_3^{m/n} - x_3^{n/n}, t_p - t_{q-1}\right)$$
$$- \boldsymbol{T}^{\beta}\left(x_1^{m/n} - x_1^{n/n-}, x_2^{m/n} - x_2^{n/n-}, x_3^{m/n} - x_3^{n/n}, t_p - t_q\right)$$
$$- \boldsymbol{T}^{\beta}\left(x_1^{m/n} - x_1^{n/n-}, x_2^{m/n} - x_2^{n/n+}, x_3^{m/n} - x_3^{n/n}, t_p - t_{q-1}\right)$$
$$+ \boldsymbol{T}^{\beta}\left(x_1^{m/n} - x_1^{n/n-}, x_2^{m/n} - x_2^{n/n+}, x_3^{m/n} - x_3^{n/n}, t_p - t_q\right)$$
$$- \boldsymbol{T}^{\beta}\left(x_1^{m/n} - x_1^{n/n+}, x_2^{m/n} - x_2^{n/n-}, x_3^{m/n} - x_3^{n/n}, t_p - t_{q-1}\right)$$
$$+ \boldsymbol{T}^{\beta}\left(x_1^{m/n} - x_1^{n/n+}, x_2^{m/n} - x_2^{n/n-}, x_3^{m/n} - x_3^{n/n}, t_p - t_q\right)$$
$$+ \boldsymbol{T}^{\beta}\left(x_1^{m/n} - x_1^{n/n+}, x_2^{m/n} - x_2^{n/n+}, x_3^{m/n} - x_3^{n/n}, t_p - t_{q-1}\right)$$
$$- \boldsymbol{T}^{\beta}\left(x_1^{m/n} - x_1^{n/n+}, x_2^{m/n} - x_2^{n/n+}, x_3^{m/n} - x_3^{n/n}, t_p - t_q\right). \tag{102}$$

### 6.3.3. Triangular Fault Element in a Three-Dimensional Medium (Figure 11c)

*Tada* (2006) obtained analytical expressions for all stress components that should arise, at an arbitrary location and time in a 3D homogeneous medium, in response to a homogeneous unit slip rate which takes place (1) either in the $x_1$- or the $x_2$-direction, (2) on a triangular patch on the $x_1x_2$-plane, and (3) beginning at time $t = 0$ and lasting perpetually. Let $\boldsymbol{T}^{\beta}(x_1, x_2, x_3, t)$ denote such a stress response tensor corresponding to slip in the $x_{\beta}$-direction.

The stress response tensor $\boldsymbol{\sigma}^{mp/\beta nq}$, corresponding to a similar slip that occurs over a finite time step $q$ on a triangular fault element $n$, is obtained by taking the difference of two similar expressions of $\boldsymbol{T}^{\beta}$, relevant to the projection of fault element $n$ onto the $x_1{}^n x_2{}^n$-plane. If fault element $n$ is elevated from the $x_1{}^n x_2{}^n$-plane by a height of $x_3{}^{n/n}$ in the $x_3{}^n$-direction, the corresponding stress response tensor $\boldsymbol{\sigma}^{mp/\beta nq}$, at location $\boldsymbol{x}^{m/n} = (x_1{}^{m/n}, x_2{}^{m/n}, x_3{}^{m/n})$ and time $t_p$, is related to $\boldsymbol{T}^{\beta}$ by

$$\boldsymbol{\sigma}^{mp/\beta nq} = \boldsymbol{T}^{\beta}\left(x_1^{m/n}, x_2^{m/n}, x_3^{m/n} - x_3^{n/n}, t_p - t_{q-1}\right)$$
$$- \boldsymbol{T}^{\beta}\left(x_1^{m/n}, x_2^{m/n}, x_3^{m/n} - x_3^{n/n}, t_p - t_q\right). \tag{103}$$

## 6.4. Numerical Modeling Studies in the Literature

Simulation studies on the dynamics of nonplanar faults using the BIEM began with the pioneering work of *Koller et al.* (1992), who presented preliminary results for a kinking fault in the 2D antiplane mode. This was followed by the works listed here:

1. *Yamashita and Fukuyama* (1996) and *Kame and Yamashita* (1997) on the dynamic interactions of 2D, antiplane, mutually parallel but noncoplanar faults

2. *Seelig and Gross* (1997, 1999a, 1999b), *Kame and Yamashita* (1999a, 1999b, 2003), *Kame et al.* (2003), *Ando et al.* (2004, 2007), *Bhat et al.* (2004, 2007), *Fliss et al.* (2005), and *Ando and Yamashita* (2007) on the dynamics of 2D, in-plane, and nonplanar faults

3. *Aochi et al.* (2000a, 2000b, 2002, 2003, 2005, 2006), *Aochi and Fukuyama* (2002), *Aochi and Madariaga* (2003), and *Aochi and Olsen* (2004) on the dynamics of 3D nonplanar faults using rectangular elements

4. *Fukuyama and Mikumo* (2006) on the dynamics of 3D nonplanar faults using triangular mesh elements

## 7. NUMERICAL STABILITY

The last problem we should be aware of in the BIEM modeling of fault dynamics is that of numerical stability. It is widely recognized that time-marching BIEM schemes are often susceptible to *numerical instability*, which manifests itself as noise of an oscillatory nature that emerges on top of the solution at a certain stage during its time evolution (e.g., *Koller et al.*, 1992; *Siebrits and Crouch*, 1994). The noise, which begins as a mild, undulating pattern of highs and lows that alternate in time and space, builds up fairly quickly and finally blows up to infinity.

The origin and nature of numerical instability are still under investigation (e.g., *Peirce and Siebrits*, 1996, 1997; *Birgisson et al.*, 1999; *Frangi and Novati*, 1999; *Tada and Madariaga*, 2001). Different authors have suppressed the onset of instability by introducing an artificial dissipation term (e.g., *Koller et al.*, 1992; *Yamashita and Fukuyama*, 1996) or a slip weakening friction law (e.g., *Andrews*, 1994; *Fukuyama and Madariaga*, 1998). Such measures did help to delay the emergence of instability but failed to provide a fundamental remedy. This implies that instability cannot be attributed alone to the inadequate modeling of the physics of rupture—unstable oscillations persisted even when appropriate account was taken of the finite length scale, which is an essential feature of fault friction laws (e.g., *Rice*, 1993). It is also suggested that instability does not depend on specific slip histories (e.g., *Peirce and Siebrits*, 1996, 1997; *Tada and Madariaga*, 2001).

The emergence of instability tends to be conditioned by the choice of grid parameters, particularly the CFL number (Section 4.4). In any numerical run, therefore, one has to carefully choose the CFL number so as to avoid the buildup of numerical instability. In the literature, the CFL number is usually defined as $c\Delta t/L$, where $L$ is the spatial grid size, but the reader is reminded that it is here defined by $c\Delta t/\Delta x$, where $\Delta x$ denotes the minimum distance from the collocation point of a certain mesh element to the surface occupied by the nearest element and equals $L/2$ in the specific case of linear or square-shaped elements.

It is computationally costly to set the CFL number in excess of unity (1.0), because an implicit solution scheme must then be used instead of the efficient explicit scheme. On the other hand, too small values of the CFL number are often prone to numerical instability. The question thus lies in finding out appropriate ranges of the CFL number that should lie in between.

In 2D modeling, the tolerable range of the CFL number tends to be fairly wide in the antiplane problem but appears fairly limited in the in-plane case. *Tada and Madariaga* (2001) conducted exhaustive numerical tests with a planar fault model to determine the acceptable range of grid parameters. When time is collocated at the endpoint of each step, as we have recommended in this chapter, they showed the BIEM scheme is apparently highly stable in the antiplane case for any CFL number $c_T \Delta t / \Delta x$ up from about 0.6 to at least 3.0, where the values have been adjusted to follow the definition of this chapter. In the in-plane case, by contrast, the scheme was stable only within narrow ranges in the neighborhood of $c_T \Delta t / \Delta x = 0.5$ and 1.0 ($c_L \Delta t / \Delta x = \sqrt{3}/2$ and $\sqrt{3}$) when $c_L = \sqrt{3} \, c_T$ was assumed. Seelig and co-workers, in the meantime, have recommended the range $1 \leq c_L \Delta t / \Delta x \leq 2$ in the modeling of 2D in-plane cracks (e.g., *Seelig and Gross*, 1997).

In 3D modeling using rectangular mesh elements, the choice of the CFL number made on a trial-and-error basis by *Fukuyama and Madariaga* (1998, 2000) was $c_T \Delta t / \Delta x = 0.5$ ($c_L \Delta t / \Delta x = \sqrt{3}/2$), whereas Aochi and coworkers preferably set the ratio $c_L \Delta t / \Delta x$ equal or sufficiently close to 1.0 (*Aochi et al.*, 2000a, 2000b, 2002, 2005, 2006; *Aochi and Fukuyama*, 2002; *Aochi and Madariaga*, 2003; *Aochi and Olsen*, 2004), where the values have again been recast in line with the definition of this chapter. *Yokoyama et al.* (2007) found that, when $c_L \Delta t / \Delta x$ was set at 0.95, their numerical scheme using triangular fault elements remained stable for at least several hundred time steps.

## 8. RELATED TOPICS

### 8.1. Fracture Criterion

Numerical modeling (not necessarily with the BIEM) of faults whose geometry evolves with time is divided into two classes: *kinematic* and *dynamic*. In kinematic modeling, the history of fault geometry evolution, or of rupture propagation, is prescribed beforehand. In dynamic modeling, the fault is made to evolve spontaneously, as part of the numerical solution, according to a given *fracture criterion*—a criterion that determines whether the tip of a crack should advance by one grid forward at a given time step or, more generally, whether a given, unruptured part of a fault should break or not under the influence of stress.

Ideally, the choice of the fracture criterion should be based on rigorous consideration of the fracture energy. The use of a criterion based on the stress intensity factor (SIF) (e.g., *Seelig and Gross*, 1997, 1999a, 1999b) is recommendable from this viewpoint. In practical applications, however, it is fairly customary to substitute this with a more easily implementable fracture criterion that follows the classical law of friction—a location on a fault should break when the shear traction there has exceeded a certain coefficient of static friction times the normal traction.

The use of this simpler criterion has been justified by the argument that it is approximately equivalent to the more rigorous, SIF-based fracture criterion as long as the spatial grid size remains small and fixed (e.g., *Das and Aki*, 1977; *Koller et al.*, 1992). When rupture involves no variation in the fault-normal traction, as in the case of an isolated planar fault, this criterion reduces to an even more convenient form, which depends alone on a yield stress, or a threshold shear traction value for rupture to occur.

A major shortcoming of this simplifying approach, however, is that the coefficient of static friction (or the yield stress) should be set differently for different grid sizes used—coarser grid spacing leads to smaller stress concentration on the grid next to the crack tip, and so the yield stress should be lowered accordingly to allow the rupture to propagate.

Another difficulty lies in the fact that, under this criterion, the velocity of rupture propagation tends to grow quickly and saturate at a limiting value as the slipping region enlarges. This happens because, for a fixed grid spacing, an expanding fault produces ever larger stress concentration on the grid next to the crack tip. When the fracture is in-plane (mode II), the modeled rupture propagation can even accelerate beyond the S-wave velocity if the yield stress is set too low (e.g., *Andrews*, 1976; *Das and Aki*, 1977; *Fukuyama and Madariaga*, 2000), although, in nature, such supershear propagation velocities are not very common.

Widely accepted remedies to these shortcomings are yet to appear. Sufficient caution should therefore be taken in using this type of fracture criterion.

## 8.2. Formulation in the Fourier and Laplace Domains

In this chapter, regularized boundary integral equations of fault dynamics have been formulated in the *time domain*, but similar boundary integral equations have also been presented in the Fourier *frequency domain* (e.g., *Budiansky and Rice*, 1979; *Nishimura and Kobayashi*, 1989; *Zhang and Achenbach*, 1989) and in the *Laplace domain* (e.g., *Sládek and Sládek*, 1984). However, the time-domain representation is more advantageous than the frequency- and Laplace-domain counterparts, because the latter can only deal with the transient response of a stationary fault that does not evolve with time.

Time- and frequency-domain formulations using the Fourier transform in space have also been presented (e.g., *Geubelle and Rice*, 1995; *Perrin et al.*, 1995; *Bouchon and Streiff*, 1997; *Lapusta et al.*, 2000). This class of formulation, however, is most suited to planar fault problems, and its application to complicated nonplanar fault geometries has yet to appear.

## 8.3. Displacement Discontinuity BIEM

Throughout this chapter, we have represented the traction on the fault as a spatiotemporal convolution of the slip (displacement discontinuity) and an appropriate integration kernel (see Equation 56). A number of studies (e.g., *Kostrov*, 1966; *Das and Aki*, 1977; *Das*, 1980; *Andrews*, 1985, 1994; *Das and Kostrov*, 1987) have adopted an alternative approach, in which, on the contrary, the displacement discontinuity on the fault is given as a spatiotemporal convolution of the traction on itself and an appropriate integration kernel:

$$\Delta u(\mathbf{x}, t) = \int_\Gamma dS(\xi) \int_0^t d\tau \, \hat{K}'(\mathbf{x}, t - \tau; \xi, 0) \, T(\xi, \tau) \quad (\mathbf{x} \in \Gamma). \tag{104}$$

The kernel, appearing in this alternative formulation, has the advantage of being free of hypersingularities. The rigorous form of this alternative expression, however, is known only for isolated planar faults.

## 8.4. Fault Opening

This chapter has so far dealt with problems of shear (mode II and mode III) fracture alone. With minimal modifications, however, it is possible to adapt the same theoretical framework to problems that include fault opening (*mode I* fracture) (e.g., *Zhang*, 1991, 2002; *Zhang and Gross*, 1993; *Geubelle and Rice*, 1995; *Seelig and Gross*, 1997, 1999a, 1999b). In the notation of our chapter, fault opening corresponds to a nonzero slip component $\Delta u_2(\xi, \tau)$ in the 2D case and a nonzero $\Delta u_3(\xi, \tau)$ in the 3D problem setting.

Displacement discontinuity on an open fault may involve components of both shear slipping and fault-normal opening. The occurrence of shear slip on a fault generally influences both shear and normal tractions on itself, whereas the same can be said of fault-normal opening. The boundary integral Equation 56 should therefore be rewritten in the present case as

$$T(\mathbf{x}, t) = T^0(\mathbf{x}) + \int_{\Gamma_{closed}+\Gamma_{open}} dS(\xi) \int_0^t d\tau \, \hat{K}^{T/shear}(\mathbf{x}, t - \tau; \xi, 0) \, \Delta \dot{u}^{shear}(\xi, \tau)$$
$$+ \int_{\Gamma_{open}} dS(\xi) \int_0^t d\tau \, \hat{K}^{T/normal}(\mathbf{x}, t - \tau; \xi, 0) \, \Delta \dot{u}^{normal}(\xi, \tau) \quad (\mathbf{x} \in \Gamma) \tag{105}$$

$$N(\mathbf{x}, t) = N^0(\mathbf{x}) + \int_{\Gamma_{closed}+\Gamma_{open}} dS(\xi) \int_0^t d\tau \, \hat{K}^{N/shear}(\mathbf{x}, t - \tau; \xi, 0) \, \Delta \dot{u}^{shear}(\xi, \tau)$$
$$+ \int_{\Gamma_{open}} dS(\xi) \int_0^t d\tau \, \hat{K}^{N/normal}(\mathbf{x}, t - \tau; \xi, 0) \, \Delta \dot{u}^{normal}(\xi, \tau) \quad (\mathbf{x} \in \Gamma), \tag{106}$$

where $N(x, t)$ accounts for the fault-normal traction and $N^0(x)$ accounts for its initial values. The symbols $\hat{K}^{T/\text{shear}}$, $\hat{K}^{T/\text{normal}}$, $\hat{K}^{N/\text{shear}}$, and $\hat{K}^{N/\text{normal}}$ each denote the generalized integration kernel that relates (1) shear slip to shear traction, (2) fault-normal opening to shear traction, (3) shear slip to normal traction, and (4) fault-normal opening to normal traction, respectively (the symbol $\hat{K}$, appearing in the main body of the present chapter, corresponds to $\hat{K}^{T/\text{shear}}$). Note that Equations 105 and 106 remain valid even when more than one fault coexist.

On an open patch of a fault, both the shear and normal tractions should be equal to zero. On a closed patch, by contrast, the shear traction $T(x, t)$ should follow a friction law, with a general, conceptual representation as in Equation 12. The boundary conditions that should be imposed are, therefore,

$$T(\mathbf{x}, t) = F\left(\Delta u^{\text{shear}}(\xi, \tau), \Delta \dot{u}^{\text{shear}}(\xi, \tau)\right) \quad (\mathbf{x} \in \Gamma_{\text{closed}}) \tag{107}$$

$$T(\mathbf{x}, t) = N(\mathbf{x}, t) = 0 \quad (\mathbf{x} \in \Gamma_{\text{open}}), \tag{108}$$

where possible dependence of Equation 107 on $N(x, t)$ has again been dropped for the sake of simplicity.

If the CFL condition (Equation 69) is satisfied, Equations 105 and 106 are discretized to

$$
\begin{aligned}
T_{mp} = T_m^0 &- \frac{\mu}{2c_T} V_{mp}^{\text{shear}} + \sum_n \sum_{q=1}^{p-1} V_{nq}^{\text{shear}} K_{m,p-q/n0}^{T/\text{shear}} \\
&+ \sum_l \sum_{q=1}^{p-1} V_{lq}^{\text{shear}} K_{m,p-q/l0}^{T/\text{shear}} + \sum_l \sum_{q=1}^{p-1} V_{lq}^{\text{normal}} K_{m,p-q/l0}^{T/\text{normal}}
\end{aligned}
\tag{109}
$$

$$
\begin{aligned}
T_{kp} = T_k^0 &- \frac{\mu}{2c_T} V_{kp}^{\text{shear}} + \sum_n \sum_{q=1}^{p-1} V_{nq}^{\text{shear}} K_{k,p-q/n0}^{T/\text{shear}} \\
&+ \sum_l \sum_{q=1}^{p-1} V_{lq}^{\text{shear}} K_{k,p-q/l0}^{T/\text{shear}} + \sum_l \sum_{q=1}^{p-1} V_{lq}^{\text{normal}} K_{k,p-q/l0}^{T/\text{normal}}
\end{aligned}
\tag{110}
$$

$$
\begin{aligned}
N_{kp} = N_k^0 &- \frac{\mu c_L}{2c_T^2} V_{kp}^{\text{normal}} + \sum_n \sum_{q=1}^{p-1} V_{nq}^{\text{shear}} K_{k,p-q/n0}^{N/\text{shear}} \\
&+ \sum_l \sum_{q=1}^{p-1} V_{lq}^{\text{shear}} K_{k,p-q/l0}^{N/\text{shear}} + \sum_l \sum_{q=1}^{p-1} V_{lq}^{\text{normal}} K_{k,p-q/l0}^{N/\text{normal}},
\end{aligned}
\tag{111}
$$

where the subscripts $m$ and $n$ are supposed to run over closed fault patches and the subscripts $k$ and $l$ are supposed to run over open fault patches. Note that the coefficients of instantaneous response, in the second term on the right-hand side, are different for shear slipping and for fault-normal opening.

The combined set of Equations 109 to 111 should be solved in a time-marching manner under the boundary conditions

$$T_{mp} = F(D_{mp}^{\text{shear}}, V_{mp}^{\text{shear}}) \quad (\mathbf{x}_m \in \Gamma_{\text{closed}}(t_p)) \tag{112}$$

$$T_{kp} = N_{kp} = 0 \quad (\mathbf{x}_k \in \Gamma_{\text{open}}(t_p)), \tag{113}$$

for three classes of unknowns, namely $V_{mp}^{\text{shear}}$ (shear slip rate on closed fault patches), $V_{kp}^{\text{shear}}$ (shear slip rate on open fault patches), and $V_{kp}^{\text{normal}}$ (fault-normal opening rate on open fault patches).

For the special case of an isolated planar fault, the occurrence of shear slip has no influence on the normal traction on the fault, whereas fault-normal opening has no influence on the shear traction. The fault-opening counterpart of the $L(x_1, t)$ function in Section 5.2, namely the normal traction response to a homogeneous unit opening rate on a semi-infinite fault in a 2D medium, is given in the work of *Tada and Madariaga* (2001). The fault-opening counterpart of the $L_{\alpha/\beta}(x_1, x_2, t)$ functions in Section 5.3, or the normal traction response to a homogeneous unit opening rate on a quadrantal fault in a 3D medium, was first published by *Aochi* (1999) and was later rewritten in a simpler form by *Tada* (2005). Using these $L$ functions, it is possible to calculate the discrete integration kernels $K_{k,p-q/l0}^{N/\text{normal}} = K_{kp/lq}^{N/\text{normal}}$ for planar 2D or 3D faults, according to the procedures described in Sections 5.1 and 5.3.

When the fault is nonplanar, the discrete integration kernels $K_{k,p-q/l0}^{T/\text{normal}}$, $K_{k,p-q/l0}^{N/\text{shear}}$, and $K_{k,p-q/l0}^{N/\text{normal}}$ should be calculated through procedures that are similar to the ones described in Sections 6.2 and 6.3. For this purpose, we need the fault-opening counterparts of the stress response tensors $T^i(x_1, x_2, t)$ that were explained in Section 6.3. Their analytical expressions are available for all three typical cases mentioned in that section—see *Tada and Madariaga* (2001) for a semi-infinite fault in a 2D medium (Figure 11a), *Aochi* (1999) for a quadrantal fault in a 3D medium (Figure 11b; later rewritten in simpler forms by *Tada*, 2005, 2006), and *Tada* (2006) for a triangular fault in a 3D medium (Figure 11c).

## 8.5. Faults in a Half-Space

With relatively minor modifications, the BIEM outlined in this chapter can be extended to account for problems of faults/cracks embedded in finite media (e.g., *Seelig and Gross*, 1999b; *Zhang*, 2002). In seismological applications, however, problems of faults embedded in a semi-infinite medium (half-space), with a flat and boundless free surface, command greater interest because of their implications for the dynamic rupture processes of shallow, damaging earthquakes.

The theory presented in this chapter can be adapted to faults in a semi-infinite medium if only we have the time-domain representation of the Green's functions of elastodynamics for a homogeneous half-space. Unlike the full space case, however, such Green's functions are hardly known in a simple, closed form, and for this reason, the treatment of faults in a half-space remains a highly challenging task. When the fault is vertical and of the strike-slip type, the presence of the free surface can be substituted, within the framework of the full-space BIEM, by placing a reflection image of the fault on the other side of the free surface (*Aochi and Fukuyama*, 2002; *Fukuyama et al.*, 2002; *Aochi and Madariaga*, 2003; *Aochi et al.*, 2003; *Fukuyama and Mikumo*, 2006). This image method, however, does not work for faults that dip at nonvertical angles or have dip-slip components.

A footstep in this challenging direction was marked by *Zhang and Chen* (2006a, 2006b, 2009) and *Chen and Zhang* (2006), who proposed a set of elaborate numerical techniques to cope with planar faults in a 3D, homogeneous half-space using the BIEM. To consolidate the path opened up by their pioneering work, follow-up studies are much awaited.

Another possible approach to this problem is to treat the free surface as if it were a huge open fault in an infinite medium of the nature discussed in Section 8.4, interacting with the shear fault in question (Figure 12a). In this approximation, horizontal displacement on the free surface should be seen

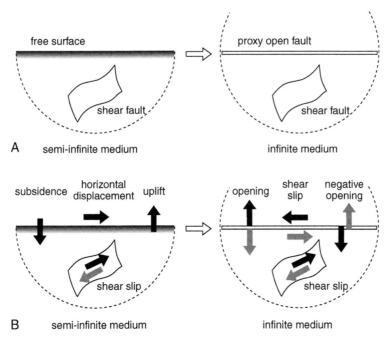

**FIGURE 12** (a) Modeling the presence of a free surface as if it were a huge open fault in an infinite medium that interacts with the shear fault under consideration. (b) Subsidence, horizontal displacement, and uplift on the free surface should each correspond to opening, shear slipping, and negative opening on the proxy open fault.

as shear slipping on the proxy open fault, and vertical subsidence on the free surface as fault-normal opening (Figure 12b). Vertical uplift on the free surface should correspond to negative opening—although such behavior would normally mean unphysical penetration of material on both sides and should therefore be prohibited, it should be tolerated in the present case where the open fault is but a convenient, imaginary substitute for a free surface. One major advantage of this proposed approach is the ability to deal with free surfaces of any arbitrary geometry—they need not necessarily be planar.

## 8.6. Galerkin Method

In Section 4.2, we illustrated how a boundary integral equation, in the general form of Equation 56, can be discretized and solved on the basis of the collocation method. The basic idea consists of determining $N_X \times N_T$ discrete values of the unknown slip function in such a way that the discrepancy (or *residual*) between the left- and right-hand sides of Equation 56 may vanish at a finite number of $N_X \times N_T$ selected combinations (collocation points) of the spatial location and time.

Although the collocation method has customarily been used in most previous studies on fault dynamics using the BIEM, work by *Goto and Bielak* (2008) relied on an alternative approach, the *Galerkin method*. In this method, it is not the values of the residual function at selected points, but its weighted integrals over appropriate domains, that are equated to zero, and the weight functions, used in evaluating those weighted residuals, are set identical to the base functions used to discretize the unknown function.

Let us denote the base functions of spatial discretization by $\phi_i(\mathbf{x})$ ($i = 1, 2, \ldots, N_X$) and express both the traction and slip rate in terms of their linear combinations:

$$T(\mathbf{x}, t) = \sum_{i=1}^{N_X} \phi_i(\mathbf{x}) \, T_i(t) \tag{114}$$

$$\Delta \dot{u}(\mathbf{x}, t) = \sum_{i=1}^{N_X} \phi_i(\mathbf{x}) \, \Delta \dot{u}_i(t). \tag{115}$$

Substituting these into Equation 56, and integrating both sides under a weight of $\phi_j(\mathbf{x})$, we have

$$\sum_{i=1}^{N_X} T_i(t) \int_\Gamma dS(\mathbf{x}) \, \phi_i(\mathbf{x}) \, \phi_j(\mathbf{x}) = \sum_{i=1}^{N_X} T_i^0 \int_\Gamma dS(\mathbf{x}) \, \phi_i(\mathbf{x}) \, \phi_j(\mathbf{x})$$
$$+ \sum_{i=1}^{N_X} \int_0^t d\tau \int_\Gamma dS(\mathbf{x}) \int_\Gamma dS(\xi) \, \hat{K}(\mathbf{x}, t - \tau; \xi, 0)$$
$$\phi_i(\xi) \, \phi_j(\mathbf{x}) \, \Delta \dot{u}_i(\tau) \quad (j = 1, 2, \ldots, N_X),$$
$$\tag{116}$$

which finally reduces to a matrix equation of the form

$$
\sum_{i=1}^{N_X} T_i(t)\Phi_{ij} = \sum_{i=1}^{N_X} T_i^0 \Phi_{ij} + \sum_{i=1}^{N_X} \int_0^t d\tau \, \Delta \dot{u}_i(\tau) \, K_{ij}(t - \tau) \quad (j = 1, 2, \ldots, N_X).
$$

(117)

*Goto and Bielak* (2008) stated that, for the problem of a planar, 2D antiplane fault, the use of the Galerkin method provided better numerical accuracy than the traditional collocation method for the same grid size, although this duly came at the expense of increased computation time.

## 9. CONCLUSION

In this chapter, we reviewed the most central concepts and procedures of the boundary integral equation method (BIEM) as applied to the modeling of fault dynamics. Advantages of the BIEM over other numerical methods include (1) smaller dimension of the problem to be solved (though this does not necessarily mean computational efficiency), (2) easy adaptability to problems that have complex boundaries (fault geometries), and (3) suitability to problems that involve unbounded domains. On the other hand, the difficulty of dealing with inhomogeneous media should be counted as one of its principal disadvantages.

Further methodological refinements and more extensive numerical implementations of the BIEM are much awaited so that the method can remain a powerful tool of modeling in the study of earthquake rupture dynamics.

## ACKNOWLEDGMENTS

Reviews by two anonymous referees and editorial comments by Eiichi Fukuyama have helped to improve the manuscript substantially.

## REFERENCES

Achenbach, J. D., (1973), *Wave Propagation in Elastic Solids*, North Holland, Amsterdam.

Aki, K. and P. G. Richards, (2002), *Quantitative Seismology, 2^{nd} Edition*, University Science Books, Sausalito.

Ando, R. and T. Yamashita, (2007), Effects of mesoscopic-scale fault structure on dynamic earthquake ruptures: Dynamic formation of geometrical complexity of earthquake faults, *J. Geophys. Res.*, **112**, B09303, doi:10.1029/2006JB004612.

Ando, R., T. Tada, and T. Yamashita, (2004), Dynamic evolution of a fault system through interactions between fault segments, *J. Geophys. Res.*, **109**, B05303, doi:10.1029/2003JB002665.

Ando, R., N. Kame, and T. Yamashita, (2007), An efficient boundary integral equation method applicable to the analysis of non-planar fault dynamics, *Earth Planets Space*, **59**, 363-373.

Andrews, D. J., (1976), Rupture velocity of plane strain shear cracks, *J. Geophys. Res.*, **81**(32), 5679-5687.

Andrews, D. J., (1985), Dynamic plane-strain shear rupture with a slip-weakening friction law calculated by a boundary integral method, *Bull. Seismol. Soc. Am.*, **75**(1), 1-21.

Andrews, D. J., (1994), Dynamic growth of mixed-mode shear cracks, *Bull. Seismol. Soc. Am.*, **84**, 1184-1198.

Aochi, H., (1999), Theoretical studies on dynamic rupture propagation along a 3D non-planar fault system, *D.Sc. thesis*, the University of Tokyo, Tokyo.

Aochi, H. and E. Fukuyama, (2002), Three-dimensional nonplanar simulation of the 1992 Landers earthquake, *J. Geophys. Res.*, **107**(B2), 2035, doi:10.1029/2000JB000061.

Aochi, H. and S. Ide, (2004), Numerical study on multi-scaling earthquake rupture, *Geophys. Res. Lett.*, **31**, L02606, doi:10.1029/2003GL018708.

Aochi, H. and R. Madariaga, (2003), The 1999 İzmit, Turkey, earthquake: Nonplanar fault structure, dynamic rupture process, and strong ground motion, *Bull. Seismol. Soc. Am.*, **93**(3), 1249-1266.

Aochi, H. and K. B. Olsen, (2004), On the effects of non-planar geometry for blind thrust faults on strong ground motion, *Pure Appl. Geophys.*, **161**, 2139-2153.

Aochi, H., E. Fukuyama, and M. Matsu'ura, (2000a), Selectivity of spontaneous rupture propagation on a branched fault, *Geophys. Res. Lett.*, **27**(22), 3635-3638.

Aochi, H., E. Fukuyama, and M. Matsu'ura, (2000b), Spontaneous rupture propagation on a nonplanar fault in 3-D elastic medium, *Pure Appl. Geophys.*, **157**, 2003-2027.

Aochi, H., R. Madariaga, and E. Fukuyama, (2002), Effect of normal stress during rupture propagation along nonplanar faults, *J. Geophys. Res.*, **107**(B2), 2038, doi:10.1029/2001JB000500.

Aochi, H., R. Madariaga, and E. Fukuyama, (2003), Constraint of fault parameters inferred from nonplanar fault modeling, *Geochem. Geophys. Geosyst.*, **4**(2), 1020, doi:10.1029/2001GC000207.

Aochi, H., O. Scotti, and C. Berge-Thierry, (2005), Dynamic transfer of rupture across differently oriented segments in a complex 3-D fault system, *Geophys. Res. Lett.*, **32**, L21304, doi:10.1029/2005GL024158.

Aochi, H., M. Cushing, O. Scotti, and C. Berge-Thierry, (2006), Estimating rupture scenario likelihood based on dynamic rupture simulations: The example of the segmented Middle Durance fault, southeastern France, *Geophys. J. Int.*, **165**, 436-446.

Beskos, D. E., (1987), Boundary element methods in dynamic analysis, *Appl. Mech. Rev.*, **40**, 1-23.

Beskos, D. E., (1997), Boundary element methods in dynamic analysis: Part II (1986-1996), *Appl. Mech. Rev.*, **50**, 149-197.

Bhat, H. S., R. Dmowska, J. R. Rice, and N. Kame, (2004), Dynamic slip transfer from the Denali to Totschunda faults, Alaska: Testing theory for fault branching, *Bull. Seismol. Soc. Am.*, **94**, S202-S213.

Bhat, H. S., M. Olives, R. Dmowska, and J. R. Rice, (2007), Role of fault branches in earthquake rupture dynamics, *J. Geophys. Res.*, **112**, B11309, doi:10.1029/2007JB005027.

Birgisson, B., E. Siebrits, and A. P. Peirce, (1999), Elastodynamic direct boundary element methods with enhanced numerical stability properties, *Int. J. Numer. Meth. Engng.*, **46**, 871-888.

Bouchon, M. and F. J. Sánchez-Sesma, (2007), Boundary integral equations and boundary elements methods in elastodynamics, *Adv. Geophys.*, **48**, 157-189.

Bouchon, M. and D. Streiff, (1997), Propagation of a shear crack on a nonplanar fault: A method of calculation, *Bull. Seismol. Soc. Am.*, **87**, 61-66.

Budiansky, B. and J. R. Rice, (1979), An integral equation for dynamic elastic response of an isolated 3-D crack, *Wave Motion*, **1**, 187-192.

Burridge, R., (1969), The numerical solution of certain integral equations with non-integrable kernels arising in the theory of crack propagation and elastic wave diffraction, *Phil. Trans. Roy. Soc. Lond., Ser. A*, **265**, 353-381.

Chen, X. and H. Zhang, (2006), Modelling rupture dynamics of a planar fault in 3-D half space by boundary integral equation method: An overview, *Pure Appl. Geophys.*, **163**, 267-299.

Cochard, A. and R. Madariaga, (1994), Dynamic faulting under rate-dependent friction, *Pure Appl. Geophys.*, **142**(3-4), 419-445.

Cochard, A. and R. Madariaga, (1996), Complexity of seismicity due to highly rate-dependent friction, *J. Geophys. Res.*, **101**(B11), 25321-25336.

Das, S., (1980), A numerical method for determination of source time functions for general three-dimensional rupture propagation, *Geophys. J. Roy. Astr. Soc.*, **62**, 591-604.

Das, S. and K. Aki, (1977), A numerical study of two-dimensional spontaneous rupture propagation, *Geophys. J. Roy. Astr. Soc.*, **50**, 643-668.

Das, S. and B. V. Kostrov, (1987), On the numerical boundary integral equation method for three-dimensional dynamic shear crack problems, *Trans. ASME J. Appl. Mech.*, **54**, 99-104.

Eringen, A. C. and E. S. Şuhubi, (1975), *Elastodynamics* (2 vols), Academic Press, New York.

Fliss, S., H. S. Bhat, R. Dmowska, and J. R. Rice, (2005), Fault branching and rupture directivity, *J. Geophys. Res.*, **110**, B06312, doi:10.1029/2004JB003368.

Frangi, A. and G. Novati, (1999), On the numerical stability of time-domain elastodynamic analyses by BEM, *Comp. Meth. Appl. Mech. Engng.*, **173**, 403-417.

Fukuyama, E. and R. Madariaga, (1998), Rupture dynamics of a planar fault in a 3D elastic medium: Rate- and slip-weakening friction, *Bull. Seismol. Soc. Am.*, **88**(1), 1-17.

Fukuyama, E. and R. Madariaga, (2000), Dynamic propagation and interaction of a rupture front on a planar fault, *Pure Appl. Geophys.*, **157**, 1959-1979.

Fukuyama, E. and T. Mikumo, (2006), Dynamic rupture propagation during the 1891 Nobi, central Japan, earthquake: A possible extension to the branched faults, *Bull. Seismol. Soc. Am.*, **96**(4A), 1257-1266.

Fukuyama, E. and K. B. Olsen, (2002), A condition for super-shear rupture propagation in a heterogeneous stress field, *Pure Appl. Geophys.*, **159**, 2047-2056.

Fukuyama, E., C. Hashimoto, and M. Matsu'ura, (2002), Simulation of the transition of earthquake rupture from quasi-static growth to dynamic propagation, *Pure Appl. Geophys.*, **159**, 2057-2066.

Fukuyama, E., T. Mikumo, and K. B. Olsen, (2003), Estimation of the critical slip-weakening distance: Theoretical background, *Bull. Seismol. Soc. Am.*, **93**, 1835-1840.

Geubelle, P. H. and J. R. Rice, (1995), A spectral method for three-dimensional elastodynamic fracture problems, *J. Mech. Phys. Solids*, **43**, 1791-1824.

Goto, H. and J. Bielak, (2008), Galerkin boundary integral equation method for spontaneous rupture propagation problems: SH-case, *Geophys. J. Int.*, **172**, 1083-1103.

Ida, Y., (1972), Cohesive force across the tip of a longitudinal-shear crack and Griffith's specific surface energy, *J. Geophys. Res.*, **77**, 3796-3805.

Kame, N. and K. Uchida, (2008), Seismic radiation from dynamic coalescence, and the reconstruction of dynamic source parameters on a planar fault, *Geophys. J. Int.*, **174**, 696-706.

Kame, N. and T. Yamashita, (1997), Dynamic nucleation process of shallow earthquake faulting in a fault zone, *Geophys. J. Int.*, **128**, 204-216.

Kame, N. and T. Yamashita, (1999a), A new light on arresting mechanism of dynamic earthquake faulting, *Geophys. Res. Lett.*, **26**(13), 1997-2000.

Kame, N. and T. Yamashita, (1999b), Simulation of the spontaneous growth of a dynamic crack without constraints on the crack tip path, *Geophys. J. Int.*, **139**(2), 345-358.

Kame, N. and T. Yamashita, (2003), Dynamic branching, arresting of rupture and the seismic wave radiation in self-chosen crack path modelling, *Geophys. J. Int.*, **155**, 1042-1050.

Kame, N., J. R. Rice, and R. Dmowska, (2003), Effects of pre-stress state and rupture velocity on dynamic fault branching, *J. Geophys. Res.*, **108**(B5), 2265, doi:10.1029/2002JB002189.

Koller, M. G., M. Bonnet, and R. Madariaga, (1992), Modelling of dynamical crack propagation using time-domain boundary integral equations, *Wave Motion*, **16**, 339-366.

Kostrov, B. V., (1966), Unsteady propagation of longitudinal shear cracks (translated from Russian), *PMM J. Appl. Math. Mech.*, **30**, 1241-1248.

Lapusta, N., J. R. Rice, Y. Ben-Zion, and G. Zheng, (2000), Elastodynamic analysis for slow tectonic loading with spontaneous rupture episodes on faults with rate- and state-dependent friction, *J. Geophys. Res.*, **105**, 23765-23789.

Nishimura, N. and S. Kobayashi, (1989), A regularized boundary integral equation method for elastodynamic crack problems, *Comp. Mech.*, **4**, 319-328.

Peirce, A. and E. Siebrits, (1996), Stability analysis of model problems for elastodynamic boundary element discretizations, *Numer. Meth. Partial Diff. Eq.*, **12**, 585-613.

Peirce, A. and E. Siebrits, (1997), Stability analysis and design of time-stepping schemes for general elastodynamic boundary element models, *Int. J. Numer. Meth. Engng.*, **40**, 319-342.

Perrin, G., J. R. Rice, and G. Zheng, (1995), Self-healing slip pulse on a frictional surface, *J. Mech. Phys. Solids*, **43**, 1461-1495.

Rice, J. R., (1993), Spatio-temporal complexity of slip on a fault, *J. Geophys. Res.*, **98**, 9885-9907.

Seelig, Th. and D. Gross, (1997), Analysis of dynamic crack propagation using a time-domain boundary integral equation method, *Int. J. Solids Structures*, **34**, 2087-2103.

Seelig, Th. and D. Gross, (1999a), On the interaction and branching of fast running cracks—A numerical investigation, *J. Mech. Phys. Solids*, **47**, 935-952.

Seelig, Th. and D. Gross, (1999b), On the stress wave induced curving of fast running cracks—A numerical study by a time-domain boundary element method, *Acta Mech.*, **132**, 47-61.

Siebrits, E. and S. L. Crouch, (1994), Two-dimensional elastodynamic displacement discontinuity method, *Int. J. Numer. Meth. Engng.*, **37**, 3229-3250.

Sládek, V. and J. Sládek, (1984), Transient elastodynamic three-dimensional problems in cracked bodies, *Appl. Math. Modelling*, **8**, 2-10.

Tada, T., (2005), Displacement and stress Green's functions for a constant slip-rate on a quadrantal fault, *Geophys. J. Int.*, **162**(3), 1007-1023.

Tada, T., (2006), Stress Green's functions for a constant slip rate on a triangular fault, *Geophys. J. Int.*, **164**(3), 653-669.

Tada, T. and R. Madariaga, (2001), Dynamic modelling of the flat 2-D crack by a semi-analytic BIEM scheme, *Int. J. Numer. Meth. Engng.*, **50**, 227-251.

Tada, T. and T. Yamashita, (1997), Non-hypersingular boundary integral equations for two-dimensional non-planar crack analysis, *Geophys. J. Int.*, **130**(2), 269-282.

Tada, T., E. Fukuyama, and R. Madariaga, (2000), Non-hypersingular boundary integral equations for 3-D non-planar crack dynamics, *Comp. Mech.*, **25**, 613-626.

Yamashita, T. and E. Fukuyama, (1996), Apparent critical slip displacement caused by the existence of a fault zone, *Geophys. J. Int.*, **125**, 459-472.

Yokoyama, K., T. Tada, and Y. Shinozaki, (2007), Simulation of dynamic source processes on an arbitrarily shaped fault (Japanese abstract), *Summ. Tech. Papers Ann. Meeting, Arch. Inst. Japan*, **B-1**, 255-256.

Zhang, Ch., (1991), A novel derivation of non-hypersingular time-domain BIEs for transient elastodynamic crack analysis, *Int. J. Solids Structures*, **28**, 267-281.

Zhang, Ch., (2002), A 2D hypersingular time-domain traction BEM for transient elastodynamic crack analysis, *Wave Motion*, **35**, 17-40.

Zhang, Ch. and J. D. Achenbach, (1989), A new boundary integral equation formulation for elastodynamic and elastostatic crack analysis, *Trans. ASME J. Appl. Mech.*, **56**, 284-290.

Zhang, Ch. and D. Gross, (1993), A non-hypersingular time-domain BIEM for 3-D transient elastodynamic crack analysis, *Int. J. Numer. Meth. Engng.*, **36**, 2997-3017.

Zhang, H. and X. Chen, (2006a), Dynamic rupture on a planar fault in three-dimensional half space—I. Theory, *Geophys. J. Int.*, **164**, 633-652.

Zhang, H. and X. Chen, (2006b), Dynamic rupture on a planar fault in three-dimensional half-space—II. Validations and numerical experiments, *Geophys. J. Int.*, **167**, 917-932.

Zhang, H. and X. Chen, (2009), Equivalence of the Green's function for a full-space to the direct-wave contributions for a half-space and a layered half-space, *Bull. Seismol. Soc. Am.*, **99**(1), 454-461.

## Chapter 10

# Dynamic Rupture Propagation of the 1995 Kobe, Japan, Earthquake

Eiichi Fukuyama

*National Research Institute for Earth Science and Disaster Prevention, Tsukuba, Japan*

The dynamic rupture propagation of the 1995 Kobe earthquake ($M_w$ 6.9) is simulated using the boundary integral equation method to investigate the physical condition of earthquake dynamic rupture. From several numerical experiments, a small fault is found to be required beneath the Akashi Kaikyo Bridge, which is located near the hypocenter and between the Nojima and Suma faults. The existence of this small fault was originally proposed by *Koketsu et al.* (1998, *Earth Planets Space*). In the present physical modeling, it plays an important role in making the rupture propagate bilaterally along both faults. In addition, the principal stress direction should be rotated 30° clockwise from the tectonic stress direction in this region. This suggests that an aseismic slip before the earthquake might occur in order to disturb the stress field in this region from the global stress field.

## 1. INTRODUCTION

The 1995 Kobe, Japan, earthquake ($M_w$ 6.9) (the formal name is the Hyogo-ken Nanbu earthquake, which was given by the Japanese government) is one of the best-analyzed earthquakes in the world in its rupture process. Because it occurred beneath one of the most crowded regions in Japan, it caused more than 6400 fatalities. The hypocenter was located beneath the Akashi Strait (Figure 1). Its focal mechanism was of the strike-slip type. The seismicity in the aftershock region had been inactive for tens of years until this mainshock occurred. The previous large earthquake in this region was the 1596 Keicho-Fushimi earthquake (M 7.5) (e.g., *Usami*, 1967, 1996), which is believed to have ruptured northeastward of the Kobe earthquake between the area lying south of Kyoto city and north of Osaka city along the Arima-Takatsuki Tectonic Line.

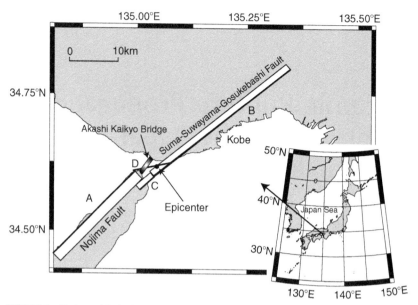

**FIGURE 1**  Fault model obtained by *Koketsu et al.* (1998). The locations of the Akashi Kaikyo Bridge (shaded rectangle) and the epicenter of the Kobe earthquake are shown. A and B are the Nojima Fault and Suma-Suwayama-Gosukebashi Fault, respectively. C is the southwest extension of the Suma Fault, and D is the jointed fault between the Nojima Fault and Suma Fault. Thick line of the rectangle indicates the updip edge of the fault. Epicenter of the 1995 Kobe earthquake is shown as a solid circle.

Four foreshocks (M 1.5-M 3) were observed near the hypocenter 11 hours before the mainshock (*Katao et al.*, 1997). The depth of the hypocenter was rather deep, about 17 km (*Katao et al.*, 1997). Beneath the hypocentral region, low velocity and high Poisson ratio anomaly was found, which might have triggered this earthquake as well as the foreshocks (*Zhao et al.*, 1996; *Zhao and Negishi*, 1998; *Okada et al.*, 2007).

The focal mechanisms of aftershocks estimated by P-wave first motion polarities are wide variations of reverse, strike-slip, and normal faults (*Katao et al.*, 1997; *Yamanaka et al.*, 2002). Similar features were reported for the 1989 Loma Prieta earthquake (*Beroza and Zoback*, 1993). This can be interpreted to mean that complete stress drop occurred during the mainshock because the effective stress was low in the source region. This low effective stress situation might be caused by the existence of high pore pressure water. In contrast, *Fukuyama et al.* (2003) reported that for the 2000 western Tottori earthquake ($M_w$ 6.6) that occurred north of the Kobe region along the coast of the Japan Sea, the effective stress state was high, suggesting pore pressure might be low in this source region because most aftershock focal mechanisms were similar to that of the mainshock.

There are several kinematic source models for this earthquake estimated from strong motion, teleseismic, and geodetic data. *Kikuchi and Kanamori*

(1996) analyzed the teleseismic waveforms and identified three subevents. The first subevent corresponds to the surface trace that appeared along the Nojima Fault on the Awaji Island (*Lin et al.*, 1995; *Nakata and Yomogida*, 1995; *Nakata et al.*, 1995). The second and third subevents correspond to the rupture beneath the Kobe city, where no surface faults appeared (*Hirata et al.*, 1996; *Katao et al.*, 1997).

*Hashimoto et al.* (1996) estimated the coseismic slip distribution using the continuous GPS data, campaign-type GPS survey, and leveling data. They assumed six segmented subfaults, each of which has uniform slip on it. There exists a step at the Akashi Strait, where the rupture initiated.

*Ide et al.* (1996) and *Sekiguchi et al.* (1996, 2000) inverted the strong ground motion data to obtain the kinematic model of the 1995 Kobe earthquake. *Horikawa et al.* (1996), *Wald* (1996), and *Yoshida et al.* (1996) inverted strong ground motion, teleseismic waveforms, and geodetic data. Although there are some differences in these models in detail, we could capture a common feature that large slip occurred at the shallow part of the Nojima Fault and beneath the Kobe city.

There are some dynamic rupture models (*Ide and Takeo*, 1997; *Bouchon et al.*, 1998; *Day et al.*, 1998; *Spudich et al.*, 1998; *Song and Beroza*, 2004, *Tinti et al.*, 2005). *Ide and Takeo* (1997) and *Song and Beroza* (2004) are based on the single fault model by *Ide et al.* (1996). *Bouchon et al.* (1998) used the model prepared by *Sekiguchi et al.* (1996). *Day et al.* (1998) and *Tinti et al.* (2005) used the model prepared by *Wald* (1996). *Spudich et al.* (1998) used the model prepared by *Yoshida et al.* (1996). All of the preceding computations of dynamic rupture were done by projecting the corresponding slip distribution onto a single planar fault. Thus, in all models, planar fault geometry is assumed.

Because the fault strike direction is significantly different between the Nojima (fault *A*) and Suma-Suwayama-Gosukebashi (fault *B*) faults with different fault dips (fault *A* dips to the east and fault *B* dips to the west; see Figure 1) and there exists a step beneath the Akashi Strait, the hypocentral region, it should be quite important to take these features into account in the dynamic rupture modeling because the dynamic rupture propagation is sensitive to such geometrical irregularity. Because the rupture initiated at the joint of these two fault segments, a detailed fault structure around this joint is crucial for constructing the dynamic fault model.

When the Kobe earthquake occured, the Akashi Kaikyo Bridge was under construction near the epicenter. To measure and monitor the locations of the two pillars and two anchorages of the bridge, GPS observation had been conducted before the earthquake. *Yamagata et al.* (1996) obtained the coseismic displacements of these four locations. Because the Akashi Kaikyo Bridge is located just above the hypocenter, this GPS data should include important information on the detailed faulting behavior near the hypocenter.

Using the Akashi Kaikyo Bridge GPS data presented earlier, as well as the leveling data along the Awaji Island, which had not been used by *Yoshida et al.* (1996), *Koketsu et al.* (1998) revised the model of *Yoshida et al.* (1996) by adding two small fault segments (fault *C* and fault *D* in Figure 1). *Koketsu et al.* (1998) obtained a reasonable fit both to the new data sets as well as to the data that *Yoshida et al.* (1996) used.

In this chapter, several numerical experiments are conducted by computing dynamic rupture propagations of the 1995 Kobe earthquake based on the fault geometry by *Koketsu et al.* (1998) under various stress conditions. These experiments help us estimate the initial stress field in this region.

## 2. COMPUTATION METHOD

To compute the dynamic rupture of earthquakes, we need the fault geometry and fault constitutive relation on the fault surface as boundary conditions and the initial stress field applied to the fault as initial conditions (e.g., *Fukuyama*, 2003, 2007). Fault geometry includes fault step, fault jog, fault branch, and change of dip angle along strike directions. These features have long been recognized in field observations of exhumed faults (e.g., *Yeats et al.*, 1997; *Snoke et al.*, 1998), but it was difficult to include these features in the earthquake rupture modeling until recently.

Because nonplanar fault geometry is one of the important aspects to be taken into account, the boundary integral equation method (BIEM) has been extensively developed recently (*Koller et al.*, 1992; *Cochard and Madariaga*, 1994; *Fukuyama and Madariaga*, 1995, 1998; *Tada and Yamashita*, 1997; *Aochi*, 1999; *Kame and Yamashita*, 1999; *Aochi et al.*, 2000; *Tada et al.*, 2000; *Tada*, 2005, 2006). For the details of the BIEM, please refer to *Tada* (2009).

In the BIEM, the boundaries (i.e., fault surfaces) are the only area to be discretized and modeled in the computation. Because the Green's functions employed in the BIEM formulation take care of the response outside the boundary, we do not have to compute explicitly the behavior outside the boundaries. Therefore, introduction of complicated fault geometry becomes feasible. On the other hand, it is difficult to introduce the heterogeneity of materials surrounding the fault (such as layered structure and bimaterial interfaces) because the corresponding Green's function becomes complicated. In the present modeling, because we will focus on the primary stage of the dynamic rupture propagation of the Kobe earthquake, it is reasonable to assume that the dynamic rupture propagations are computed on a nonplanar fault system embedded in the homogeneous elastic materials.

For the simulation of the Kobe earthquake, BIEM is applied with triangular elements (*Fukuyama et al.*, 2002; *Fukuyama and Mikumo*, 2006; *Tada*, 2006; *Fukuyama*, 2007) because the fault model requires a 3D feature to reproduce the complicated fault geometry. Concerning the kernels of triangular elements in the BIEM, theoretical formulations for nonplanar faults derived by *Tada et al.* (2000) and the corresponding discretized form for triangular elements derived by *Tada* (2006) are employed here.

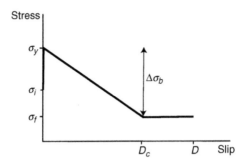

**FIGURE 2**  Slip weakening friction law used for the computation of dynamic rupture propagation. $\sigma_y$, $\sigma_i$, and $\sigma_f$ are yield stress, initial stress, and residual stress, respectively. $D_c$ and $D$ are slip weakening distance and total slip, respectively. $\Delta\tau_b$ stands for breakdown stress drop.

As a constitutive relation on the fault, a simple linear slip weakening constitutive law (e.g., *Andrews*, 1976) is applied, which is defined by three parameters: strength (or yielding stress) ($\sigma_y$), frictional stress (or residual stress) ($\sigma_f$), and slip weakening distance ($D_c$), as shown in Figure 2. According to the Coulomb law, $\sigma_y$ is given by $\mu_s \times \sigma_n$, where $\mu_s$ is the coefficient of static friction and $\sigma_n$ is the normal stress on the fault. Similarly, $\sigma_f$ is given by $\mu_d \times \sigma_n$, where $\mu_d$ is a coefficient of dynamic friction. Because we consider that a uniform triaxial stress field is applied remotely to the entire fault system, $\sigma_n$ and shear stress $\sigma_s$ are computed from the triaxial stress tensor. Then, $\sigma_s$ is used for the initial stress, and $\sigma_n$ is used for the computation of $\sigma_y$ and $\sigma_f$. Although a uniform triaxial stress field is assumed remotely, $\sigma_s$, $\sigma_y$, and $\sigma_f$ on the fault vary depending on the orientation of fault strike and dip. Slip direction is assumed to be parallel to the initial shear traction direction on the fault. These variations of the parameters cause the heterogeneous rupture on the fault (e.g., *Aochi and Fukuyama*, 2002; *Aochi and Madariaga*, 2003).

The discretized kernel of the boundary integral equation enables us to compute the shear stress at arbitrary time and space on the fault by convolving the past slip history. The constitutive relation governs the rupture and provides the relation between shear stress and slip (or slip velocity) at every time on the fault. Thus, by solving these two equations simultaneously with the initial condition (initial shear stress), both slip and stress distributions can be estimated at each time step. Therefore, spatiotemporal variations of shear stress and slip are obtained.

## 3. FAULT MODEL

To initiate a spontaneous rupture, a fault model without geometrical discontinuity (i.e., fault step) at the initiation point is required. However, most kinematic fault models (e.g., *Hashimoto et al.*, 1996; *Horikawa et al.*, 1996; *Sekiguchi et al.*, 1996; *Wald*, 1996; *Yoshida et al.*, 1996) have fault step at the hypocenter, which makes the rupture initiation difficult there.

*Koketsu et al.* (1998) proposed a new model (Figure 1) that took into account the GPS data at the piers of the Akashi Kaikyo Bridge located just above the hypocenter. They improved the model by *Yoshida et al.* (1996) by adding two small faults and could successfully fit the Akashi Kaikyo Bridge GPS data and the leveling data along Awaji Island in addition to what *Yoshida et al.* (1996) used. The model by *Koketsu et al.* (1998) consists of the Nojima Fault (*A*), the Suma-Suwayama-Gosukebashi Fault (*B*), the southern extension of the Suma Fault with small offset (*C*), and a vertical strike-slip fault beneath the Akashi Kaikyo Bridge (*D*) connecting fault *A* and fault *B* (Figure 1). Based on the model by *Koketsu et al.* (1998), a fault model is constructed for the dynamic rupture modeling as shown in Figure 3 and Table 1. It consists of 1440 triangular elements whose average area is 0.75 km$^2$. The hypocenter is located at the joint between fault *B* and fault *D*, as shown in Figure 3.

Because this earthquake is of the strike-slip type, the intermediate principal stress ($\sigma_2$) is assumed to be vertical. The magnitude of $\sigma_2$ is computed from the lithostatic stress level. The mean stress, the average of the maximum

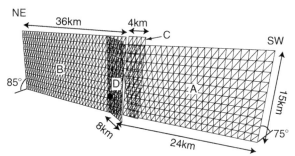

**FIGURE 3** Fault model used in this computation viewed from southwest. A, B, C, and D correspond to the faults in Figure 1. Their detailed parameters are shown in Table 1.

**TABLE 1** Geometry of Four Subfaults

| Fault Name | Length (km) | Width (km) | Strike (N deg. E) | Dip (deg.) |
|---|---|---|---|---|
| A | 24 | 15 | 45 | 75 |
| B | 36 | 15 | 232 | 85 |
| C | 4 | 15 | 232 | 85 |
| D | 8 | 15 | 78 | 90 |

($\sigma_1$) and minimum ($\sigma_3$) principal stresses, is assumed to equal to $\sigma_2$. Finally, the magnitudes of $\sigma_1$ and $\sigma_3$ are adjusted to fit the amount of slip estimated from the waveform inversion. Then one free parameter, the direction of $\sigma_1$, remains, which we need to estimate. This parameter controls the propagation of rupture along the fault segments, because the amount of strength and stress drop is controlled by the strike direction with respect to the $\sigma_1$ direction.

At the hypocenter, initial stress is raised to the corresponding strength to make the rupture initiate and propagate outward. Because the hypocenter is located at the joint between fault $B$ and fault $D$, the initiation region, where the initial stress is raised, is defined as the fault elements whose distance from the hypocenter is less than 1 km. This nucleation zone size is taken from the kinematic inversion result of initial rupture of this earthquake by *Shibazaki et al.* (2002), which also roughly coincides with the $L_c$ parameter defined by *Andrews* (1976) as follows:

$$L_c = \frac{8}{\pi} \frac{\mu(\lambda + \mu)}{\lambda + 2\mu} \frac{G}{(\sigma_0 - \sigma_r)^2} \tag{1}$$

where $\lambda$ and $\mu$ are the Lamé constants and $G$ is fracture energy.

## 4. COMPUTATION RESULTS

Two cases are considered here; one is that $\sigma_1$ direction is east-west, which is consistent with the tectonic stress direction in this region (*Okada and Ando*, 1979; *Tsukahara and Kobayashi*, 1991) (Case 1). The other is that $\sigma_1$ oriented to N120°E, 30° rotated clockwise from the tectonic stress direction in this region (Case 2). Case 2 is set after several trial computations. In both cases, the magnitudes of principal stresses $\sigma_1$, $\sigma_2$, and $\sigma_3$ are assumed to be 75, 50, and 25 MPa, respectively. P- and S-wave velocities and rigidity ($\mu$) are assumed to be 6000 m/s, 3500 m/s, and 30 GPa, respectively. $\mu_s$, $\mu_d$, and $D_c$ are assumed to be 0.2, 0.05, and 0.6 m, respectively. $\mu_s$ is set after several trial computations, and the rupture could not propagate if the $\mu_s$ value were larger. It might be rather small compared to a typical friction of rocks (0.6, e.g., *Byerlee*, 1978), but some clay materials have a low coefficient of friction (~0.2, e.g., *Paterson and Wong*, 2005, p. 170), which the fault might be composed of. A low total stress field suggested by the variation of aftershock focal mechanisms (*Katao et al.*, 1997) also supports this small coefficient of friction.

In Case 1, the rupture did not propagate bilaterally at the hypocenter but it propagated northeast (fault $B$) first, then with some time delay, it propagated southwest (fault $A$), as shown in Figure 4. This is because the initial stress was not accumulated sufficiently on fault $D$ (Figure 4), which makes the rupture propagate along fault $B$ and fault $C$ first. Once the rupture starts to propagate on fault $B$ and fault $C$, the rupture along fault $D$ becomes difficult because of the shadow effect (e.g., *Aochi et al.*, 2000; *Poliakov et al.*,

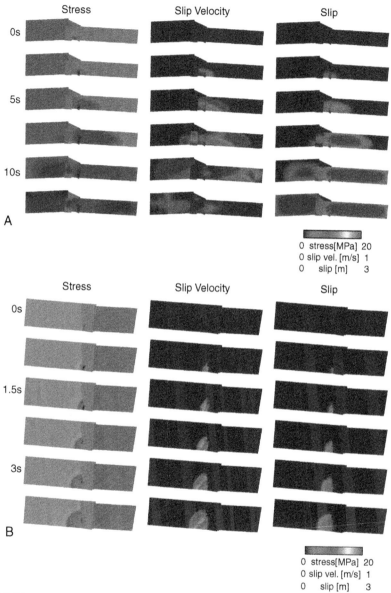

**FIGURE 4**  Snapshots of stress, slip velocity, and slip of the dynamic rupture propagation for Case 1. (a) Views from the east. (b) Views from the west for the beginning of the rupture. (**See Color Plate 24.**)

2002; *Kame et al.*, 2003). The shadow effect occurs when two parallel faults are aligned and one of them starts to rupture, then the other fault is suppressed to propagate rupture. Although the rupture started to propagate on fault *A* after about 5 s, the rupture could not propagate on fault *D* (Figure 4), which is inconsistent with the model by *Koketsu et al.* (1998). In addition, it is believed that the rupture propagated along fault *A* first then along fault B (e.g., *Kikuchi and Kanamori*, 1996; *Yoshida et al.*, 1996), which again is not consistent with the computation result of Case 1.

To improve the fit of the computed rupture propagation to the kinematic model, the stress field applied to the fault need to be modified. After several trial computations changing the $\sigma_1$ direction by 5-degree intervals, a suitable direction of principal stress is found when the $\sigma_1$ direction is rotated 30° clockwise. This stress field is consistent with the result observed by *Yamashita et al.* (2004), who reported that the $\sigma_1$ direction was rotated 30° clockwise after the earthquake. *Yamashita et al.* (2004) estimated the preseismic stress field from the postseismic stress by the *in situ* hydraulic fracturing stress measurements (*Ikeda et al.*, 2001; *Tsukahara et al.*, 2001) with the stress drop distribution estimated by the waveform inversion analysis (*Yoshida et al.*, 1996). This rotation results in more accumulation of the initial shear stress on the joint fault segment *D*, and thus the fault *D* could rupture first.

Figure 5 shows the computation result of Case 2. The rupture first propagated along fault *D* then propagated bilaterally to both fault *A* and *B*. This rupture propagation pattern is similar to that of the kinematic fault models (e.g., *Yoshida et al.*, 1996; *Koketsu et al.*, 1998). In this case, a time lag is observed between the rupture initiation and the main rupture propagation along the Nojima and Suma faults. This feature has been pointed out by the analysis of the strong motion seismometers installed at the Akashi Kaikyo Bridge (*Megawati et al.*, 2001). It is also consistent with the kinematic modeling of the initial source process by *Shibazaki et al.* (2002) who inverted the initial part of P-waveform data.

## 5. DISCUSSION AND CONCLUSION

The rupture process of the 1995 Kobe earthquake has been modeled using the boundary integral equation method with triangular elements. Referring to the geometry of the kinematic fault model by *Koketsu et al.* (1998), the dynamic rupture was successfully reproduced, especially on its rupture propagation pattern, using physically reasonable parameters. In this modeling, the detailed fault structure in the hypocentral region is introduced, which is found to be crucial for reproducing the rupture propagation similar to that of the kinematic model. The simulation result suggests that a small fault connecting Nojima Fault and Suma Fault is required to make the rupture propagate bilaterally, and the stress orientation is not consistent with that of the tectonic stress in this region but is rotated 30° clockwise to make the shear stress accumulate on the small fault.

This rotation of the initial stress field from the global stress pattern could be explained by the preslip. If the preslip occurs before the earthquake, the

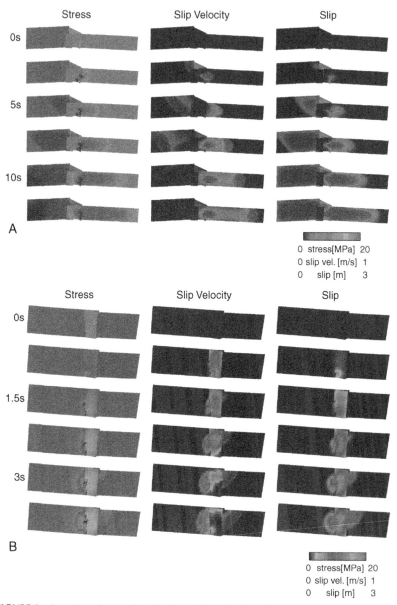

FIGURE 5    Snapshots of stress, slip velocity, and slip of the dynamic rupture propagation for Case 2. (a) Views from the east. (b) Views from the west for the beginning of the rupture. (See Color Plate 25.)

initial stress field should be contaminated by the stress change caused by the preslip. The stress rotation due to the stress change of the preslip becomes predominant if the total stress level of the background stress is low. As the variation of the focal mechanisms of aftershocks was reported to be large

(*Katao et al.*, 1997), total stress could be low. In this situation, a small amount of preslip can cause the rotation of the initial stress field.

Even if the fault geometry, constitutive law on the fault, and applied stress field around the fault are *a priori* known, to reproduce the dynamic rupture propagation similar to real earthquakes, we need additional information on how the rupture initiates, which was *a priori* assumed in the present study. There are several models on the initiation of the rupture (e.g., *Dieterich*, 1992; *Shibazaki and Matsu'ura*, 1992; *Ide and Aochi*, 2005). These models are based on planar faults. As discussed before, because detailed geometry of the fault system with applied stress controls the rupture, more detailed structures of the fault near the hypocenter are required with the loading process of the stress applied. If the rupture grows with increasing the fracture energy as *Ide and Aochi* (2005) proposed, very fine structure near the hypocenter becomes important during the initiation process of the dynamic rupture. However, it is difficult and almost impossible to obtain the detailed fault structure near the hypocenter before the earthquake.

The dynamic rupture computations introduced in this study can be applied for the prediction of the rupture scenario of future disastrous earthquakes (e.g., *Aochi et al.*, 2006; *Fukuyama et al.*, 2009). Once the rupture scenarios are prepared, we can compute the strong ground motions that we will suffer during future earthquakes. How the ground shakes during the earthquake is very important information for the public because they can understand *a priori* what kind of damages will be suffered and thus they can prepare against the future earthquake disasters. For this purpose, the hypocenter location is sensitive to the further rupture scenario depending on the initial stress applied to the fault in the present study. Even if a realistic modeling of the initiation process seems difficult at present, we will be able to prepare several scenarios based on the computation of the dynamic rupture by assuming several possible hypocenter locations (*Fukuyama et al.*, 2009). Using possible rupture scenarios, we can compute the strong ground motions using 3D velocity structure and obtain the average image of ground motion distribution and that of the worst, which will be useful for the prevention of earthquake disasters.

## ACKNOWLEDGMENTS

Comments by Hideo Aochi were very helpful.

## REFERENCES

Andrews, D. J., (1976), Rupture velocity of plane strain shear cracks, *J. Geophys. Res.*, **81**(32), 5679-5687.

Aochi, H., (1999), Theoretical studies on dynamic rupture propagation along a 3D non-planar fault system, *D.Sc. thesis*, the University of Tokyo, Tokyo.

Aochi, H. and E. Fukuyama, (2002), Three-dimensional nonplanar simulation of the 1992 Landers earthquake, *J. Geophys. Res.*, **107**(B2), 2035, doi:10.1029/2000JB000061.

Aochi, H., M. Cushing, O. Scotti, and C. Berge-Thierry, (2006), Estimating rupture scenario like-lihood based on dynamic rupture simulations: The example of the segmented Middle Durance fault, southeastern France, *Geophys. J. Int.*, **165**, 436-446.

Aochi, H., E. Fukuyama, and M. Matsu'ura, (2000), Selectivity of spontaneous rupture propaga-tion on a branched fault. *Geophys. Res. Lett.*, **27**(22), 3635-3638.

Aochi, H. and R. Madariaga, (2003), The 1999 Izmit, Turkey, earthquake: Nonplanar fault struc-ture, dynamic rupture process, and strong ground motion, *Bull. Seismol. Soc. Am.*, **93**(3), 1249-1266.

Beroza, G. C. and M. D. Zoback, (1996), Mechanism diversity of the Loma Prieta aftershocks and the mechanics of mainshock-aftershock interaction, *Science*, **259**, 210-213.

Bouchon, M., H. Sekiguchi, K. Irikura, and T. Iwata, (1998), Some characteristics of the stress field of the 1995 Hyogo-ken Nanbu (Kobe) earthquake, *J. Geophys. Res.*, **103**(B10), 24271-24282.

Byerlee, J. D., (1978), Friction of rocks, *Pure Appl. Geophys.*, **116**, 615-626.

Cochard, A. and R. Madariaga, (1994), Dynamic faulting under rate-dependent friction, *Pure Appl. Geophys.*, **142**(3-4), 419-445.

Day, S. M., G. Yu, and D. J. Wald, (1998), Dynamic stress changes during earthquake rupture, *Bull. Seismol. Soc. Am.*, **88**(2), 512-522.

Dieterich, J. H., (1992), Earthquake nucleation on faults with rate- and state-dependent strength, *Tectonophys.*, **211**, 115-134.

Fukuyama, E., (2003), Numerical modeling of earthquake dynamic rupture: Requirements for realistic modeling, *Bull. Earthq. Res. Inst., Univ. Tokyo.*, **78**, 167-174.

Fukuyama, E., (2007), Fault structure, stress, friction and rupture dynamics of earthquakes, In: *Advances in Earth Sciences*, edited by Sammons, P. R. and J. M. T. Thompson, Imperial Col-lege Press, London, 109-131.

Fukuyama, E., W. L. Ellsworth, F. Waldhauser, and A. Kubo, (2003), Detailed fault structure of the 2000 western Tottori, Japan, earthquake sequence, *Bull. Seismol. Soc. Am.*, **93**(4), 1468-1478.

Fukuyama, E., C. Hashimoto, S. Aoi, R. Ando, and M. Matsu'ura, (2009), A physics-based simu-lation of the 2003 Tokachi-oki, Japan, earthquake toward strong motion prediction, *submitted to Bull. Seismol. Soc. Am.*

Fukuyama, E. and R. Madariaga, (1995), Integral equation method for plane crack with arbitrary shape in 3D elastic medium, *Bull. Seismol. Soc. Am.*, **85**(2), 614-628.

Fukuyama, E. and R. Madariaga, (1998), Rupture dynamics of a planar fault in a 3D elastic medium: Rate- and slip-weakening friction, *Bull. Seismol. Soc. Am.*, **88**(1), 1-17.

Fukuyama, E. and T. Mikumo, (2006), Dynamic rupture propagation during the 1891 Nobi, cen-tral Japan, earthquake: A possible extension to the branched faults, *Bull. Seismol. Soc. Am.*, **96**(4A), 1257-1266.

Fukuyama, E., T. Tada, and B. Shibazaki, (2002), Three dimensional dynamic rupture propagation on a curved/branched fault based on boundary integral equation method with triangular ele-ments, *EOS Trans.*, AGU, **83**(47), Fall Meet. Suppl., NG62A-0930.

Hashimoto, M., T. Sagiya, H. Tsuji, Y. Hatanaka, and T. Tada, (1996), Co-seismic displacements of the 1995 Hyogo-ken Nanbu earthquake, *J. Phys. Earth*, **44**, 255-279.

Hirata, N., S. Ohmi, S. Sakai, K. Katsumata, S. Matsumoto, T. Takanami, A. Yamamoto, T. Iidaka, T. Urabe, M. Sekine, T. Ooida, F. Yamazaki, H. Katao, Y. Umeda, M. Nakamura, N. Seto, T. Matsushima, H. Shimizu, and Japanese University Group of the Urgent Joint Obser-vation for the 1995 Hyogo-ken Nanbu earthquake, (1996), Urgent joint observation of after-shocks of the 1995 Hyogo-ken Nanbu earthquake, *J. Phys. Earth*, **44**, 317-328.

Horikawa, H., K. Hirahara, Y. Umeda, M. Hashimoto, and F. Kusano, (1996), Simultaneous inversion of geodetic and strong-motion data for the source process of the Hyogo-ken Nanbu, Japan, earthquake, *J. Phys. Earth*, **44**, 455-471.

Ide, S. and H. Aochi, (2005), Earthquakes as multiscale dynamic ruptures with heterogeneous fracture surface energy, *J. Geophys. Res.*, **110**, B11303, doi:10.1029/2004JB003591.

Ide, S. and M. Takeo, (1997), Determination of constitutive relations of fault slip based on seismic wave analysis, *J. Geophys. Res.*, **102**(B12), 27379-27391.

Ide, S., M. Takeo, and Y. Yoshida, (1996), Source process of the 1995 Kobe earthquake: Determination of spatio-temporal slip distribution by Bayesian modeling, *Bull. Seismol. Soc. Am.*, **86**(3), 547-566.

Ikeda, R., Y. Iio, and K. Omura, (2001), In situ stress measurements in NIED boreholes in and around the fault zone near the 1995 Hyogo-ken Nanbu earthquake, Japan, *Island Arc*, **10**(3-4), 252-260.

Kame, N. and T. Yamashita, (1999), Simulation of the spontaneous growth of a dynamic crack without constraints on the crack tip path, *Geophys. J. Int.*, **139**(2), 345-358.

Kame, N., J. R. Rice, and R. Dmowska, (2003), Effects of prestress state and rupture velocity on dynamic fault branching, *J. Geophys. Res.*, **108**(B5), 2265, doi:10.1029/2002JB002189.

Katao, H., N. Maeda, Y. Hiramatsu, Y. Iio, and S. Nakano, (1997), Detailed mapping of focal mechanisms in/around the 1995 Hyogo-ken Nanbu earthquake rupture zone, *J. Phys. Earth*, **45**, 105-119.

Kikuchi, M. and H. Kanamori, (1996), Rupture process of the Kobe, Japan, earthquake of Jan. 17, 1995, determined from teleseismic body waves, *J. Phys. Earth*, **44**, 429-436.

Koketsu, K., S. Yoshida, and H. Higashihara, (1998), A fault model of the 1995 Kobe earthquake derived from the GPS data on the Akashi Kaikyo Bridge and other datasets, *Earth Planets Space*, **50**, 803-811.

Koller, M. G., M. Bonnet, and R. Madariaga, (1992), Modelling of dynamical crack propagation using time-domain boundary integral equations, *Wave Motion*, **16**, 339-366.

Lin, A., H. Imiya, S. Uda, K. Iinuma, T. Misawa, T. Yoshida, Y. Abematsu, T. Wada, and K. Kawai, (1995), Investigation of the Nojima earthquake fault occurred on Awaji Island in the southern Hyogo Prefecture earthquake, *J. Geography*, **104**, 113-126 (in Japanese with English abstract).

Megawati, K., H. Higashihara, and K. Koketsu, (2001), Derivation of near-source ground motions of the 1995 Kobe (Hyogo-ken Nanbu) earthquake from vibration records of the Akashi Kaikyo Bridge and its implications, *Eng. Struct.*, **23**, 1256-1268.

Nakata, T., K. Yomogida, J. Odaka, T. Sakamoto, K. Asahi, and N. Chida, (1995), Surface fault ruptures associated with the 1995 Hyogoken-Nanbu earthquake, *J. Geography*, **104**, 127-142 (in Japanese with English abstract).

Nakata, T. and K. Yomogida, (1995), Surface fault characteristics of the 1995 Hyogoken-Nanbu earthquake, *J. Nat. Disas. Sci*, **16**, 1-9.

Okada, A. and M. Ando, (1979), Active faults and earthquakes in Japan. *Kagaku*, **49**, 158-169 (in Japanese).

Okada, T., A. Hasegawa, J. Suganomata, D. Zhao, H. Zhang, and C. Thurber, (2007), Imaging the source area of the 1995 southern Hyogo (Kobe) earthquake (M7.3) using double-difference tomography, *Earth Planet. Sci. Lett.*, **253**, 143-150.

Paterson, M. S. and T.-F. Wong, (2005), *Experimental Rock Deformation—The Brittle Fields*, 2nd ed., Springer, NewYork, 347 pp.

Poliakov, A. N. B., R. Dmowska, and J. R. Rice, (2002), Dynamic shear rupture interactions with fault bends and off-axis secondary faulting, *J. Geophys. Res.*, **107**(B11), 2295, doi:10.1029/2001JB000572.

Sekiguchi, H., K. Irikura, T. Iwata, Y. Kakehi, and M. Hoshiba, (1996), Minute locating of faulting beneath Kobe and the waveform inversion of the source process during the 1995 Hyogo-ken Nanbu, Japan, earthquake using strong ground motion records, *J. Phys. Earth*, **44**, 473-487.

Sekiguchi, H., K. Irikura, and T. Iwata, (2000), Fault geometry at the rupture termination of the 1995 Hyogo-ken Nanbu Earthquake, *Bull. Seismol. Soc. Am.*, **90**(1), 117-133.

Shibazaki, B., and M. Matsu'ura, (1992), Spontaneous processes for nucleation, dynamic propagation, and stop of earthquake rupture, *Geophys. Res. Lett.*, **19**(12), 1189-1192.

Shibazaki, B., Y. Yoshida, M. Nakamura, M. Nakamura and H. Katao, (2002), Rupture nucleations in the 1995 Hyogo-ken Nanbu earthquake and its large aftershocks, *Geophys. J. Int.*, **149**, 572-588

Snoke, A. W., J. Tullis, and V. R. Todd, (eds.) (1998), *Fault-related Rocks: A Photographic Atlas*, Princeton University Press, New Jersey, 617 pp.

Song, S. G. and G. C. Beroza, (2004), A simple dynamic model for the 1995 Kobe, Japan earthquake, *Geophys. Res. Lett.*, **31**, L18613, doi:10.1029/2004GL020557.

Spudich, P., M. Guatteri, K. Otsuki, and J. Minagawa, (1998), Use of fault striations and dislocation models to infer tectonic shear stress during the 1995 Hyogo-ken Nanbu (Kobe) earthquake, *Bull. Seismol. Soc. Am.*, **88**(2), 413-427.

Tada, T., (2005), Displacement and stress Green's functions for a constant slip-rate on a quadrantal fault, *Geophys. J. Int.*, **162**(3), 1007-1023.

Tada, T., (2006), Stress Green's functions for a constant slip rate on a triangular fault, *Geophys. J. Int.*, **164**(3), 635-669.

Tada, T., (2009), Boundary integral equation method for earthquake rupture dynamics, In: *Fault-Zone Properties and Earthquake Rupture Dynamics*, edited by E. Fukuyama, Elsevier, *International Geophysics Series*, **94**, 217-267.

Tada, T., E. Fukuyama, and R. Madariaga, (2000), Non-hypersingular boundary integral equations for 3-D non-planar crack dynamics, *Comp. Mech.*, **25**, 613-626.

Tada, T. and T. Yamashita, (1997), Non-hypersingular boundary integral equations for two-dimensional non-planar crack analysis, *Geophys. J. Int.*, **130**(2), 269-282.

Tinti, E., P. Spudich, and M. Cocco, (2005), Earthquake fracture energy inferred from kinematic rupture models on extended faults, *J. Geophys. Res.*, **110**, B12303, doi:10.1029/2005JB003644.

Tsukahara, H. and Y. Kobayashi, (1991), Crustal stress in the central and western parts of Honshu, Japan, *Zisin*, **44**, 221-231 (in Japanese with English abstract).

Tsukahara, H., R. Ikeda, and K. Yamamoto, (2001), In situ stress measurements in a borehole close to the Nojima Fault, *Island Arc*, **10**(3-4), 261-265.

Usami, T., (1967), Descriptive table of major earthquakes in and near Japan which were accompanied by damages, *Bull. Earthq. Res. Inst., Univ. Tokyo*, **44**(4), 1571-1622 (in Japanese with English abstract).

Usami, T., (1996), *Materials for Comprehensive List of Destructive Earthquakes in Japan* (Revised and Enlarged Edition), University of Tokyo Press, Tokyo, p. 497.

Wald, D. J., (1996), Slip history of the 1995 Kobe, Japan, earthquake determined from strong motion, teleseismic, and geodetic data, *J. Phys. Earth*, **44**, 489-503.

Yamagata, M., M. Yasuda, A. Nitta, and S. Yamamoto, (1996), Effects on the Akashi Kaikyo bridge, *Soils and Foundations*, special issue, 179-187.

Yamanaka, H., Y. Hiramatsu, and H. Katao, (2002), Spatial distribution of atypical aftershocks of the 1995 Hyogo-ken Nanbu earthquake, *Earth Planets Space*, **54**, 933-945.

Yamashita, F., E. Fukuyama, and K. Omura, (2004), Estimation of fault strength: Reconstruction of stress before the 1995 Kobe earthquake, *Science*, **306**, 261-263.

Yeats, R. S., K. Sieh, and C. R. Allen, (1997), *The Geology of Earthquakes*, Oxford University Press, New York, 568 pp.

Yoshida, S., K. Koketsu, B. Shibazaki, T. Sagiya, T. Kato, and Y. Yoshida, (1996), Joint inversion of near- and far-field waveforms and geodetic data for the rupture process of the 1995 Kobe earthquake, *J. Phys. Earth*, **44**, 437-454.

Zhao, D., H. Kanamori, H. Negishi, and D. Wiens, (1996), Tomography of the source area of the 1995 Kobe earthquake: Evidence for fluids at the hypocenter?, *Science*, **274**, 1891-1894.

Zhao, D. and H. Negishi, (1998), The 1995 Kobe earthquake: Seismic image of the source zone and its implications for the rupture nucleation, *J. Geophy. Res.*, **103**(B5), 996-986.

## CHAPTER 2:

| | |
|---|---|
| ER-Mapper | name of the commercial software for processing digital remote sensing data |
| ETM | Enhanced Thematic Mapper |
| ICD-CSB | Institute of Crustal Dynamics, China Seismological Bureau |
| IKONOS | name of the commercial earth observation satellite |
| InSAR | Interferometric Synthetic Aperture Radar |
| IRIS-DMC | Data Management Center, Incorporated Research Institutions for Seismology |
| Landsat TM | Landsat Thematic Mapper |
| QuickBird | name of the commercial earth observation satellite |
| SPOT | Systememe Probatoire de l'Observation de la Terre |
| spyder | name of the software that provides online near real-time data service. |
| SRTM | Shuttle Radar Topography Mission |
| SBQP | Seismological Bureau of Qinghai Province |

## CHAPTER 3:

| | |
|---|---|
| BSR | Bottom Simulating Reflector |
| CRISP | Costa Rica Seismogenesis Project |
| GPS | Global Positioning System |
| InSAR | Interferometric Synthetic Aperture Radar |
| IODP | Integrated Ocean Drilling Project |
| LFE | low-frequency earthquake |
| Moho | Mohorovicic discontinuity |
| $M_w$ | moment magnitude |
| NantroSEIZE | Nankai Trough Seismogenic Zone Experiment |
| ODP Leg | Ocean Drilling Project Leg |
| OBH | ocean bottom hydrophone |
| OBS | ocean bottom seismometer |
| OOST | out of sequence thrust |
| PT | pressure-temperature |
| SAFOD | San Andreas Observatory at Depth |
| SSE | slow slip event |
| SVU | Sestola-Vidiciatico Unit |
| UHP | ultra high-pressure |

## CHAPTER 4:

| ATF | Alto Tiberiana normal fault |
|-----|-----|
| $CO_2$ | carbon dioxide |
| IWL | Interconnected Weak Layer |
| LANF | low-angle normal fault |
| LBM | Load-Bearing Microstructure |
| Ma | Million years |
| SEM | Scanning Electron Microscope |
| XRD | X-Ray Diffraction |
| ZF | Zuccale Fault |
| $\mu s$ | static friction coefficient |

## CHAPTER 5:

| 3D | three-dimensional |
|-----|-----|
| BSE-FE-SEM | Back Scatter Electron-Field Emission Scanning Electron Microscope |
| FE-SEM | Field Emission Scanning Electron Microscope |
| GPS | Global Positioning System |
| InSAR | Interferometric Synthetic Aperture Radar |
| LIDAR | Light Detection And Ranging |
| Ma | Million years |
| $M_L$ | local magnitude |
| $M_w$ | moment magnitude |
| R&S friction | rate- and state-dependent friction |
| SEM | Scanning Electron Microscope |
| XRF | X-Ray Fluorescence |

## CHAPTER 6:

| 1D | one-dimensional |
|-----|-----|
| 2D | two-dimensional |

## CHAPTER 7:

| 2D | two-dimensional |
|-----|-----|
| 3D | three-dimensional |

## CHAPTER 8:

| 1D | one-dimensional |
|-----|-----|
| 2D | two-dimensional |
| 3D | three-dimensional |
| DEM | Discrete Element Method |

## CHAPTER 9:

| | |
|---|---|
| BEM | Boundary Element Method |
| BIEM | Boundary Integral Equation Method |
| CFL number | Courant-Friedrichs-Lewy number, $c\Delta t/\Delta x$ |
| FDM | Finite Difference Method |
| FEM | Finite Element Method |
| SIF | Stress Intensity Factor |

## CHAPTER 10:

| | |
|---|---|
| BIEM | Boundary Integral Equation Method |
| GPS | Global Positioning System |
| $M_w$ | moment magnitude |
| M | magnitude |

# International Geophysics Series

EDITED BY

**RENATA DMOWSKA**
*Division of Applied Science*
*Harvard University*
*Cambridge, Massachusetts*

**DENNIS HARTMANN**
*Department of Atmospheric Sciences*
*University of Washington*
*Seattle, Washington*

**H. THOMAS ROSSBY**
*Graduate School of Oceanography*
*University of Rhode Island*
*Narragansett, Rhode Island*

*Out of print

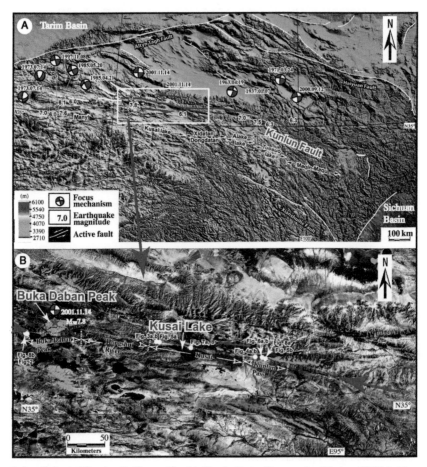

**Color Plate 1 (Chapter 2, Figure 1)** (a) Shuttle Radar Topography Mission (SRTM, 90-m resolution) color-shaded relief map showing the topographic features and distribution of the major active faults in the northern Tibet plateau and (b) Landsat TM image showing the 2001 coseismic surface rupture (indicated by red arrows) in the study area. (a) The Kunlun Fault is divided into six main segments, from the west to east: Kusai Lake, Xidatan-Dongdatan, Alake Lake, Tuosuo Lake, and Maqing-Maqu segments, respectively (*Seismological Bureau of Qinghai Province and Institute of Crustal Dynamics*, 1999). Red lines indicate the Kunlun Fault. Star indicates the epicenter of the 2001 earthquake. Beach balls show the focal mechanisms of large historic earthquakes of M ≥ 6.0 from 1904 to 2001 (*International Seismological Centre*, 2001). (b) The 2001 Kunlun surface rupture (indicated by red arrows) is divided into four main segments, from the west to east: Buka Daban Peak, Hongshui River, Kusai Lake, and Kunlun Pass segment, respectively (*Lin et al.*, 2003). Solid white cycles show the locations of figures (Figures 2 to 8). (Part a modified from Lin and Guo, 2008b.)

**Color Plate 2 (Chapter 2, Figure 2)** Representative example of extensional cracks that occurred in the western termination area of the 2001 surface rupture zone. See Figure 1b for detail location.

**Color Plate 3 (Chapter 2, Figure 3)** Fault outcrop of the 2001 rupture zone. (a) Red clay sediments (left site) are bounded with the glacial deposits composed of unconsolidated sand-gravel (right site) by fault along which the 2001 coseismic ruptures occurred. (b) Numerous striations are developed on the fault plane shown in (a), which indicate a major horizontal movement. See Figure 1b for detail location.

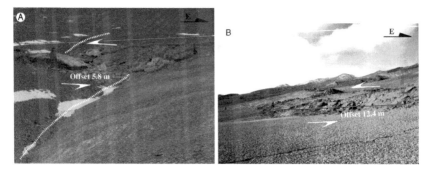

**Color Plate 4 (Chapter 2, Figure 4)** Typical outcrops of the 2001 surface ruptures. (a) Gully was offset 5.8 m. (b) Alluvial fan was offset about 12.4 m. The south-facing and north-facing fault scarps are developed on these alluvial fans, which are interpreted as a scissoring structure caused by pure strike-slip faulting (*Lin et al.*, 2003). See Figure 1b for detail location.

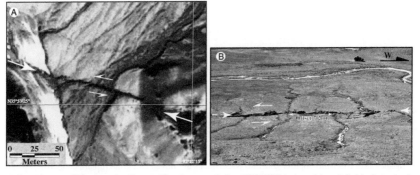

**Color Plate 5 (Chapter 2, Figure 5)** 1-m-resolution IKONOS image (a) and field photograph (b) showing typical topographic features of strike-slip offset on gullies along the 2001 coseismic surface ruptures (indicated by large white arrows). The river channel (white part in the left side of the image) is displaced by 2.3 m during the 2001 earthquake (a). See Figure 1b for detail location.

**Color Plate 6 (Chapter 2, Figure 6)** 1-m-resolution IKONOS image (a) and field photograph (b) showing typical right-stepping echelon ruptures. The geometric pattern indicates a left-lateral movement of the 2001 rupture zone. See Figure 1b for detail location.

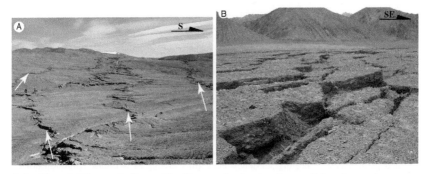

**Color Plate 7 (Chapter 2, Figure 7)** Photographs showing the parallel surface rupture zones (indicated by white arrows) (a) and extensional ruptures distributed in a wide zone (b). See Figure 1b for detail location.

**Color Plate 8 (Chapter 2, Figure 8)** Photographs showing the mole track structures developed along the 2001 coseismic rupture zone. (a) Mole tracks are linked along the rupture zone. (b) Typical angular-type of mole track developed on the unconsolidated alluvial deposits. See Figure 1b for detail location.

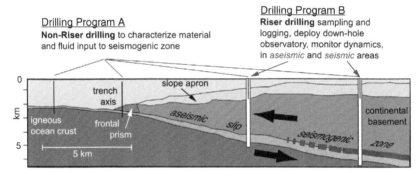

**Color Plate 9 (Chapter 3, Figure 1)** Example of the proposed drilling plan of the Costa Rica Seismogenesis Project (CRISP). Program A consists of shallow holes to be drilled with the conventional, nonriser drill technique. During Program A data will be collected, as rock composition and fluid regime, that will be used during Program B preparation. Program B plan is to drill the deep holes with riser technique, capable of controlling pressure changes.

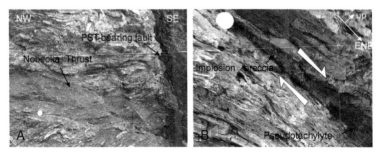

**Color Plate 10 (Chapter 3, Figure 4)** (a) Photograph of the 20-cm-thick Nobeoka Thrust fault core and pseudotachylyte-bearing subsidiary fault. (b) Close-up of Nobeoka Thrust subsidiary fault zone characterized by a dilational jog with implosion breccias and pseudotachylyte. (Pictures taken during a field trip led by G. Kimura in 2006.)

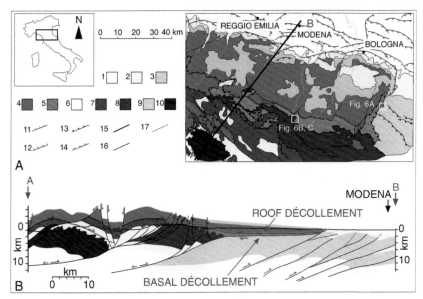

**Color Plate 11 (Chapter 3, Figure 5)** (a) Schematic geological map of the Northern Apennines with its geographical location shown in the inset. In the map, there is the location of outcrops photographed in Figures 6a, 6b, and 6c. Key: 1. Quaternary deposits; 2. late Miocene-Pleistocene marine deposits; 3. forearc slope deposits; 4. oceanic units of the late Cretaceous-early Eocene accretionary prism, European plate; 5. Sestola-Vidiciatico tectonic unit-subduction channel; 6. Mesozoic carbonate units of the Adria plate; 7. late Oligocene–early Miocene (Aquitanian) trench turbidites of the Adria plate; 8. early Miocene (Aquitanian-Burdigalian) foredeep turbidites of the Adria plate; 9. middle Miocene-late Miocene (Langhian-Messinian) foredeep turbidites of the Adria plate; 10. metamorphic continental units of the Adria plate; 11. normal faults; 12. normal faults (subsurface); 13. thrust faults and overthrusts; 14. thrust faults (subsurface); 15. strike-slip faults; 16. high-angle faults of unknown displacement (subsurface); 17. lithological boundaries. (b) Geological cross section of the Northern Apennines as marked on map (A-B). The thickness of the SVU is about 500 m, slightly decreasing toward the NE. (modified from *Vannucchi et al.*, 2008)

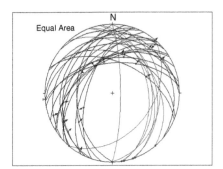

**Color Plate 12 (Chapter 3, Figure 6d)** Stereographic projection (equal area lower hemisphere) of the structures characterizing the décollement of Figure 6b and 6c. Faults, black circles, with kinematic indication of normal or reverse movement; foliation, red circles. All the structures show a consistent movement direction toward NE.

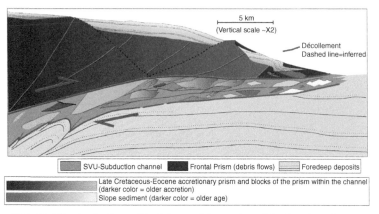

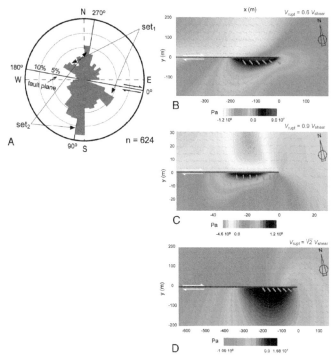

**Color Plate 13 (Chapter 3, Figure 7)** Scheme of the erosive subduction channel.

**Color Plate 14 (Chapter 5, Figure 5)** (a) Area-weighted rose diagram showing the orientation of the injection veins filled by pseudotachylyte from 28 fault segments of the Gole Larghe Fault Zone. The fractures are measured clockwise from the east side of the fault (see Figure 3c). Most fractures are toward the south and oriented at about 30° and 85° from the main fault. (b-d) Numerical models of the tensile stress field (positive is tension, negative compression) close to rupture tip for three different rupture velocities (b: $V_{rupt} = 0.6\ V_{shear}$; c: $V_{rupt} = 0.9\ V_{shear}$; d: $V_{rupt} = \sqrt{2}V_{shear}$). The fracture tip is shown as a black thick line and viewed from above. The fault is dextral, the rupture is propagating eastward and the wall rocks are under tension in the southern side in all models. The planes of maximum tensile stress are indicated by thin black segments: the planes in the southern side and near to the rupture tip are evidenced in orange. For $V_{rupt} = 0.9$ $V_{shear}$ (Figure 3c), the planes of maximum tension are oriented at about 85° from the main fault, consistently with the most common orientation of the fractures observed in the Gole Larghe Fault. All figures are from *Di Toro et al.* (2005a).

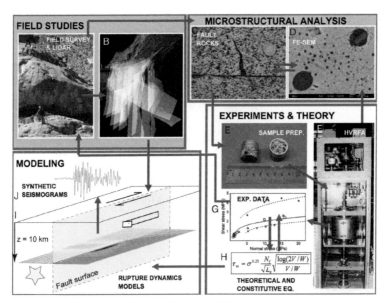

**Color Plate 15 (Chapter 5, Figure 11)** A multidisciplinary approach to the study of pseudotachylyte-bearing fault networks: synoptic view (see the text for an explanation). The Back Scatter Electron-Field Emission-Scanning Electron Microscope (BSE-FE-SEM) image in Figure 11d is courtesy of INGV.

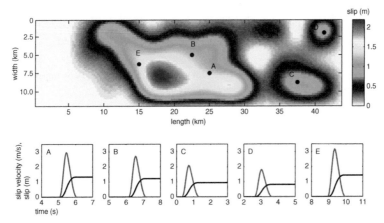

**Color Plate 16 (Chapter 7, Figure 3)** Upper panel: Slip distribution of kinematic model by *Hartzell and Heaton* (1983) for the 1979 Imperial Valley earthquake. Bottom panel: Slip velocity (gray lines) and slip (black lines) time histories for five subfaults as indicated by the capital letters above.

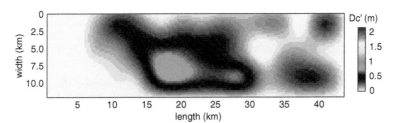

**Color Plate 17 (Chapter 7, Figure 6)** $D_c'$ distribution for the 1979 Imperial Valley earthquake inferred from the kinematic model of *Hartzell and Heaton* (1983).

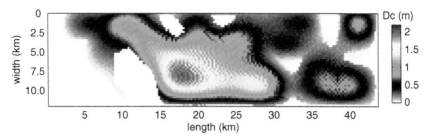

**Color Plate 18 (Chapter 7, Figure 8)** $D_c$ distribution for the 1979 Imperial Valley earthquake using the *Hartzell and Heaton* (1983) kinematic model as a boundary condition to compute traction history.

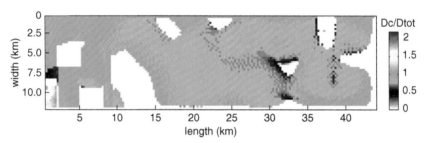

**Color Plate 19 (Chapter 7, Figure 10)** $D_c/D_{tot}$ distribution for the 1979 Imperial Valley earthquake using the *Hartzell and Heaton* (1983) kinematic model.

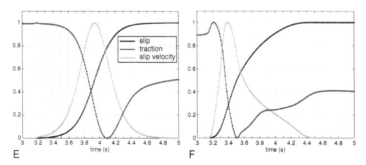

**Color Plate 20 (Chapter 7, Figure 11e and f)** Normalized time histories of slip, slip velocity, and dynamic traction calculated with $f_1$ (*tanh* function) (e) and $f_2$ (Yoffe function) (f) for the same target point.

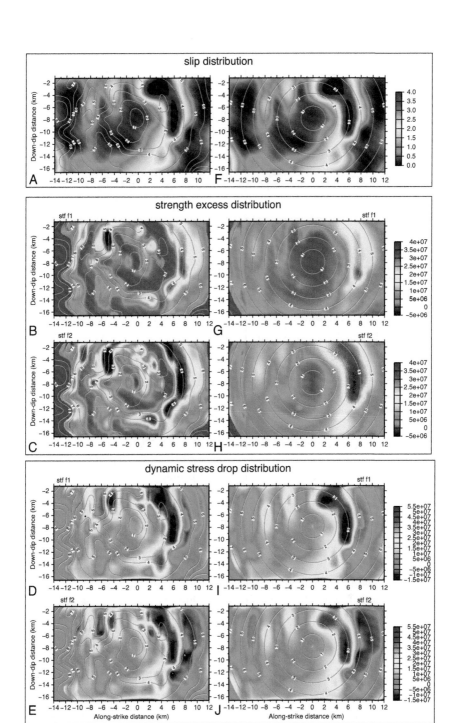

**Color Plate 21 (Chapter 7, Figure 12)** Distribution of slip (a, f), strength excess (b, c, g, j), and dynamic stress drop (d, e, i, l) on the fault plane retrieved for the two source time function $f_1$ and $f_2$ and for heterogeneous (left panels) and constant (right panels) rupture velocity models. (panels a-e are from *Piatanesi et al.*, 2004)

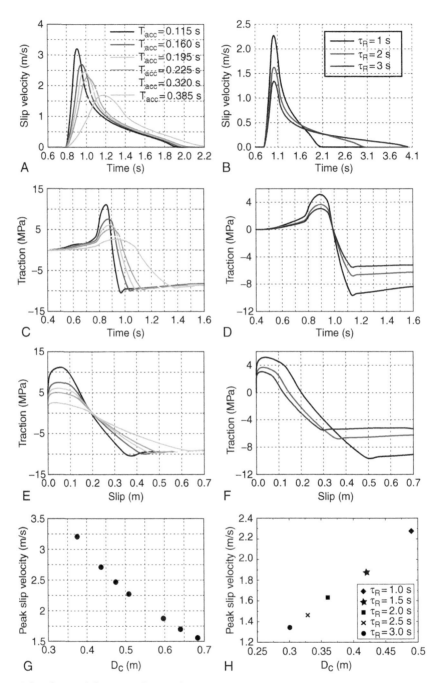

**Color Plate 22 (Chapter 7, Figure 13)** The dynamic traction changes (panels c, d, e, and f) for source models having uniform slip of 1 m, constant rupture velocity (2.0 km/s), and slip velocity time histories represented by the Yoffe function (panels a and b). Left panels: Calculations obtained for different $T_{acc}$ values and constant rise time (1.0 s). Right panels: Calculations for different rise times and a constant $T_{acc}$ (0.225 s). (modified from *Tinti et al.*, 2005b)

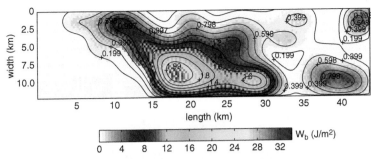

**Color Plate 23 (Chapter 7, Figure 14)** $W_b$ distribution (J/m$^2$) of the 1979 Imperial Valley earthquake using the *Hartzell and Heaton* (1983) kinematic model. Black lines and numbers represent the contour of the slip distribution shown in Figure 3.

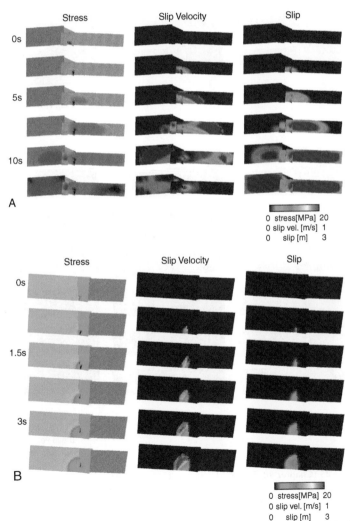

**Color Plate 24 (Chapter 10, Figure 4)** Snapshots of stress, slip velocity and slip of the dynamic rupture propagation for Case 1. (a) Views from the east. (b) Views from the west for the beginning of the rupture.

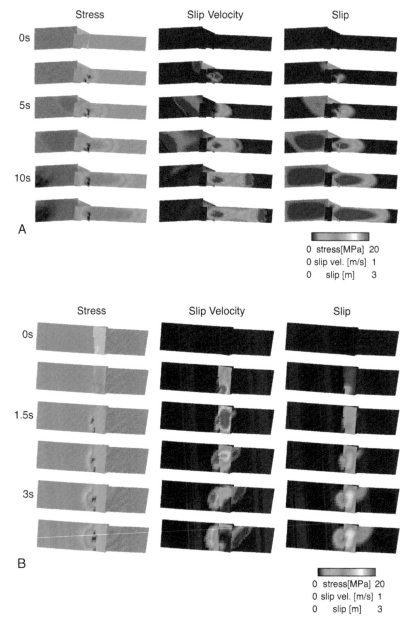

**Color Plate 25 (Chapter 10, Figure 5)**   Snapshots of stress, slip velocity, and slip of the dynamic rupture propagation for Case 2. (a) Views from the east. (b) Views from the west for the beginning of the rupture.

Printed and bound by CPI Group (UK) Ltd, Croydon, CR0 4YY

03/10/2024

01040412-0007